FUELS AND FUEL-ADDITIVES

FUELS AND FUEL-ADDITIVES

S. P. Srivastava
Jenő Hancsók

Published by John Wiley & Sons, Inc., Hoboken, New Jersey
Published simultaneously in Canada

For general information on our other products and services or for technical support, please contact our
Customer Care Department within the United States at (800) 762–2974, outside the United States at
(317) 572–3993 or fax (317) 572–4002.

Wiley also publishes its books in a variety of electronic formats. Some content that appears in print may
not be available in electronic formats. For more information about Wiley products, visit our web site at
www.wiley.com.

Library of Congress Cataloging-in-Publication Data

Srivastava, S.P., 1940–
    Fuels and fuel-additives / S.P. Srivastava, Jenő Hancsók.
      pages   cm
    Includes bibliographical references and index.
    ISBN 978-0-470-90186-1 (cloth)
1. Motor fuels–Additives.   I. Hancsók, Jenő. 1947–   II. Title.
    TP343.H336 2013
    665.5′3827–dc23
                                                 2013023527

Printed in the United States of America

10 9 8 7 6 5 4 3 2 1

Contents

Petroleum based fuels are being used for over 100 years and the specifications of such fuels have evolved to meet the changing demands of the users. New processes have been used to convert maximum refinery streams into useful distillate fuels of acceptable quality at reasonable profits. Technically many products can be conveniently used as fuels, such as methanol, ethanol, other alcohols, gasoline, diesel, gas oil, dimethyl ether, natural gas, liquefied petroleum gas, compressed natural gas, coal derived liquid fuels, bio fuels, hydrogen and many others. However, engine technologies have developed around gasoline and diesel fuels over others. The pricing of the crude oil also favors petroleum fuels in the engines. Whenever, crude oil prices go up, alternate fuels are extensively discussed and investigated. Based on the current oil availability and prices, the use of petroleum based fuels and lubricants will continue in the current century. However, alternate fuels will find their place wherever, cost benefit analysis permits or regulations force their use. According to an IFP study, gasoline demand seems to be static and is likely to decrease in future, whereas diesel and kerosene markets will grow by 5%. It is an indication that in future, the use of hydro-conversion technologies will increase in their refineries and FCC throughput will decrease.

Environmental considerations, regulation of emission norms, energy efficiency and new engine technologies during the last twenty years have been responsible for dramatic changes in the fuels and additive quality. The dynamics of these changes and their interrelationship need to be properly understood, since the subject has now become quite complex. The present book on the "Fuel and Fuel-additives" is a unique effort to bring out these aspects. It discuss the science and technology involved in the production and application of modern conventional and alternate fuels, and fuel additives. Additives can be incorporated into fuels to improve a product's properties or to introduce new properties. Generally, they are produced synthetically and are used in low concentrations (1–500 mg/kg) in the finished product. A separate chapter on fuel additives has been incorporated, providing complete details of chemical additives used in oil industry.

This book has been jointly authored by an oil industry professional Dr. S. P. Srivastava and a chemical engineering professor Dr. Jenő Hancsók; combining both industrial and fundamental experience in their respective fields.

The book discusses the production of gasoline, diesel, aviation turbine fuel, and marine fuels, from both crude oil and alternative sources and discusses the main properties of these products in a simple language. Related environmental issues, fuel quality up gradation and application of fuel additives have been discussed in greater details.

Dr. Jenő Hancsók is grateful to Zoltán Varga, and Zoltán Eller, as well as to his postgraduate students, for their assistance in preparing the manuscript.

This book would thus be useful to all those engaged in the teaching, research, application and marketing of petroleum products.

S. P. Srivastava and Jenő Hancsók

October 2013

CHAPTER 1

Petroleum-Based Fuels – An Outlook

1.1 INTRODUCTION

Petroleum-based fuels have been used to power automotive vehicles and industrial production for well over 100 years [1–2]. Petroleum is one of the most important fuels derived fossil energy sources.

Currently, global annual energy consumption is about 12.2×10^9 tons of crude oil. Energy consumption is expected to increase to 17.5×10^9 tons of oil by 2035 [3–9]. Southeast Asia's energy demand alone will expand by about 75% by 2030 based on the strong economic growth trends in China and India [8–11]. The reserves of oil, gas, and coal that we depend on are therefore declining, and oil production is becoming ever more expensive, and causing significant environmental impact as well.

The industrial sector uses more energy than any other end-user sector, and currently it consumes about half of the world's total delivered energy [7]. Huge amounts of energy are consumed in manufacturing, mining, and construction, mainly by processing and assembly equipment but also by air conditioning and lighting. Worldwide, industrial energy consumption is expected to grow by 1.75×10^9 tons from 2010 to 2030, while transportation by about 0.6×10^9 tons and other energy consumption by about 0.8×10^9 tons during the same time period [7].

Industrial energy demand varies across countries depending on the level and mixes of economic activity and technological development. About 90% of the increase in world energy consumption is projected to occur in the non-OECD countries, where rapid economic growth is taking place. The key countries—Brazil, Russia, India, and China—will account for more than two-thirds of the growth of non-OECD industrial energy use by 2030 [7,9]. The transportation sector follows the industrial sector in world energy use, and it is of particular interest worldwide, as extensive improvements are being continually made in the quality of engine fuels.

To comply with climate change regulations, the energy sector is required to limit the long-term concentration of greenhouse gases to 450 ppm (mg/kg) of carbondioxide equivalent in the atmosphere so that the global temperature rise can be contained

Fuels and Fuel-Additives, First Edition. S. P. Srivastava and Jenő Hancsók.
© 2014 John Wiley & Sons, Inc. Published 2014 by John Wiley & Sons, Inc.

to about $2\,°C$ above the pre-industrial level [12]. In order for this target to be met, energy-related carbon dioxide emissions need to fall to 26.4 gigatonnes (ca. 26.4×10^9 tons) by year 2050 from the level at 28.9 Gt in 2009 [7,13]. Even given this outlook, fossil fuel demand will peak by year 2020.

The projected growth of energy consumption is based on the fast increase in the world population and in the standard of living. The world population was estimated to increase to about 7.03 billion (7.03×10^9) by April 2012 [14]. The fastest population growth rate (about 1.8%) were witnessed during the 1950s and then for a longer time period during the 1960s and 1970s. At this rate the world population is expected to reach about 9 billion (9×10^9) by year 2040 [4,14].

In North America and Western Europe, the automobile population has been growing roughly in parallel to the human population growth. But in the developing world, the automobile population growth is becoming almost exponential, due to effect of faster economic growth [15].

Globally, the number of vehicles on the road may reach 1 billion (10^9) by 2011 [15]. The growth is being fueled primarily by the rapidly expanding Asian market, which will see 5.7% average compound annual growth in vehicles in operation in the next three years. Asia will account for more than 23% (231 million vehicles) of global vehicles in use by 2011 [15]. Thus every seventh person in the world will have a vehicle by 2011. Europe and the Americas will account for 34% and 36% of the global share of automobiles by 2011, respectively. The Americas and Western Europe will continue to see approximately 1.3% and 2.0% compound annual growth in the next three years respectively, while Eastern Europe's vehicle population growth rate is forecasted to be 4.3% [15].

With the growth in the number of vehicles, especially passanger cars with internal combustion engines, fuels consumption has gone up significantly [9,16,17]. This has had a deleterious effect on the environment.

A large part of energy consumption is in form of engine fuels. Fuels for internal combustion engines produced from primarily sources are composed of combustionable molecules. Heat energy is a derivative of fuel's oxidation, which is converted to kinetic energy. Different gas, liquid, and solid (heavy diesel fuel, which is solid below $20\,°C$) products are usable as engine fuels [8,9,18–20]. These fuels are classified as crude oil based—namely gasoline, diesel fuels, and any other gas and liquid products [18–21]—and non-crude oil based—namely natural gas based fuels—compressed natural gas (CNG) and dimethyl-ether—biofuels, like methanol, ethanol, any other alcohols and different mixtures of them; biodiesel; biogas oil (mixtures of iso- and n-paraffins from natural trygliceries). Liquefied petroleum gases (LPG), which can be crude oil or natural gas based, and hydrogen are derivatives from different fuel sources [22–38].

Over the years fuel specifications have evolved considerably to meet the changing demands of engine manufacturers and consumers [20,39–42]. Both engines [43–45] and fuels [9,20,39,41,42] have been improved due to environmental and energy efficiency considerations. New processes have been developed to convert maximum refinery streams into useful fuels of acceptable quality at reasonable refinery margins [8,9,46,47].

Gasoline and diesel fuels have been preferred [20] in the development of engine technology. The price of crude oil is also often at a level that makes petroleum–based fuels in engines desirable for economic reasons [46,47,53–54]. Whenever crude oil prices do rise, the issue of alternative fuels comes up [9,22–27,48–51], but the discussions and investigations get dropped out soon after crude oil prices settle down [28]. The oil crises of the 1970s and 2008 reflect this tendency. However, oil is not going to last forever, and it is also not going to be exhausted in the near future [5,7,9,10]. So, while the use of petroleum-based fuels and lubricants may continue in the current century, it is likely that a significant decrease will occur after crude oil usage peaks [3,7,10,11,52].

The application of alternative fuels will find a place wherever cost–benefit analyses permit or wherever regulations force their use. (The use of compressed natural gas in all of New Delhi's city transportation vehicles is an example. This was decreed by the Supreme Court of India and is now being enforced in the other cities of India as well.)

While world gasoline demand is expected to be static and possibly to decrease in future, the consumption of diesel and kerosene, the rail and water transport fuel, will likely expand 1,3% to 15% by 2030 [9,16,17]. With the demand for heavy fuel oil expected to decrease [9,16,17], heavy fuel oil is being converted into lighter products such as LPG, gasoline, and diesel [53–55].

The world demand for middle distillate fuel, mainly diesel oil and heating fuel, will grow faster than that for any other refined oil products toward 2030 [9,56]. Globally, new car fleets are shifting to diesel from gasoline, and therefore the demand for middle distillates will grow and account for about 60% of the expected 20 million barrels per day (bpd) (2.66×10^6 t/day) rise in global oil production by 2030. In 2008, the difference in demand between gasoline and diesel was around 3 million bpd (0.4×10^6 t/day). By 2020, the projected gas oil/diesel demand is 6.5 million bpd (0.9×10^6 t/day), higher than for gasoline, and by 2030, the difference exceeds 9 million bpd (1.2×10^6 t/day). The expected global demand for diesel and gas oil (mainly used for heating) will grow to 34.2 million bpd (4.5×10^6 t/day) by 2030 from 24.5 million bpd (3.3×10^6 t/day) in 2008. Gasoline demand will rise to about 25.1 million bpd (2.9×10^6 t/day) by 2030 from 21.4 million bpd (2.5×10^6 t/day) in 2008. Jet fuel/kerosene demand will rise to 8.1 million bpd (1.0×10^6 t/day) from 6.5 million bpd (0.8×10^6 t/day), while residual fuel demand, used as a refinery feedstock and as marine fuel, will fall to 9.4 million bpd from 9.7 million bpd.

The United States accounts for most of the world's gasoline demand, whereas in Europe the demand for diesel is increasing [4,9,42] due to the rising number of diesel vehicles [15]. India and several other Asian countries also consume more diesel fuel than gasoline [4,49]. With the development of more fuel-efficient diesel vehicles, the demand for diesel will increase significantly and its use might have to be restricted to meet future demand. However, gasoline-powered engines seem still to be favorable for hybrid vehicles.

Among the primary energy carriers, in absolute terms, coal demand will increase to the highest rate, followed by gas and oil over the projected time period [4]. Nevertheless, oil still remains the largest single fuel source by 2030, even if its share drops from the present 34% to about 30% (in 2010 ca. 92.3% and in 2030 ca. >85% fuels from petroleum) [4,11].

The projected worldwide crude oil demand reflects the leading role of crude oil in engine fuel production. The global total crude oil demand in 2011 was 4.05×10^9 t, and this is expected to increase to 4.4×10^9 t/year by 2015, and to 5.25×10^9 t/year by 2030 [25–27]. Non-OECD countries contribute to the increase by more than 90%; China and India account for more than 50% alone [4,49].

Among the alternative fuels, the biofuels (e.g., first-generation ethanol and biodiesel, second-generation bioethanol from lignocelluloisic materials and the biogas oil blends of iso- and n-paraffins processed from natural triglycerides) [40,51,57,58] will become an increasingly important unconventional source of the liquid fuel supply, likely reaching around 5.9 million barrels per day by 2030. Particularly strong growth in biofuel consumption is projected for the United States [58], where the production of biofuels could increase from 0.3 million barrels per day of the 2006 level to 1.9 million barrels per day by 2030 [58,59]. The Energy Independence and Security Act passed in 2007 has made the use of biofuels compulsory in the United States. Sizable increases in biofuels production are projected for other regions as well: the OECD countries with 10% biofuel energy by 2020 [60], non-OECD countries in Asia [49], and Central and South America [58]. The total biofuel production of the world was ca. 60 million tons oil equivalent. For bioethanol, in 2012, Middle and South America: ca. 12.1 Million tons oil equivalent, Europe and Eurasia: ca. 2.2 million tons oil equivalent and North America: ca. 25.7 million tons oil equivalent and production of other countries was ca. 2.0 million tons oil equivalent. For Biodiesel, Middle and South America: ca. 4.5 million tons oil equivalent, Europe and Eurasia: 7.6 million tons oil equivalent and North America: ca. 2.8 million tons oil eq. (BP Statistical review 2013), and production of other countries was ca. 3.1 million tons oil equivalent.

1.2   ENVIRONMENTAL ISSUES

Environmental pollution from fossil fuel use has long been a major government concern, and over the past 20 to 30 years, attempts have been made to control pollution by improving the quality of fuels and lubricants. The combustion of fossil fuels leads to the formation of CO_2, CO, unburned hydrocarbons, NO_x, SO_x, soot, and particulate matter. Liquid fuels, by nature, are volatile and produce volatile organic compounds (VOC), as emitted especially by gasoline. From time to time different countries have attempted to enforce regulations to minimize these harmful emissions. Their legislators have prompted major inter industry cooperation to improve fuel quality, lubricant quality, and engine/vehicle designs (Figure 1.1) [39].

Environmental issues thus drive the development of modern fuels, engines, and lubricants. The advances in these industries are interrelated, although biofuel development has the strongest linkages to environmental concerns, binding government legislation, engine development, exhaust treatment catalysts, tribology, and the fuel economy [39,60].

Greenhouse gas emission, global warming, and climate change are other important issues related to fuels and lubricants. About 10,000 years passed between the last

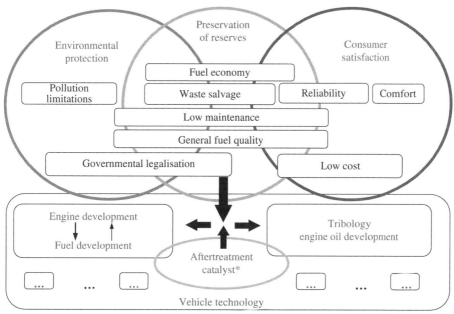

*Including particulate filter

FIGURE 1.1   The mechanism of the development of vehicles and fuels

Ice Age and the Industrial Revolution. During that time period the atmospheric CO_2 level varied by only about 5% [12,13]. From the start of the Industrial Revolution to 2030, in about 150 years, the amount of atmospheric CO_2 will have doubled [7,12,13]. World carbon dioxide emissions are projected to rise from 28.8 billion (28.8×10^9) metric tons in 2007 to 33.1 billion (33.1×10^9) metric tons by 2015 and 40.4 billion (40.4×10^9) metric tons by 2030—an increase of 39% over the examined period [7,12,13]. The only way to reduce carbon dioxide emissions is to reduce the consumption of hydrocarbon fuels and/or improve the energy efficiency of engines and equipment using hydrocarbons. The biggest single contributor to the rise in greenhouse gases is the burning of fossil fuels. Since in hydrocarbon fuels the CO_2 emission is proportional to the amount of energy produced, a reduction in energy consumption will reduce the CO_2 emission as well [12,13]. Improving energy efficiency is the first step in reducing carbon dioxide emission. This calls for a combined engineering effort to introduce more efficient fuels, better power systems, and new materials and processes.

Through coordinated action, it may be possible to lower the long-term concentration of greenhouse gases in the atmosphere to around 450 ppm (mg/kg) of the carbondioxide equivalent. This would correspond to the global temperature goal of environmentalists of not exceeding the 2 °C rise of the pre-industral period temperatures. To meet this target, energy-related carbon dioxide emissions must be lowered to around 26.4 gigatonnes (26.4×10^9 tons) by 2030 from the 28.8 gigatonnes (28.8×10^9 tons) of 2007 [7].

1.3   CLASSIFICATION OF FUELS

Engine fuels can be any liquid or gaseous hydrocarbons used for the generation of power in an internal combustion engine. There are several materials that can be used in the internal combustion engine either alone or blended as a component. These materials are classified as follows [19]:

- Drivetrains:
  - Otto engines (gasolines, PB, CNG, ethanol, etc.)
  - Diesel engines (diesel gas oils, CNG, dimethyl-ether, etc.)
- Origin:
  - Produced from exhaustible energy carriers,
  - Produced from renewable energy carriers (biofuels based on biomass)
- Number of feedstock resources:
  - One resource (e.g., fatty acid methyl esters from only triglyceride and fatty acid containing feedstocks)
  - Multiple resources (e.g., ethanol; from sugar crops, from crops containing starch, lignocellulose, hydration of ethylene)

Alternative fuels are those fuels that are other than gasoline or gas oil derived from petroleum. The main types of motor fuels are shown in Figure 1.2. [61].

The choice of fuel to use depends on the engine design, availability of the energy source, environmental protection issues, energy policy, safety technology, human biology, the aftertreater catalytic system, lubricants, additives, economy, traditions, and so forth [62,106]. The fuel industry categorizes the different types of fuels as follows:

Gasoline   A volatile mixture of liquid hydrocarbons generally containing small amount of additives suitable for use as a fuel in a spark-ignition internal combustion engine.

Unleaded gasoline   Any gasoline to which no lead have been intentionally added and which contains not more than 0.013 gram lead per liter (0.05 g lead/US gal).

E85 fuel   A blend of ethanol and hydrocarbons in gasoline with 75–85% of ethanol. E85 fuel ethanol must meet the most recent standard of a region or country.

M85 fuel   A blend of methanol and hydrocarbons where the methanol is nominally 70% to 85%.

Racing gasoline   A special automotive gasoline that is typically of lower volatility, has a narrower boiling range, a higher antiknock index, and is free of significant amounts of oxygenates. It is designed for use in racing vehicles, which have high compression engines.

Aviation gasoline   A fuel used in an aviation spark-ignition internal combustion engine.

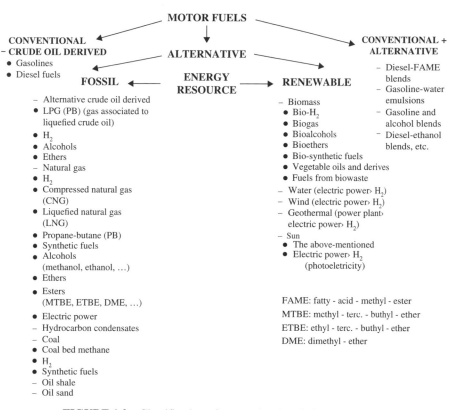

FIGURE 1.2 Classification of conventional and alternative fuels

Petroleum gases (LPG) Gas phase hydrocarbons, mainly C_3 and in low quantity C_4. Their quality is determined by the country or regional standards.

Compressed natural gas (CNG) Predominantly methane compressed at high pressures suitable as fuel in internal combustion engine.

Aviation turbine fuel A refined middle distillate suitable for use as a fuel in an aviation gas turbine engine.

Diesel fuel A middle distillate from crude oil commonly used in internal combustion engines where ignition occurs by pressure and not by electric spark.

Low or ultra-low sulfur diesel (ULSD) Diesel fuel with less than 50 and 10 mg/ kg respectively.

Biodiesel A fuel based on mono-alkyl esters of long-chain fatty acids derived from vegetable oils or animal fats. Biodiesel containing diesel gas oil is a blend of mono-alkyl esters of long chain fatty acids and diesel gas oil from petroleum. A term B100 is used to describe neat biodiesel used for heating, which does not contain any mineral oil based diesel fuel.

Biogas oil   Mixture of iso-paraffins and normal-paraffins, produced by catalytic hydrogenation of triglyceride-containing feedstocks.

Ethanol/gasoil(/biodiesel) Emulsions   A fuel that contains at a minimum 80% diesel gas oil. Stability of the emulsion is assured with additives and sometimes with biodiesel too.

Bunker oil   Used for marine ships.

Some *other alternate fuels* are less dispersed, such as dimethyl ether (DME) and hydrogen. These fuels are discussed in chapter 4.

Among the engine fuels, *fuel oil belongs to the energy products of the oil industry*. Fuel oils can be refined middle distillates (heating oils), heavy distillates, or residues after atmospheric and vacuum distillation, and a blend of these is also suitable for use as a fuel for heating or power generation.

There are several standards for fuels. This book mainly concentrates on those fuels that require additives to improve the performance characteristics of the base fuel.

REFERENCES

1. Favennec J.-P. (2001). *Petroleum Refining: Refinery Operation and Management*. Editions Technip, Paris.
2. Maugeri L. (2006). *The Age of Oil: The Mythology, History, and Future of the World's Most Controversial Resource*. Praeger, New York.
3. Hirsch, R. L. (2008). Elements of the World's energy future. Af&V Conference, May 12.
4. *BP Statistical Review of World Energy*, June 2010 and 2011.
5. Radler, M., Bell. L. (2010). US energy demand set for slim expansion in 2010. *Oil Gas J.*, 108(2), 20–29.
6. Radler, M., Bell, L. (2011). US energy demand growth slows on economic sluggishness. *Oil Gas J.*, 109(14), 24–38.
7. *Shell Energy Scenarios to 2050*, Shell International BV, 2011.
8. Europia (European Petroleum Industry Association), *Annual Report* (2010). Wood Mackenzie, Cambridge, UK.
9. Europia (European Petroleum Industry Association), *White Paper on EU Refining* (2010). Wood Mackenzie, Cambridge, UK.
10. Brower D. (2011). An alarming global energy outlook. *Petrol. Econ.*, 78(10), 4–7.
11. Thinnes B. (2012). BP's energy outlook to 2030. *Hydrocarb. Proc.*, 91(3), 17–18.
12. European Environmental Agency (2007). Greenhouse gas emission trends and projections in Europe 2007: Tracking progress towards Kyoto targets. Report N85, Copenhagen.
13. International Energy Agency (2011). CO_2 emissions from fuel combustion. IEA, Paris.
14. US Cencus Bureau (2012), Washington, DC.
15. *Shell PKW-Szenarien bis 2030 (Fakten, Trends und Handlungsoptionen für nachhaltige Auto-Mobilitat)*, Shell Deutschland Oil GmbH, Hamburg, 2009.
16. 2011 HPI Market Data (2011). Executive summary. *Hydrocarb. Proc.*, 90(1), 21–26.

17. Radler, M., Bell, L. (2010). US energy demand set for slim expansion in 2010, *Oil Gas J.*, 108(2), 20–29.

18. Hancsók, J. (1997). Korszerű Motor- és Sugárhajtómű Üzemanyagok: I. Motorbenzinek. Veszprém, University of Pannonia, Hungary.

19. Hancsók, J. (1999). Korszerű Motor- és Sugárhajtómű Üzemanyagok: II. Dízelgázolajok. Veszprém, University of Pannonia, Hungary.

20. Lucas, G. A. (2000). *Modern Petroleum Technology*. Wiley, Chichester.

21. Hancock, E. G. (1985). *Technology of Gasoline*. Society of Chemical Industry, Oxford.

22. Cammack, R., Frey, M., Robson, R., eds. (2001). *Hydrogen as a Fuel*. Taylor & Francis, London.

23. Mittelbach, M., Remschmidt, C., eds. (2004). *Biodiesel the Comprehensive Handbook*. Mittelbach, Graz.

24. Olah, G. A., Goeppert, A., Prakash, S., eds. (2006). *Beyond Oil and Gas: The Methanol Economy*, Wiley-VCH Verlag, Weinheim.

25. Sunggyu, L., Speight, J. G., Loyalka, S. K., eds. (2007). *Handbook of Alternative Fuels Technologies*. Taylor & Francis, London.

26. Kamm, B., Gruber, P. R., Kamm, M., eds. (2006). *Biorefineries—Industrial Processes and Products*. Wiley-VCH Verlag, Weinheim.

27. Minteer, S. (2006). *Alcoholic Fuels*. CRC Press, Taylor & Francis, London.

28. McGowan, T. F. (2009). *Biomass and Alternative Fuel Systems*. Wiley, Hoboken, NJ.

29. US Government (2008). *21st Century Complete Guide to Natural Gas Vehicles—NGVs, Compressed Natural Gas (CNG), Liquefied Natural Gas (LNG), Autogas, Engines, Infrastructure* (CD-ROM). Progressive Management, February 1.

30. Thrän, D. (2012). European biomethane potentials. Workshop on Biomethane Trade, Brussels, February 21.

31. IGU—International Gas Union. (2005). *Global Opportunities for Natural Gas as a Transportation Fuel for Today and Tomorrow*. Fornetu, Norway.

32. The petroleum economist. *Energy Magazine: LNG Rev.*, 2011.

33. DOE/NETL (2009). *Affordable, Low-Carbon Diesel Fuel from Domestic Coal and Biomass*, January 14. Government Printing Office, Washington, DC.

34. Bartis, T. J., Camm, F., Ortiz, S. D., eds. (2008). *Producing Liquid Fuels from Coal Prospects and Policy Issues*.

35. Toman, M., Curtright, E. A., Ortiz, S. D., Darmstadter, J., Shannon, B., eds. (2008). *Unconventional Fossil-Based Fuels*.

36. Honkanen, S. (2011). Hydrotreated vegetable oil, comparative analyses of automobile fuel sources. 20th World Petroleum Congress, December, Doha Quatar.

37. Hancsók, J., Krár, M., Magyar, Sz., Boda, L., Holló, A., Kalló, D. (2007). Investigation of the production of high cetane number biogasoil from pre-hydrogenated vegetable oils over Pt/HZSM-22/Al$_2$O$_3$. *Microporous Mesoporous Materials*, 101(1–2), 148–152.

38. Hancsók, J., Varga, Z., Kovács, S., Kasza, T. (2011). Comparing the biodiesel and biogasoil production from different natural triglycerides.*Proceedings of 2nd Eurpoean Conference of Chemical Engineering (ECCE'11)*. December 10–12, Puerto de La Cruz, Tenerife, Spain.

39. Hancsók, J., Kasza, T. (2011). The importance of isoparaffins at the modern-engine fuel production. *Proceedings of 8th International Colloquium Fuels*, January 19–20, Stuttgart/Ostfildern, Germany.

40. Röj, A. (2011). Requirements on present and future fuels and fuel quality—An automotive view (VOLVO). *Proceedings of 5th International Biodiesel Conference*, October 6–7, Berlin, Germany.

41. Szalkowska, U. (2009). Fuel quality—global overview. *Proceedings of 7th International Colloquium Fuels—Mineral Oil Based and Alternative Fuels*, January 14–15, Stuttgart/ Ostfildern.

42. *World Wide Fuel Charter*, 4th edn, January 18–19, 2006. Cambodia.

43. Basshuysen, R., Schäfer F., eds. (2004). *Internal Combustion Engine Handbook*. SAE International, Troy, MI.

44. Xin Q. (2011). *Diesel Engine System Design*. Woodhead Publishing, Oxford.

45. Langen, P., Nehse, C. W. (2010). Vision 2000—The combustion engine on a dead end road, 31st International, Vienna Motor Symposium, April, 29–30, Vienna.

46. Energy Information Administration, 2010. and 2011.

47. International Energy Agency (2011). *Oil Market Report*. IEA, Paris.

48. Beddoes, L. (2011). EU energy pathways to 2050: What role for the EU oil refining industry? *Proceedings of Interfaces'11*, September 28–30.

49. Woo, C. (2012). Transportation and alternative fuels in Asia. *Hydrocarb. Proc.*, 91(2), 44–48.

50. Block, M. (2011). Opportunities in the biodiesel industry for bio-jet, European Biodiesel Conference, ACI, Rotterdam, June 15–16.

51. Blommel, P. (2011). Conversion of carbohydrates to renewable diesel. European Biodiesel Conference, ACI, Rotterdam, June 15–16.

52. Towler, B. (2011). World peak oil production still years away. *Oil Gas J.*, 109(18), 90–97.

53. Weber, C. Yeung, S. (2011). Trend sin opportunity crude processing, *PTQ*, (3), 5–13.

54. Wedlake, D. A., Elliot, J.N. (2011). Paying residues. *Hydrocarb. Eng.*, 16(6), 35–38.

55. Mothagi, M., Shree, K., Krishnamurthy, S. (2010). Consider new methods for bottom of the barrel processing—Part 1. *Hydrocarb Proc.*, 89(2), 35–40.

56. Robertson, H. (2011). Biofuel here to stay, *Petrol. Econ.*, 78(8), 34–35.

57. Rabobank Industry Note 303. (2012). The future of ethanol; Brazilian and US perspectives. February 2012. Utrecht.

58. Rabobank Industry Note 308. (2012). Recalculating the route; Can the EU meet its renewable energy targets for land transport by 2020? Utrecht.

59. Flach, B., Lieberz, S., Bendz, K., Dahlbacka, B. (2011). *EU-27 Annual Biofuels Report*. Gain Report Number NL1013, Brussels.

60. Hancsók, J., Auer, J., Baladincz, J., Kocsis, Z., Bartha, L., Bubálik, M., Molnár, I. (2005) Interactions between modern engine oils and reformulated fuels. *Petrol. Coal*, 47(2), 55–64.

61. Hancsók, J. (2004). *Korszerű motor és sugárhajtómű üzemanyagok. III. Alternatív motorhajtóanyagok*. Veszprém University Press, Veszprém.

CHAPTER 2

Emission Regulation of Automotive Vehicles and Quality of Automotive Fuels

Four parameters influence the emissions from automotive vehicles: engine design, vehicle design, fuel, and lubricant quality. Automotive emissions can be regulated directly (by emission standards) or indirectly by directives (predetermined data for 5–20 years), fuel quality standards, government regulations, engine oil specifications such as low-SAPS (SAPS: sulphated ash, phosphorus and sulfur), and regulation on utilization of aftertreatment catalysts [1].

Over the last two decades tighter emission norms and strict fuel quality specifications have been worked out to improve all four parameters. Currently, emission standards on fuel quality and vehicle technology follow the year 2000 recommendations of the Auto Oil Programmes in Europe, the United States, and Japan [2–7].

2.1 DIRECT REGULATION OF EMISSIONS

Compounds that contaminate the environment are formed in internal combustion engines by the oxidation of hydrocarbon-type fuels. The key contaminants are carbon-dioxide, carbon-monoxide, hydrocarbons, nitrogen-oxides, particles, and sulfuric compounds—SO_2—aldehydes. Their damaging effects manifest as acid rain, the greenhouse effect, destruction of ground and atmospheric ozone layers, respiratory diseases, and soil and water pollution, among other deleterious effects to life on earth [1].

Recently the contribution of carbon dioxide emissions from deforestation has attained importance, since the removal of CO_2 consuming trees contributes further to the carbon dioxide imbalance in the atmosphere. Carbon dioxide molecules trap energy from the sun (the greenhouse effect) and thus cause global warming. Current estimates of anthropogenic carbon dioxide show the increases to be about

Fuels and Fuel-Additives, First Edition. S. P. Srivastava and Jenő Hancsók.
© 2014 John Wiley & Sons, Inc. Published 2014 by John Wiley & Sons, Inc.

0.5% per annum [8–11]. Emission of carbon dioxide from vehicles can only be reduced by lowering the consumption of fuels from fossil energy sources.

Reductions in the emissions of CO_2, CO, NO_x and, particle matter, for example, have been possible with engine design changes combined with fuel quality upgrading. A major advancement in engine hardware has been the development of a variable geometry turbocharger with inter cooling, ultra high injection pressures of 1500 bars and higher (common rail injection technology), electronic fuel injection and control, a multi-valve with swirl and variable valve timing, an exhaust gas recycle with temperature management, onboard diagnostics (OBD), catalytic exhaust converters (aftertreatment systems), lean de-NO_x catalyst and regenerative diesel particulate filters [1,12,13]. Newer technologies are still emerging and being tested to reduce fuel consumption, emissions and improve engine performance.

About 21.5% of the anthropogenic carbon dioxide emission in 2009 can be attributed to the transport sector [8–11]. To comply with Kyoto Protocol, the European Union has voluntarily decided to cut down CO_2 emission from the transport sector by one-third. Key to reducing CO_2 emissions is to cut fuel consumption, since the final combustion products of hydrocarbons are always carbon dioxide and water. It has been estimated that a vehicle consuming 3 L/100 km will achieve a 25% CO_2 reduction [2–4].

The European Parliament voted to adopt a Regulation on CO_2 for cars based on a proposal by the EU Commission in December 2008. Some highlights of the text follow [14]:

Limit value curve: The fleet average to be achieved by all **new** cars registered in the European Union is 130 grams per kilometer (g/km). A limit value curve is used to allow heavier cars higher emissions than lighter cars, while preserving the overall fleet average.

Phasing-in of requirements: In 2012, 65% of each manufacturer's newly registered cars must comply, on average, with the limit value curve set by the legislation. This will rise to 75% in 2013, 80% in 2014, and 100% from 2015 onward.

Lower penalty payments for small excess emissions until 2018: A manufacturer whose fleet exceeds the CO_2 limit value in any year after 2012 will have to pay an excess emissions premium for each registered car. This premium will amount to €5 for exceeding the first g/km, €15 for the second g/km, €25 for the third g/km, and €95 for each subsequent g/km. After 2019, the cost of exceeding the first g/km of will be €95.

Long-term target: A target of 95 g/km is specified for the year 2020. The modalities for reaching this target and the implementation details, including the excess emissions premium, will be defined in a review by 2013.

Eco-innovations: The test procedures used for vehicle type approval are outdated. Revised standards are to be completed by 2014. To demonstrate the CO_2-reducing effects of innovative technologies, an interim procedure grants manufacturers a maximum of 7 g/km of emission credits on average for their fleet if they equip vehicles with innovative technologies, based on independently verified data.

Many countries have introduced their own measures to control automotive exhaust emissions. Some important decisions are provided in the following pages.

2.1.1 Emission Standards in Europe

The stages are typically referred to as Euro 1, Euro 2, Euro 3, Euro 4, Euro 5, and Euro 6 fuels for light duty vehicle standards. The corresponding series of standards for heavy duty vehicles use roman numbers (Euro I, II, III, IV, V, and VI).

The legal framework consists of a series of Directives, each amendment to the 1970 Directive 70/220/EEC. A summary of the standards and the relevant EU directives providing the definition of the standard is provided below:

Euro 1 (1993): For passenger cars—91/441/EEC (EEC: European Economic Community)

Also for passenger cars and light trucks—93/59/EEC.

Euro 2 (1996): For passenger cars—94/12/EC (and 96/69/EC)

Euro 3 (2000): For any vehicle—98/69/EC

Euro 4 (2005): For any vehicle—98/69/EC (and 2002/80/EC)

Euro 5 (2008/9) and Euro 6 (2014): For light passenger and commercial vehicles—2007/715/EC

These limits supersede the original Directive on emission limits 70/220/EEC.

In the area of fuels, the 2003/30/EC Directive [15] required that 5.75% of all transport fossil fuels (petrol and diesel) should be replaced by biofuels by December 31, 2010, with an intermediate target of 2% by the end of 2005. However, MEPS (Members of the European Parliament) have since voted to lower this target in the wake of new scientific evidence about the sustainability of biofuels and the impact on food prices. In a vote in Strasbourg, the European Parliament's environment committee supported a plan to reduce the EU target for renewable sources in transport to 4% by 2015 and a thorough review to bring the target to the 8–10% mark by 2020 [16].

Emission Standards for Passenger Cars [17–19] Emission standards for passenger cars and light commercial vehicles (vehicle categories M_1 and N_1, respectively) are summarized in the Tables 2.1 through 2.4. Since the Euro 2 stage, EU regulations have introduced different emission limits for diesel and gasoline vehicles. Diesels have more stringent CO standards but are allowed higher NO_x. Gasoline vehicles are exempted from PM standards through the Euro 4 stage, but vehicles with direct injection engines will be subject to a limit of 0.005 g/km for Euro 5 and Euro 6.

Euro 5/6 regulations introduce PM mass emission standards, numerically equal to those for diesel and gasoline cars with DI engines. All dates listed in the tables refer to new type approvals. The EC Directives also specify a second date one year later (unless indicated otherwise) that applies to first registration of existing and previously type-approved vehicle models.

TABLE 2.1 European emission standards for passenger cars (category M$_1$): gasoline

Emission Stage	Implementation Date	Emission (g/km)				
		CO	HC	HC+NO$_x$	NO$_x$	PM
Euro 1	July 1992	2.72 (3.16)	—	0.97(1.13)	—	—
Euro 2	January 1996	2.2	—	0.50	—	—
Euro 3[a]	January 2000	2.3	0.20	—	0.15	—
Euro 4[b]	January 2005	1.0	0.10	—	0.08	—
Euro5[c]	September 2009[d]	1.0	0.10[e]	—	0.06	0.005[f, g]
Euro 6[c]	September 2014	1.0	0.10[e]	—	0.06	0.005[f, g]

Note: At the Euro 1–4 stages, passenger vehicles >2500 kg were type approved as category N$_1$ vehicles. Values in brackets are in conformity with production (COP) limits. The dash means no regulation.

[a] Euro 3 stage—80,000 km or 5 years, whichever occurs first.
[b] Euro 4 stage—100,000 km or 5 years, whichever occurs first.
[c] Euro 5/6 stage—in-service conformity: 100,000 km or 5 years; durability testing of pollution control devices for type approval: 160,000 km or 5 years (whichever occurs first).
[d] 2011.01 for all models.
[e] NMHC = 0.068 g/km (NMHC: nonmethane hydrocarbon).
[f] Applicable only for vehicles using DI engines.
[g] 0.0045 g/km using the PMP measurement procedure (PMP: Particulate Measurement Program).

Emission Standards for Lorries and Buses [17–19]   Whereas for passenger cars, the standards are defined in g/km, these are defined by engine power, g/kWh for Lorries (trucks), and are therefore in no way comparable. Tables 2.5 and 2.6 provide summaries of the emission standards and their implementation dates. Dates in the tables refer to new type approvals; the dates for all type approvals are in most cases one year later (EU type approvals are valid for longer than one year). The official category name is heavy-duty (HD) diesel engines, which generally includes trucks and buses.

2.1.2 US (EPA) Emission Standards [20,21]

In the United States, emissions standards are managed by the Environmental Protection Agency (EPA) (Tables 2.7 through 2.11). The state of California has special dispensation to promulgate more stringent vehicle emissions standards, and other states may choose to follow either the national or Californian standards (Tables 2.15 through 2.25). California's emissions standards are set by the California Air Resources Board, known locally by its acronym CARB.

Clean Fuel Fleet Program   Table 2.12 shows a voluntary Clean Fuel Fleet (CFF) emission standard. It is a federal standard applied to 1998 to 2003 model year engines, both CI and SI, over 8500 lbs GVWR. In addition to the CFF standard, vehicles must meet applicable conventional standards for other pollutants.

TABLE 2.2 European emission standards for passenger cars (category M_1): Diesel

Emission Stage	Implementation Date	Emission (g/km)				
		CO	HC+NO_x	NO_x	PM	PN* (#/km)
Euro 1	July 1992	2.72(3.16)	0.97(1.13)	—	0.14(0.18)	—
Euro 2—IDI**	January 1996	1.00	0.70	—	0.08	—
Euro 2—DI***	January 1999[a]	1.00	0.90	—	0.10	—
Euro 3	January 2000	0.64	0.56	0.50	0.05	—
Euro 4	January 2005	0.50	0.30	0.25	0.025	—
Euro 5[a]	September 2009[b]	0.50	0.23	0.18	0.005[d]	—
Euro 5[b]	September 2011[c]	0.50	0.23	0.18	0.005[d]	6.0×10^{11}
Euro 6	September 2014	0.50	0.17	0.08	0.005[d]	6.0×10^{11}

Note: At the Euro 1–4 stages, passenger vehicles > 2500 kg were type approved as category N_1 vehicles. Values in brackets are in conformity with production (COP) limits. * PN—Particle number limit, ** IDI—indirect injection, *** DI—direct injection. The dash means no regulation.

[a] Until 1999.09.30 (after that date DI engines must meet the IDI limits).

[b] 2011.01 for all models.

[c] 2013.01 for all models.

[d] 0.0045 g/km using the PMP measurement procedure.

TABLE 2.3   EU emission standards for light commercial vehicles (category N_1): gasoline

Category	Stage	Date	Emission (g/km)				
			CO	HC	HC+NO_x	NO_x	PM
Class I	Euro 1	October 1994	2.72	—	0.97	—	—
(≤1305 kg)	Euro 2	January 1998	2.2	—	0.50	—	—
	Euro 3	January 2000	2.3	0.20	—	0.15	—
	Euro 4	January 2005	1.0	0.10	—	0.08	—
	Euro 5	September 2009[a]	1.0	0.10[c]	—	0.06	0.005[f,g]
	Euro 6	September 2014	1.0	0.10[c]	—	0.06	0.005[f,g]
Class II	Euro 1	October1994	5.17	—	1.40	—	—
(1305–1760 kg)	Euro 2	January 1998	4.0	—	0.65	—	—
	Euro 3	January 2001	4.17	0.25	—	0.18	—
	Euro 4	January 2006	1.81	0.13	—	0.10	—
	Euro 5	September 2010[b]	1.81	0.13[d]	—	0.075	0.005[f,g]
	Euro 6	September 2015	1.81	0.13[d]	—	0.075	0.005[f,g]
Class III	Euro 1	October1994	6.90	—	1.70	—	—
(1760–3500 kg)	Euro 2	January 1998	5.0	—	0.80	—	—
	Euro 3	January 2001	5.22	0.29	—	0.21	—
	Euro 4	January 2006	2.27	0.16	—	0.11	—
	Euro 5	September 2010[b]	2.27	0.16[e]	—	0.082	0.005[f,g]
	Euro 6	September 2015	2.27	0.16[e]	—	0.082	0.005[f,g]

Note: For Euro 1 and 2 the category N_1 reference mass classes were class I≤1250 kg, class II 1250–1700 kg, class III>1700 kg. The dash means no regulation.
[a] 2011.01 for all models.
[b] 2012.01 for all models.
[c] NMHC=0.068 g/km.
[d] NMHC=0.090 g/km.
[e] NMHC=0.108 g/km.
[f] Applicable only to vehicles using DI engines.
[g] 0.0045 g/km using the PMP measurement procedure.

Model Year 2004 and Later   In October 1997 the EPA adopted new emission standards for model year 2004 and later heavy-duty diesel truck and bus engines. These standards reflect the provisions of the Statement of Principles (SOP) signed in 1995 by the EPA, California ARB, and the manufacturers of heavy-duty diesel engines. The goal was to reduce NO_x emissions of highway heavy-duty engines to levels approximately 2.0 g/bhpv·h beginning in 2004. Manufacturers have the flexibility to certify their engines to one of the two options shown in Table 2.13. All emission standards other than NMHC and NO_x applying to 1998 and later model year heavy duty engines will continue at their 1998 levels.

EPA established revised useful engine lives, with significantly extended requirements for the heavy heavy-duty diesel engine service class, as follows:

LHDDE—110,000 miles or 10 years

MHDDE—185,000 miles or 10 years

HHDDE—435,000 miles or 10 years or 22,000 hours of travel

TABLE 2.4 EU emission standards for light commercial vehicles (category N_1): Diesel

Category	Stage	Date	Emission (g/km)				
			CO	HC+NO$_x$	NO$_x$	PM	PN* (#/km)
Class I (<1305 kg)	Euro 1	October 1994	2.72	0.97	—	0.14	—
	Euro 2—IDI**	January 1998	1.0	0.70	—	0.08	—
	Euro 2—DI***	January 1998[a]	1.0	0.90	—	0.10	—
	Euro 3	January 2000	0.64	0.56	0.50	0.05	—
	Euro 4	January 2005	0.50	0.30	0.25	0.025	—
	Euro 5a	September 2009[b]	0.50	0.23	0.18	0.005[e]	—
	Euro 5b	September 2011[d]	0.50	0.23	0.18	0.005[e]	6.0×10^{11}
	Euro 6	September 2014	0.50	0.17	0.08	0.005[e]	6.0×10^{11}
Class II (1305–1760 kg)	Euro 1	October 1994	5.17	1.40	—	0.19	—
	Euro 2—IDI**	January 1998	1.25	1.0	—	0.12	—
	Euro 2—DI***	January 1998[a]	1.25	1.3	—	0.14	—
	Euro 3	January 2001	0.80	0.72	0.65	0.07	—
	Euro 4	January 2006	0.63	0.39	0.33	0.04	—
	Euro 5a	September 2010[c]	0.63	0.295	0.235	0.005[e]	—
	Euro 5b	September 2011[d]	0.63	0.295	0.235	0.005[e]	6.0×10^{11}
	Euro 6	September 2015	0.63	0.195	0.105	0.005[e]	6.0×10^{11}
Class III (1760–3500 kg)	Euro 1	October 1994	6.9	1.70	—	0.25	—
	Euro 2—IDI**	January 1998	1.5	1.20	—	0.17	—
	Euro 2—DI***	January 1998[a]	1.5	1.60	—	0.20	—
	Euro 3	January 2001	0.95	0.86	0.78	0.10	—
	Euro 4	January 2006	0.74	0.46	0.39	0.06	—
	Euro 5a	September 2010[c]	0.74	0.350	0.280	0.005[e]	—
	Euro 5b	September 2011[d]	0.74	0.350	0.280	0.005[e]	6.0×10^{11}
	Euro 6	September 2015	0.74	0.215	0.125	0.005[e]	6.0×10^{11}

Note: At the Euro 1–4 stages, passenger vehicles >2500 kg were type approved as category N_1 vehicles. *PN—Particle number limit, **IDI—indirect injection, ***DI—direct injection. The dash means no regulation.

[a] Until 1999.09.30 (after that date DI engines must meet the IDI limits).

[b] 2011.01 for all models.

[c] 2012.01 for all models.

[d] 2013.01 for all models.

[e] 0.0045 g/km using the PMP measurement procedure.

TABLE 2.5   EU emission standards for HD diesel engines, g/kWh (Category N$_3$, >12,000 kg)

Tier	Date	Test Cycle	CO	HC	NO$_x$	PM	Smoke (m^{-1})
Euro I	1992, < 85 kW	ECE R-49	4.5	1.1	8.0	0.612	
	1992, > 85 kW		4.5	1.1	8.0	0.36	
Euro II	October 1996		4.0	1.1	7.0	0.25	
	October 1998		4.0	1.1	7.0	0.15	
Euro III	October 1999 EEVs only	ESC & ELR	1.0	0.25	2.0	0.02	0.15
	October 2000	ESC & ELR	2.1	0.66	5.0	0.100.13^a	0.8
Euro IV	October 2005		1.5	0.46	3.5	0.02	0.5
Euro V	October 2008		1.5	0.46	2.0	0.02	0.5

Note: EEV is "Enhanced environmentally friendly vehicle."
^{a} for engines of less than 0.75 dm^3 swept volume per cylinder and a rated power speed of more than 3.000 per minute.

TABLE 2.6   Emission standards for large goods vehicles: Euro norm emissions for category N$_2$, EDC (2000 and up)

Standard	Date	CO (g/kWh)	NO$_x$ (g/kWh)	HC (g/kWh)	PM (g/kWh)
Euro 0	1988–1992	12.3	15.8	2.6	None
Euro I	1992–1995	4.9	9.0	1.23	0.40
Euro II	1995–1999	4.0	7.0	1.1	0.15
Euro III	1999–2005	2.1	5.0	0.66	0.1
Euro IV	2005–2008	1.5	3.5	0.46	0.02
Euro V	2008–2012	1.5	2.0	0.46	0.02
Euro VI	Future				

The emission warranty remains at 5 years for 100,000 miles. With the exception of turbocharged and supercharged diesel-fueled engines, discharge of crankcase emissions is not allowed for any new 2004 or later model year engines.

The federal standards issued in 2004 for highway trucks are harmonized with California standards, with the intent that manufacturers can use a single engine or machine design for both markets. However, California certifications for model years 2005 to 2007 additionally require SET testing and NTE limits of 1.25 × FTP standards. California also adopted more stringent standards for MY 2004–2006 engines for public urban bus fleets.

Model Year 2007 and Later   The EPA signed emission standards for model year 2007 and later heavy-duty highway engines on December 21, 2000 (the California ARB adopted virtually identical 2007 heavy-duty engine standards in October 2001). The rule included two components: emission standards and diesel fuel quality regulations.

TABLE 2.7 EPA tier 1 emission standards for passenger cars and light-duty trucks (FTP, g/km)

Category	80,000 (km/5 years)						160,000 (km/10 years)[a]					
	THC	NMHC	CO	NO_x* Diesel	NO_x Gasoline	PM**	THC	NMHC	CO	NO_x* Diesel	NO_x* Gasoline	PM**
Passenger cars	0.26	0.16	2.1	0.6	0.25	0.05	—	0.19	2.6	0.4	0.4	0.06
LLDT, LVW <1700 kg	—	0.16	2.1	0.6	0.25	0.05	0.5	0.19	2.6	0.4	0.4	0.06
LLDT, LVW >1700 kg	—	0.2	0.3	—	0.44	0.05	0.5	0.25	3.5	0.6	0.6	0.06
HLDT, ALVW <2600 kg	0.2	—	0.3	—	0.44	—	0.5	0.29	4	0.6	0.6	0.06
HLDT, ALVW >2600 kg	0.26	—	3.1	—	0.7	—	0.5	0.35	4.6	0.96	0.96	0.08

Note: *More relaxed NO_x limits for diesels applicable to vehicles through the 2003 model year. **PM standards applicable to diesel vehicles only. Abbreviations: LVW—loaded vehicle weight (curb weight +140 kg); ALVW—adjusted LVW (the numerical average of the curb weight and the GVWR); LLDT—light light-duty truck (below 2700 kg GVWR); HLDT — heavy light-duty truck (above 2700 kg GVWR).
[a] Useful life of 200,000 km/11 years for all HLDT standards and for THC standards for LDT.

TABLE 2.8 EPA tier 1 SFTP standards

Category	NMHC+NO$_x$ (g/km) Weighted	CO (g/km) US06	SC03	Weighted
Passenger cars & LLDT, LVW < 1700 kg	0.57/1.29* (0.41/0.92*)	6.9 (5.6)	2.3 (1.9)	2.6 (2.1)
LLDT, LVW > 1700 kg	0.86 (0.6)	9.1 (7.3)	3.1 (2.4)	3.5 (2.8)
HLDT, ALVW < 2600 kg	0.9 (0.6)	10.6 (7.3)	3.5 (2.4)	4.0 (2.8)
HLDT, ALVW > 2600 kg	1.3 (0.9)	12.1 (8.3)	4.0 (2.8)	4.6 (3.1)

Note: *The more relaxed value is for diesel-fueled vehicles.

The first component of the regulation introduces new, very stringent emission standards, as follows:

PM—0.01 g/bhp-h (0.007 g/kWh)

NO$_x$—0.20 g/bhp-h (0.15 g/kWh)

NMHC—0.14 g/bhp-h (0.10 g/kWh)

The PM emission standard come into effect fully for 2007 model year of heavy-duty engine. The NO$_x$ and NMHC standards were phased in for diesel engines between 2007 and 2010. The phase-in took place gradually on a percent-of-sales basis: 50% from 2007 to 2009 and 100% in 2010 (gasoline engines are subject to these standards based on a phase-in requiring 50% compliance in 2008 and 100% compliance in 2009). A few engines meeting the 0.20 g/bhp-h NO$_x$ requirement were manufactured before 2010. In 2007 most manufacturers opted instead to meet a Family Emission Limit (FEL) of around 1.2 to 1.5 g/bhp-h NO$_x$ for most of their engines, a few manufacturers still certified some of their engines as high as 2.5 g/bhp-h NO$_x$ + NMHC.

Besides the transient FTP testing requirements, emissions certification includes:

Testing supplemental emissions (using SET) with limits equal to the FTP standards, and testing NTE (Not-to-Exceed) emissions with limits of 1.5×FTP standards for engines meeting the limit of NO$_x$ FEL of 1.5 g/bhp-h or less, and 1.25×FTP standards for engines with a NO$_x$ limit of FEL higher than 1.5 g/bhp-h.

The EPA regulation maintains the earlier crankcase emission control exception for turbocharged heavy-duty diesel for the 2007 model year engines but requires that their emissions to the atmosphere are to be added to the exhaust emissions during all testing. In this case, the deterioration of crankcase emissions must also be accounted for in exhaust deterioration factors.

The diesel fuel regulation restricted the sulfur content of on-highway diesel fuel to 15 ppm (wt) from the previous 500 ppm. Refiners were required to start producing the 15-ppm sulfur fuel as of June 1, 2006. Highway diesel fuel sold as low sulfur fuel has had to meet the 15-ppm sulfur standard starting on July 15, 2006, at the terminal

TABLE 2.9 Vehicle categories used in EPA tier 2 standards

Vehicle Category		Abbreviation	Requirements
Light-duty vehicle		LDV	Max. 8500 lb (3855 kg) GVWR
Light-duty truck		LDT	Max. 8500 lb (3855 kg) GVWR, max. 6000 lb (2721 kg) curb weight, and max. 45 ft^2 (4.18 m^2) frontal area
Light light-duty truck		LLDT	Max. 6000 lb (2721 kg) GVWR
	Light-duty truck[a]	LDT1	Max. 3750 lb (1700 kg) LVW[1]
	Light-duty truck[b]	LDT2	Min. 3750 lb (1700 kg) LVW[1]
Heavy light-duty truck		HLDT	Min. 6000 lb (2721 kg) GVWR
	Light-duty truck[c]	LDT3	Max. 5750 lb (2600 kg) ALVW[2]
	Light-duty truck[c]	LDT4	Min. 5750 lb (2600 kg) ALVW[2]
Medium-duty passenger vehicle		MDPV	Max. 10,000 lb (4535 kg) GVWR[3]

[a] LVW (loaded vehicle weight) = curb weight + 300 lb (136 kg).
[b] ALVW (adjusted loaded vehicle weight) = average of GVWR and curb weight.
[c] Manufacturers may alternatively certify engines for diesel-fueled MDPVs through the heavy-duty diesel engine regulations.

TABLE 2.10 Tier 2 emission standards (FTP 75, g/mi)

Bin#	Intermediate Life (5 years / 50,000 miles, 80,500 km)					Full Useful Life				
	NMOG*	CO	NO_x	PM	HCHO	NMOG*	CO	NO_x	PM	HCHO
Temporary bins										
11 MDPV^c						0.280	7.3	0.9	0.12	0.032
10^{a,b,d,f}	0.125 (0.160)	3.4 (4.4)	0.4	—	0.015 (0.018)	0.156 (0.230)	4.2 (6.4)	0.6	0.08	0.018 (0.027)
9^{a,b,e,f}	0.075 (0.140)	3.4	0.2	—	0.015	0.090 (0.180)	4.2	0.3	0.06	0.018
Permanent bins										
8^b	0.100 (0.125)	3.4	0.14	—	0.015	0.125 (0.156)	4.2	0.20	0.02	0.018
7	0.075	3.4	0.11	—	0.015	0.090	4.2	0.15	0.02	0.018
6	0.075	3.4	0.08	—	0.015	0.090	4.2	0.10	0.01	0.018
5	0.075	3.4	0.05	—	0.015	0.090	4.2	0.07	0.01	0.018
4	—	—	—	—	—	0.070	2.1	0.04	0.01	0.011
3	—	—	—	—	—	0.055	2.1	0.03	0.01	0.011
2	—	—	—	—	—	0.010	2.1	0.02	0.01	0.004
1	—	—	—	—	—	0.000	0.0	0.00	0.00	0.000

Note: Average manufacturer fleet NO_x standard is 0.07 g/mi (0.0435 g/km) for tier 2 vehicles. * for diesel fueled vehicle, NMOG (nonmethane organic gases) means NMHC (nonmethane hydrocarbons)

^{a}Bin deleted at end of 2006 model year (2008 for HLDTs).

^{b}The higher temporary NMOG, CO, and HCHO values apply only to HLDTs and MDPVs and expire after 2008.

^{c}An additional temporary bin restricted to MDPVs expires after model year 2008.

^{d}The optional temporary NMOG standard of 0.195 g/mi (50,000) (0.121 g/km at 80,467 km) and 0.280 g/mi (0.174 g/km) (full useful life) applies for qualifying LDT4s and MDPVs only.

^{e}The optional temporary NMOG standard of 0.100 g/mi (50,000) (0.062 g/km at 80,467 km) and 0.130 g/mi (0.080 g/km) (full useful life) applies for qualifying LDT2s only.

^{f}The 50,000 miles at (80,467 km) standard optional for diesels is certified to bins 9 or 10.

TABLE 2.11   Phase-in percentages for tier 2 requirements

Model Year	LDV/LLDT Tier 2[a]	HLDT/MDPV Tier 2[b]	HLDT/MDPV Interim Non–Tier 2[c]
2004	25		25
2005	50		50
2006	75		75
2007	100		100
2008	100	50	100
2009 and subsequent	100	100	

[a]Percentage of LDV/LLDTs that must meet tier 2 requirements.
[b]Percentage of HLDT/MDPVs that must meet tier 2 requirements.
[c]Percentage of non–tier 2 HLDT/MDPVs that must meet interim non-tier 2 fleet average NO_x requirements.

TABLE 2.12   Clean Fuel Fleet program for heavy-duty SI and CI engines (g/bhp·h, g/kWh)

Category	CO	$NMHC+NO_x$	PM	HCHO
LEV (federal fuel)		3.8 (2.8)		
LEV (California fuel)		3.5 (2.6)		
ILEV	14.4 (10.7)	2.5 (1.9)		0.050 (0.037)
ULEV	7.2 (5.4)	2.5 (1.9)	0.05 (0.04)	0.025 (0.019)
ZEV	0	0	0	0

Note: LEV—low-emissions vehicle; ILEV—inherently low-emissions vehicle; ULEV—ultra low-emissions vehicle; ZEV—zero-emission vehicle.

TABLE 2.13   EPA Emission standards for year 2004 model and later HD diesel engines (g/bhp·h, g/kWh)

Option	$NMHC+NO_x$	NMHC
1	2.4 (1.8)	n/a
2	2.5 (1.9)	0.5 (0.4)

level. For retail stations and wholesale purchasers, highway diesel fuel sold as "low sulfur fuel" has had to be the 15-ppm sulfur standard after September 1, 2006.

Refiners also took advantage of a temporary compliance option that allowed them to continue producing 500 ppm fuel in 20% of the volume of diesel fuel to be produced until December 31, 2009. In addition, refiners participated in an averaging, banking, and trading program with other refiners in their geographic areas.

Ultra low-sulfur diesel fuel requires advanced exhaust emission control technologies, such as catalytic diesel particulate filters and NO_x catalysts, in order to meet the 2007 emission standards. The EPA has estimated the cost of reducing the sulfur content of diesel fuel to increase fuel prices at approximately 4.5 to 5 cents per

TABLE 2.14   EPA emission standards for heavy-duty diesel engines (g/bhp-hr, g/kwh)

Year	HC	CO	NO$_x$	PM
		Heavy-duty diesel truck engines		
1988	1.3 (1.0)	15.5 (11.6)	10.7 (8.0)	0.60 (0.45)
1990	1.3 (1.0)	15.5 (11.6)	6.0 (4.5)	0.60 (0.45)
1991	1.3 (1.0)	15.5 (11.6)	5.0 (3.7)	0.25 (0.19)
1994	1.3 (1.0)	15.5 (11.6)	5.0 (3.7)	0.10 (0.07)
1998	1.3 (1.0)	15.5 (11.6)	4.0 (3.0)	0.10 (0.07)
		Urban bus engines		
1991	1.3 (1.0)	15.5 (11.6)	5.0 (3.7)	0.25 (0.19)
1993	1.3 (1.0)	15.5 (11.6)	5.0 (3.7)	0.10 (0.07)
1994	1.3 (1.0)	15.5 (11.6)	5.0 (3.7)	0.07 (0.05)
1996	1.3 (1.0)	15.5 (11.6)	5.0 (3.7)	0.05 (0.04)[a]
1998	1.3 (1.0)	15.5 (11.6)	4.0 (3.0)	0.05 (0.04)[a]

[a] In-use PM standard was 0.07.

TABLE 2.15   California emission standards for heavy-duty diesel engines (g/bhp-h, g/kWh)

Year	NMHC	THC	CO	NO$_x$	PM
			Heavy-duty diesel truck engines		
1987	—	1.3 (1.0)	15.5 (11.6)	6.0 (4.5)	0.60 (0.45)
1991	1.2 (0.9)	1.3 (1.0)	15.5 (11.6)	5.0 (3.7)	0.25 (0.19)
1994	1.2 (0.9)	1.3 (1.0)	15.5 (11.6)	5.0 (3.7)	0.10 (0.07)
			Urban bus engines		
1991	1.2 (0.9)	1.3 (1.0)	15.5 (11.6)	5.0 (3.7)	0.10 (0.07)
1994	1.2 (0.9)	1.3 (1.0)	15.5 (11.6)	5.0 (3.7)	0.07 (0.05)
1996	1.2 (0.9)	1.3 (1.0)	15.5 (11.6)	4.0 (3.0)	0.05 (0.04)

gallon. The EPA has also estimated that the new emission standards will raise the prices of vehicles by $1200 to $1900 (for comparison, new heavy-duty trucks typically sell for up to $150,000 and buses up to $250,000).

The emission standards for heavy-duty diesel truck and bus engines of model years 1988 to 2003 (US EPA) and 1987 to 2003 (California ARB) are summarized in Tables 2.14 and 2.15. Notice that starting with 1994 standards, the sulfur content in the certification fuel was reduced to 500 ppm wt.

Useful Life and Warranty Periods   Compliance with emission standards has to be demonstrated over the useful life of the engine. The standards adopted by the federal government and California are as follows:

LHDDE—8 years/110,000 miles (177,000 km) (whichever occurs first)
MHDDE—8 years/185,000 miles (297,700 km)
HHDDE—8 years/290,000 miles (466,700 km)

The federal useful life requirements were later increased to 10 years, with no change in the mileage numbers, for the urban bus PM standard (1994+) and for the NO_x standard (1998+). The emission warranty period is now 5 years per 100,000 miles (160,934 km) (compare 5 years/100,000 miles/3000 hours in California), which is close to the basic mechanical warranty for engines.

Cars and Light-Duty Trucks Emission Standards: California   There were three major regulatory steps in the evolution of California emission standards:

Tier 1/LEV California emission standards extended through the year 2003
LEV II California regulations phased in through model years 2004 to 2010
LEV III California regulations, proposed in 2010, to be phased-in through model
     years 2014 to 2022

The California emission standards that applied through model year 2003 were expressed using the following emission categories (detailed in Tables 2.16 to 2.25):

Tier 1
Transitional low-emission vehicles (TLEV)
Low-emission vehicles (LEV)
Ultra low-emission vehicles (ULEV)
Super ultra low-emission vehicles (SULEV)
Zero-emission vehicles (ZEV)

2.1.3   Emission Regulation in Japan [7,21]

Japanese emission standards for engines and vehicles and fuel efficiency targets are jointly developed by a number of government agencies:

Ministry of the Environment (MOE)
Ministry of Land, Infrastructure, and Transport (MLIT)
Ministry of Economy, Trade, and Industry (METI)

Engine and vehicle emission standards (Tables 2.22 to 2.25) are developed under the authority of the "Air Pollution Control Law," while fuel efficiency targets are adopted under the "Law Concerning the Rational Use of Energy" (Energy Conservation Law).

2.1.4   Emission Standards in India [22,23]

In India, a smoke test for diesel vehicles has been enforced since 1986 in various states. The first Indian emission regulations were idle emission limits that became effective in 1989. These idle emission regulations were soon replaced by mass emission limits for

TABLE 2.16 LEV Emission standards for light-duty vehicles (FTP 75, g/km)

Category	50,000 Miles (80,500 km/5 years)					100,000 Miles (160,900 km/10 years)				
	NMOG[a]	CO	NO$_x$	PM	HCHO	NMOG[a]	CO	NO$_x$	PM	HCHO
Passenger cars										
Tier 1	0.16	2.13	0.25	0.05	—	0.19	2.63	0.38	—	—
TLEV	0.078	2.125	0.25	—	0.009	0.098	2.63	0.38	0.050	0.011
LEV	0.047	2.125	0.125	—	0.009	0.056	2.63	0.188	0.050	0.011
ULEV	0.025	1.063	0.125	—	0.005	0.034	1.313	0.188	0.025	0.007
LDT1, LVW < 3750 lbs (1700 kg)										
Tier 1	0.16	2.13	0.25	0.05	—	0.19	2.63	0.38	—	—
TLEV	0.078	2.125	0.25	—	0.009	0.098	2.63	0.38	0.050	0.011
LEV	0.047	2.125	0.125	—	0.009	0.056	2.63	0.188	0.050	0.011
ULEV	0.025	1.063	0.125	—	0.005	0.034	1.313	0.188	0.025	0.007
LDT2, LVW > 3750 lbs (1700 kg)										
Tier 1	0.2	2.75	0.438	0.05	—	0.25	3.438	0.606	—	—
TLEV	0.1	2.75	0.438	—	0.011	0.125	3.438	0.563	0.063	0.014
LEV	0.063	2.75	0.25	—	0.011	0.081	3.438	0.313	0.063	0.014
ULEV	0.031	1.375	0.25	—	0.006	0.044	1.75	0.313	0.031	0.008

Abbreviations: LVW—loaded vehicle weight (curb weight + 300 lbs [136 kgs]), LDT—light-duty truck, NMOG—nonmethane organic gases, HCHO—formaldehyde.
[a] NMHC for all tier 1 standards.

TABLE 2.17 LEV emission standards for medium-duty vehicles (FTP 75, g/km)

Category	50,000 miles (80,500 km/5 years)					120,000 miles (193,100 km/10 years)				
	NMOG[a]	CO	NO_x	PM	HCHO	NMOG[a]	CO	NO_x	PM	HCHO
MDV1, 0–3750 lbs (0–1700 kg)										
Tier 1	0.156	2.125	0.25	—	—	0.225	3.125	0.344	0.05	—
LEV	0.078	2.125	0.25	—	0.009	0.113	3.125	0.375	0.05	0.014
ULEV	0.047	1.063	0.125	—	0.005	0.067	1.563	0.1875	0.025	0.0075
MDV2, 3751–5750 lbs (1701–2600 kg)										
Tier 1	0.2	2.75	0.438	—	—	0.288	4	0.613	0.063	—
LEV	0.1	2.75	0.25	—	0.011	0.144	4	0.375	0.063	0.017
ULEV	0.063	2.75	0.25	—	0.006	0.089	4	0.375	0.031	0.008
SULEV	0.031	1.375	0.125	—	0.003	0.045	2	0.188	0.031	0.004
MDV3, 5751–8500 lbs (2601–3855 kg)										
Tier 1	0.244	3.125	0.688	—	—	0.350	4.563	0.956	0.063	—
LEV	0.122	3.125	0.375	—	0.014	0.175	4.563	0.563	0.063	0.02
ULEV	0.073	3.125	0.375	—	0.007	0.104	4.563	0.563	0.031	0.01
SULEV	0.037	1.563	0.188	—	0.004	0.053	2.313	0.281	0.031	0.005
MDV4, 8501–10,000 lbs (3856–4535 kg)										
Tier 1	0.288	3.438	0.813	—	0.018	0.413	5.063	1.131	0.075	0.018
LEV	0.144	3.438	0.438	—	0.018	0.206	5.063	0.625	0.075	0.025
ULEV	0.086	3.438	0.438	—	0.009	0.123	5.063	0.625	0.038	0.013
SULEV	0.043	1.750	0.219	—	0.004	0.063	2.563	0.313	0.038	0.006
MDV5, 10,001–14,000 lbs (4536–6350 kg)										
Tier 1	0.375	4.375	1.250	—	—	0.538	6.438	1.731	0.075	—
LEV	0.188	4.375	0.625	—	0.023	0.269	6.438	0.938	0.075	0.033
ULEV	0.113	4.375	0.625	—	0.011	0.161	6.438	0.938	0.038	0.016
SULEV	0.056	2.188	0.313	—	0.006	0.081	3.250	0.438	0.038	0.008

Abbreviations: MDV—medium-duty vehicle (the maximum GVW from 8500 to 14,000 lbs (3855–4535 kg). The MDV category is divided into five classes, MDV1 ... MDV5, based on vehicle test weight. The definition of "test weight" in California is identical to the federal ALVW. NMOG—nonmethane organic gases; HCHO—formaldehyde.

[a]NMHC for all tier 1 standards.

TABLE 2.18 LEV II emission standards for passenger cars and LDVs < 8500 lbs (3855 kg, g/km)

Category	50,000 miles (80,467 km/5 years)					120,000 miles (193.121 km/10 years)				
	NMOG	CO	NO_x	PM	HCHO	$NMOG^a$	CO	NO_x	PM	HCHO
LEV	0.047	2.125	0.031	—	0.009	0.056	2.625	0.044	0.006	0.011
ULEV	0.025	1.063	0.031	—	0.005	0.034	1.313	0.044	0.006	0.007
SULEV	—	—	—	—	—	0.006	0.625	0.013	0.006	0.003

TABLE 2.19 LEV II emission standards for medium-duty vehicles, durability 120,000 miles (193,100 km, g/km)

Weight (GVW)	Category	NMOG	CO	NO_x	PM	HCHO
8500–10,000 lbs	LEV	0.122	4	0.125	0.075	0.020
(3856–4535 kg)	ULEV	0.089	4	0.125	0.038	0.010
	SULEV	0.063	2	0.063	0.038	0.005
10,001–14,000 lbs	LEV	0.144	4.563	0.250	0.075	0.025
(4536– 6350 kg)	ULEV	0.104	4.563	0.250	0.038	0.013
	SULEV	0.073	2.313	0.125	0.038	0.006

TABLE 2.20 Proposed LEV III NMOG + NO_x emission standards for LDVs, g/miles durability 150,000 miles (241,400 km)

Category	NMOG + NO_x
LEV	0.1
ULEV	0.078
ULEV70	0.044
ULEV50	0.031
SULEV	0.019
SULEV20	0.013

TABLE 2.21 Proposed LEV III NMOG + NO_x emission standards for MDVs, g/km durability 150,000 miles (241,400 km)

Weight (GVW)	Category	NMOG + NO_x
8500–10,000 lbs(3855–4535 kg)	ULEV	0.125
	SULEV	0.091
10,001–14,000 lbs(4536–6350 kg)	ULEV	0.198
	SULEV	0.125

both petrol (1991) and diesel (1992) vehicles, and were gradually tightened over the 1990s. Since the year 2000, India has been adopting European emission and fuel regulations for four-wheeled light-duty and heavy-duty vehicles. India's own emission regulations are still being applied to two- and three-wheeled vehicles. The National

TABLE 2.22   Japanese emission standards for gasoline vehicles

		PM	NO$_x$	NMHC	CO	Date in Effect
Passenger car		0.005	0.05	0.024	0.63	2009
		(New)	(0%)	(−0%)	(−0%)	
Trucks and buses	Light-weight (GVW 1.7 t or less)	0.005	0.05	0.024	0.63	2009
		(New)	(0%)	(−0%)	(−0%)	
	Middle-weight (GVW over 1.7 t~3.5 t or less)	0.007	0.07	0.024	0.63	2009
		(New)	(0%)	(−0%)	(−0%)	
	Heavy-weight (GVW over 3.5 t)	0.01	0.07	0.17	2.22	2009
		(New)	(0%)	(−0%)	(−0%)	

Note: Units for heavy-weight trucks are g/kWh; all others are g/km. Percentage reductions from 2005 standards are given in parentheses. GVW = gross vehicle weight; NMHC = nonmethane hydrocarbons.

TABLE 2.23   Japanese emission standards for diesel passenger cars (g/km)

			CO	HC	NO$_x$	PM
Vehicle Weight	Date	Test	Mean (Max.)	Mean (Max.)	Mean (Max.)	Mean (Max.)
< 1250 kg	1986	10–15 mode	2.1 (2.7)	0.40 (0.62)	0.70 (0.98)	—
	1990		2.1 (2.7)	0.40 (0.62)	0.50 (0.72)	—
	1994		2.1 (2.7)	0.40 (0.62)	0.50 (0.72)	0.20 (0.34)
	1997		2.1 (2.7)	0.40 (0.62)	0.40 (0.55)	0.08 (0.14)
	2002[a]		0.63	0.12	0.28	0.052
	2005[b]	JC08[c]	0.63	0.024[d]	0.14	0.013
	2009		0.63	0.024[d]	0.08	0.005
> 1250 kg	1986	10–15 mode	2.1 (2.7)	0.40 (0.62)	0.90 (1.26)	—
	1992		2.1 (2.7)	0.40 (0.62)	0.60 (0.84)	—
	1994		2.1 (2.7)	0.40 (0.62)	0.60 (0.84)	0.20 (0.34)
	1998		2.1 (2.7)	0.40 (0.62)	0.40 (0.55)	0.08 (0.14)
	2002[a]		0.63	0.12	0.30	0.056
	2005[b]	JC08[c]	0.63	0.024[d]	0.15	0.014
	2009[e]		0.63	0.024[d]	0.08	0.005

Note: Vehicle weight is the equivalent inertia weight (EIW) of a vehicle weighing 1265 kg.
[a] 2002.10 for domestic cars; 2004.09 for imports.
[b] Full implementation by the end of 2005.
[c] Full phase-in by 2011.
[d] Nonmethane hydrocarbons.
[e] 2009.10 for new domestic models; 2010.09 for existing models and imports.

Auto Fuel Policy announced on October 6, 2003, envisages a phased program for introducing Euro 2, 3, and 4 emission and fuel regulations by 2010. A new auto fuel policy for 2025 is currently under preparation.

India is gradually harmonizing its automotive standards with global norms, demonstrating due cognizance of prevailing national concerns. The Union Cabinet in October 2002 approved a proposal to join the World Forum for Harmonization of

TABLE 2.24 Japanese diesel emission standards for light commercial vehicles, GVW ≤ 3500 kg (≤ 2500 kg before 2005)

Vehicle Weight	Date	Test	Unit	CO Mean (Max.)	HC Mean (Max.)	NO$_x$ Mean (Max.)	PM Mean (Max.)
≤ 1700 kg	1988	10–15 mode	g/km	2.1 (2.7)	0.40 (0.62)	0.90 (1.26)	
	1993			2.1 (2.7)	0.40 (0.62)	0.60 (0.84)	0.20 (0.34)
	1997			2.1 (2.7)	0.40 (0.62)	0.40 (0.55)	0.08 (0.14)
	2002			0.63	0.12	0.28	0.052
	2005[b]	JC08[c]		0.63	0.024[d]	0.14	0.013
	2009			0.63	0.024[d]	0.08	0.005
> 1700 kg	1988	6 mode	ppm	790 (980)	510 (670)	DI: 380 (500)	—
						IDI: 260 (350)	—
	1993	10–15 mode	g/km	2.1 (2.7)	0.40 (0.62)	1.30 (1.82)	0.25 (0.43)
	1997[a]			2.1 (2.7)	0.40 (0.62)	0.70 (0.97)	0.09 (0.18)
	2003			0.63	0.12	0.49	0.06
	2005[b]	JC08[c]		0.63	0.024[d]	0.25	0.015
	2009[e]			0.63	0.024[d]	0.15	0.007

Note: Vehicle weight is gross vehicle weight (GVW).
[a] 1997 for manual transmission vehicles; 1998 for automatic transmission vehicles.
[b] Full implementation by the end of 2005.
[c] Full phase-in by 2011.
[d] Nonmethane hydrocarbons.
[e] 2009.10 for new domestic models; 2010.09 for existing models and imports.

TABLE 2.25 Japanese diesel emission standards for heavy commercial vehicles, GVW > 3500 kg (> 2500 kg before 2005)

Date	Test	Unit	CO Mean (Max.)	HC Mean (Max.)	NO$_x$ Mean (Max.)	PM Mean (Max.)
1988/89	6 mode	ppm	790 (980)	510 (670)	DI: 400 (520)	
					IDI: 260 (350)	
1994	13 mode	g/kWh	7.40 (9.20)	2.90 (3.80)	DI: 6.00 (7.80)	0.70 (0.96)
					IDI: 5.00 (6.80)	
1997[a]			7.40 (9.20)	2.90 (3.80)	4.50 (5.80)	0.25 (0.49)
2003[b]			2.22	0.87	3.38	0.18
2005[c]	JE05		2.22	0.17[d]	2.0	0.027
2009			2.22	0.17[d]	0.7	0.01

[a] 1997: GVW ≤ 3500 kg; 1998: 3500 < GVW ≤ 12,000 kg; 1999: GVW > 12,000 kg.
[b] 2003: GVW ≤ 12,000 kg; 2004: GVW > 12,000 kg.
[c] Full implementation by the end of 2005.
[d] Nonmethane hydrocarbons.

TABLE 2.26   Indian emission standards (four-wheel vehicles)

Standard	Reference	Date	Region
India 2000	Euro 1	2000	Nationwide
Bharat stage II	Euro 2	2001	NCR, Mumbai, Kolkata, Chennai
		2003.04	NCR, 10 cities[a]
		2005.04	Nationwide
Bharat stage III	Euro 3	2005.04	NCR, 10 cities[a]
		2010.04	Nationwide
Bharat stage IV	Euro 4	2010.04	NCR, 10 cities[a]

Note: NCR (national capital region) refers to Delhi.
[a] Mumbai, Kolkata, Chennai, Bengaluru, Hyderabad, Ahmedabad, Pune, Surat, Kanpur, and Agra

Vehicle Regulations (WP.29) as an Observer in the first instance. The government of India on November 18, 2009, also notified national ambient air quality standards.

Indian Emission Norms   The evolution of emission norms introduced in India over the last decades can be summarized as follows:

1991—Idle CO limits for gasoline vehicles and free acceleration smoke for diesel vehicles, mass emission norms for gasoline vehicles.

1992—Mass emission norms for diesel vehicles.

1996—Revision of mass emission norms for gasoline and diesel vehicles, requiring mandatory fitment of catalytic converter for cars in metro-cities (Delhi, Mumbai, Kolkata, Chennai) on unleaded gasoline.

1998—Cold start norms introduced.

2000—India 2000 (equivalent to Euro I), modified IDC (Indian driving cycle), Bharat stage II norms for Delhi.

2001—Bharat Stage II (equivalent to Euro II) for all metro-cities (Delhi, Mumbai, Kolkata, Chennai), emission norms for CNG and LPG vehicles.

2003—Bharat stage II (equivalent to Euro II) norms for 11 major cities.

2005—From April 1, Bharat stage III (equivalent to Euro III) norms for 11 major cities.

2010—Bharat stage III emission norms for four-wheelers for entire country and Bharat stage IV (equivalent to Euro IV) for 11 major cities. OBD (diluted Euro III).

The implementation schedule of emission standards in India is summarized in Table 2.26.

The standards listed in the table apply to all new four-wheel vehicles to be sold and registered in the respective regions. In addition the National Auto Fuel Policy has set emission requirements for interstate buses with routes originating or terminating in Delhi and in 10 other cities. For two- and three-wheelers, Bharat stage II (Euro 2)

TABLE 2.27   Indian emission standards for diesel truck and bus engines (g/kWh)

Year	Reference	CO	HC	NO_x	PM
1992	—	17.3–32.6	2.7–3.7	—	—
1996	—	11.20	2.40	14.4	—
2000	Euro I	4.5	1.1	8.0	0.36[a]
2005[b]	Euro II	4.0	1.1	7.0	0.15
2010[b]	Euro III	2.1	0.66	5.0	0.10

[a] 0.612 for engines below 85 kW.
[b] Earlier introduction in selected regions; see Table 2.25.

TABLE 2.28   Indian emission standards for light-duty diesel vehicles (g/km)

Year	Reference	CO	HC	$HC+NO_x$	PM
1992	—	17.3–32.6	2.7–3.7	—	—
1996	—	5.0–9.0	—	2.0–4.0	—
2000	Euro 1	2.72–6.90	—	0.97–1.70	0.14–0.25
2005[a]	Euro 2	1.0–1.5	—	0.7–1.2	0.08–0.17

[a] Earlier introduction in selected regions; refer Table 2.25.

has been in effect since April 1, 2005 and stage III (Euro 3) standards began to be enforced on April 1, 2008.

Trucks and Buses   Emission standards for new heavy-duty diesel engines— applicable to vehicles of GVW > 3500 kg—are listed in Table 2.27. Emissions are tested over the ECE R49 13-mode test (through the Euro II stage).

Light-Duty Diesel Vehicles   Emission standards for light-duty diesel vehicles (GVW ≤ 3500 kg) are summarized in Table 2.28. Ranges of emission limits refer to different classes (by reference mass) of light commercial vehicles; compare the EU light-duty vehicle emission standards page for details on the Euro 1 and later standards. The lowest limit in each range applies to passenger cars (GVW ≤ 2500 kg; up to six seats).

The test cycle has been ECE + EUDC for low-power vehicles (with maximum speed limited to 90 km/h). Before 2000, emissions were measured with an Indian test cycle. Engines used in light-duty vehicles can also be emission tested applying an engine dynamometer. The respective emission standards are listed in Table 2.29.

Light-Duty Gasoline Vehicles—Four-Wheel Vehicles   Emission standards for gasoline vehicles (GVW ≤ 3500 kg) are summarized in Table 2.30. The range of emission limits refers to different classes of light commercial vehicles (compare the EU light-duty vehicle emission standards page). The lowest limit in each range regards to passenger cars (GVW ≤ 2500 kg; up to six seats).

Three- and Two-Wheel Vehicles   Emission standards for three- and two-wheel gasoline vehicles are listed in Tables 2.31 and 2.32.

TABLE 2.29   Indian emission standards for light-duty diesel engines (g/kWh)

Year	Reference	CO	HC	NO$_x$	PM
1992	—	14.0	3.5	18.0	—
1996	—	11.20	2.40	14.4	—
2000	Euro I	4.5	1.1	8.0	0.36[a]
2005[b]	Euro II	4.0	1.1	7.0	0.15

[a] 0.612 for engines below 85 kW.
[b] Earlier introduction in selected regions, refer Table 2.25.

TABLE 2.30   Indian emission standards for gasoline vehicles, GVW ≤ 3500 kg (g/km)

Year	Reference	CO	HC	HC + NO$_x$
1991	—	14.3–27.1	2.0–2.9	—
1996	—	8.68–12.4	—	3.00–4.36
1998[a]	—	4.34–6.20	—	1.50–2.18
2000	Euro 1	2.72–6.90	—	0.97–1.70
2005[b]	Euro 2	2.2–5.0	—	0.5–0.7

Note: Gasoline vehicles must also meet an evaporative (SHED) limit of 2 g/test (effective in 2000).
[a] For catalytic converter fitted vehicles.
[b] Earlier introduction in selected regions; refer Table 2.25.

TABLE 2.31   Indian emission standards for three-wheel gasoline vehicles (g/km)

Year	CO	HC	HC + NO$_x$
1991	12–30	8–12	—
1996	6.75	—	5.40
2000	4.00	—	2.00
2005 (BS II)	2.25	—	2.00 (DF = 1.2)

TABLE 2.32   Emission standard for two–wheeler vehicles (g/km)

Year	CO	HC	HC + NO$_x$	
1991	12–30	8–12	—	—
1996	4.50	—	3.60	—
2000	2.00	—	2.00	—
2005(BS II)	1.50	—	1.50	(DF = 1.2)

2.1.5   Emission Standards in China [24]

Along with rapid industrialization, coal power plants and particularly the number of cars on China's roads are rapidly growing. The pollution problem became so pronounced that China enacted its first emission control on automobiles in 2000, equivalent to Euro I standard. This was upgraded to Euro II in 2005. A more stringent emission standard, National Standard III, equivalent to the Euro III standard, went into effect on July 1, 2007. Plans are for the Euro IV standard to take effect in 2010. Beijing initiated the Euro IV standard on January 1, 2008, and became the first city in China mainland to be adopted it.

2.2   INDIRECT EMISSION REGULATIONS (INTERNATIONAL STANDARDS)

Indirect regulations, as mentioned before, can be directives (medium or long term, 5 to 20 years), fuel quality standards, government regulations, or specifications on engine oils (e.g., low-SAPS lubricating oils) and utilization of aftertreatment catalysts [25,26]. In this book we consider only the harmful materials emission due to fuel quality and improvements to fuel quality. Figure 2.1 shows the effects of different fuel components on emissions.

Fuel quality in different regions of the world is determined and prescripted by formation of different action plans and directives over the middle and long term (5 to 10 years). Of course, general and specific requirements must first be formed and refined for engine fuel specifications. This runs currently.

In the United States, there are currently active the 1967 Clean Air Act [5], the 1970 EPA (Environmental Protection Agency), the Society of Automative Engineers' global Auto-Oil Program [4] and the previously mentioned CARB (California Air Resources Board) [6,27]. In Europe, Auto Oil Programmes I and II [2–4,15,16], through the Directives of the European Union [15,16,28,29], provide the general and specific aims and target data for internal combustion engines and emissions over a determined period of time.

In the European Union, the two Auto Oil Programs have set the ground work for legislation of emission limits, and thus for all quality specifications of engine fuels since 1990. The Auto Oil Programme was a conception of the European Programme on Emissions, Fuels and Engine Technologies (EPEFE) initiated in cooperation with

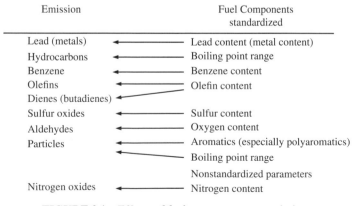

FIGURE 2.1   Effects of fuel components on emissions

the Association des Constructeurs Européan d'Automobiles (ACEA) and the European Petroleum Industry Association (EUROPIA). EU legislation concerning fuel quality results from these associations.

Also an active contributor to EU legislation is an association of European oil companies concerned with fuel industry safety, health effects, and environmental issues. This association, named CONCAWE (Conservation of Clean Air and Water in Europe), was founded 1963 to deal with fuel quality and emission standards, air and water quality, soil pollution, waste management, industrial safety and health, trade of crude oil products, and related operations of international pipelines [30].

Indisputably, then, it can be stated that the quality of engine fuels has as indirect effect on emissions of internal combustion engines, and that the critical stakeholders are vehicle and engine manufacturers, refining and/or hydrocarbon processing companies, environment protection organizations, and then partly the standardizer organizations. Indeed, the general quality requirements for modern gasolines—high energy content and environment friendly blending components with high efficiency and quality improvement additives—have evolved over many years. In the developed regions of the world fuels must have the following characteristics[31]:

- High research and motor octane number
- Balanced octane number distribution
- Be lead free
- Minimal sulfur content (≤ 50, ≤ 10 mg/kg)
- Reduced aromatic content (≤ 35 v/v%)
- Low benzene content (≤ 1.0 v/v%)
- Reduced olefin content (≤ 18.0 v/v%)
- High iso-paraffin content
- Suitable boiling point range
- Suitable vapor pressure (≤ 60 kPa)
- Be halogenic free
- Suitable oxygen content
- Economic application of alternative components (blendability with bio-components)
- Suitable levels of additives
- Compatible with engine oils
- Suitable cost
- Easy and safe handling
- Formation of relative environment friendly burning product

Similar specifications of fuels have been worked out in the United States, in the European Union, and in Japan. Concurrently, the leading vehicle and engine manufacturers consisting of members from European Automobile Manufacturers'

TABLE 2.33   Key changes in specifiations of gasolines

Properties	WWFC				
	Category 1	Category 2	Category 3	Category 4	Category 5
Sulfur content, mgS/kg maximum	1000	1000	30	10	10
Aromatic content, v/v% maximum	50.0	40.0	35.0	35.0	35
Olefinic content, v/v% maximum	—	18.0	10.0	10.0	10
Benzene content, v/v% maximum	5.0	5.0	1.0	1.0	1.0
Oxygen content, % maximum	2.7	2.7	2.7	2.7	2.7
Reid vapor pressure, kPa maximum	—	—	—	45–105[a]	45–105

Note: RFG: Reformulated gasoline, CARB: California Air Resources Board, WWFC: World Wide Fuel Charter (2006). The dash means "no prescription."
[a] depend on geological position and average temperature
WWFC - 2013, 5th edition has just been released incorporating an additional category 5. Refer to original documents for full details.

TABLE 2.34   Unleaded gasoline—WWFC category 1

Properties	Limit
'91 RON research octane number	Min. 91
'91 RON motor octane number	Min. 82
'95 RON research octane number	Min. 95
'95 RON motor octane number	Min. 85
'98 RON research octane number	Min. 98
'98 RON motor octane number	Min. 88
Oxidation stability, minutes	360
Sulfur content, mg/kg	1000
Metal content, mg/L	Nondetectable
Oxygen content, %	2.7
Aromatics content, v/v%	50.0
Benzene content, v/v%	5.0
Unwashed gum, mg/100 ml	70
Washed gum, mg/100 ml	5
Density, kg/m³	715–780
Copper corrosion	Class 1

Note: Additives must be compatible with engine oils (no increase in engine sludge/vanish deposits) and addition of ash-forming components is not allowed. Good industry practice is to reduce contamination (dust, water, other fuels, etc.).

Association (ACEA) with 13 members, the Alliance of Automobile Manufacturers (AAM) with 9 members, the Engine Manufacturers Association with 30 members, and the Japan Automobile Manufacturers Association (JAMA) with 14 members created the world wide fuel charter. There are further smaller member companies, and 14 other associate members and supporting organisations [32]. The main object is to harmonize the fuel quality specifications with the requirements of engine and vehicle manufacturing, and with the different world markets. Together, the four associations collaborated on and released in September 2006 the quality category specifications each for unleaded gasolines and diesel gas oils. They based their classifications on the emissions

TABLE 2.35   Unleaded gasoline—WWFC category 2

Properties	Limit
'91 RON research octane number	Min. 91
'91 RON motor octane number	Min. 82.5
'95 RON research octane number	Min. 95
'95 RON motor octane number	Min. 85
'98 RON research octane number	Min. 98
'98 RON motor octane number	Min. 88
Oxidation stability, minutes	480
Sulfur content, mg/kg	150
Metal content, mg/L	Nondetectable
Phosphorous content, mg/L	Nondetectable
Silicon content, mg/kg	Nondetectable
Oxygen content, %	2.7
Aromatics content, v/v%	40.0
Olefin content, v/v%	18.0
Benzene content, v/v%	2.5
Unwashed gum, mg/100 ml	70
Washed gum, mg/100 ml	5
Density, kg/m^3	715–770
Copper corrosion	Class 1

Note: Additives must be compatible with engine oils (no increase in engine sludge/vanish deposits), and addition of ash-forming components is not allowed. Good industry practice is to reduce contamination (dust, water, other fuels, etc.).

TABLE 2.36   Unleaded gasoline—WWFC category 3

Properties	Limit
'91 RON research octane number	Min. 91
'91 RON motor octane number	Min. 82.5
'95 RON research octane number	Min. 95
'95 RON motor octane number	Min. 85
'98 RON research octane number	Min. 98
'98 RON motor octane number	Min. 88
Oxidation stability, minutes	480
Sulfur content, mg/kg	30
Metal content, mg/L	Nondetectable
Phosphorous content, mg/L	Nondetectable
Silicon content, mg/kg	Nondetectable
Oxygen content, %	2.7
Aromatics content, v/v%	35.0
Olefin content, v/v%	10.0
Benzene content, v/v%	1.5
Unwashed gum, mg/100 ml	30
Washed gum, mg/100 ml	5
Density, kg/m^3	715–770
Copper corrosion	Class 1

Note: Additives must be compatible with engine oils (no increase in engine sludge/vanish deposits), and addition of ash-forming components is not allowed. Good industry practice is to reduce contamination (dust, water, other fuels, etc.).

TABLE 2.37   Unleaded gasoline—WWFC category 4

Properties	Limit
'91 RON research octane number	Min. 91
'91 RON motor octane number	Min. 82.5
'95 RON research octane number	Min. 95
'95 RON motor octane number	Min. 85
'98 RON research octane number	Min. 98
'98 RON motor octane number	Min. 88
Oxidation stability, minutes	480
Sulfur content, mg/kg	10
Metal content, mg/L	Nondetectable
Phosphorous content, mg/L	Nondetectable
Silicon content, mg/kg	Nondetectable
Oxygen content, %	2.7
Aromatics content, v/v%	35.0
Olefin content, v/v%	10.0
Benzene content, v/v%	1.0
Unwashed gum, mg/100 ml	30
Washed gum, mg/100 ml	5
Density, kg/m^3	715–770
Copper corrosion	Class 1

Note: Additives must be compatible with engine oils (no increase in engine sludge/vanish deposits), and addition of ash-forming components is not allowed. Good industry practices is to reduce contamination (dust, water, other fuels, etc.).

regulations and efficiency potentials for each market (shown in Tables 2.33 to 2.37 for gasolines):

• No regulation or low regulation of emission; basic engine and vehicle efficiency and emission formation
• Rigorous emission regulation or other market demands
• Improved emission regulation or other market demands
• Improved emission regulation, as applicable to possible NO_x converter technologies

Of course, in each country and region specifications of gasolines are different. Table 2.38 shows the main EU specified quality parameters of gasoline, compared with those in the United States (as well as in California) and in Japan.

For *diesel fuels*, the general requirements pertaining to quality are the following form [26,33]:

• Energy compliance with operational parameters of diesel engines
• Easily processed
• Small emissions of harmful substances during processing
• Availablility in high quantities of consistent quality

TABLE 2.38 Key changes in quality specifications of gasolines

| Properties | United States | | European Union | Japan | WWFC |
	RFG Phase 3 (2004/2006)	CARB Phase 3 (2009)	EN 228 (2009)		Category 4
Sulfur content, mgS/kg, maximum	300/80	30/20	10	10	10
Aromatic content, v/v% maximum	—	35	35		35
Olefinic content, v/v%, maximum	—	10	18		10
Benzene content, v/v% maximum	1.0	0,7	1.0		1.0
Oxygen content, % maximum	1.5–3.5/2.1	1.8–3.5	2.7		2.7
Reid vapor pressure, kPa maximum	44–69	41–50	45–105[a]		45–105[1]

Note: RFG: Reformulated gasoline; CARB: California Air Resources Board; WWFC: World Wide Fuel Charter (2006). The dash means "no requirement."
[a] Depend on geological position and average temperature.

TABLE 2.39   Key changes in the specifications of gasoils

Properties	WWFC Category 1	WWFC Category 2	WWFC Category 3	WWFC Category 4
Cetane number, minimum	48	51	53	55
Density, kg/m^3	820–860	820–850	820–840	820–840
Total aromatic content, %, maximum	—	25.0	20.0	15.0
Policyclic aromatic content, %, maximum	—	5.0	3.0	2.0
Sulfur content, mg/kg, maximum	2000	300	50	10

Note: WWFC: WorldWide Fuel Charter (2006). The dash means "no requirement."
WWFC - 2013, 5th edition has just been released incorporating an additional category 5. Refer to original documents for full details.

- High energy content
- Low volatility
- Vaporizable in required quantities
- Easily pumped
- Adequate viscosity
- Low sulfur content (≤ 10 mg/kg)
- Reduced aromatic content (≤ 20–25% mg/kg)
- Low polyaromatic content (≤ 2–8%)
- Not cause corrosion
- Good lubricity
- Good thermal stability
- Good chemical stability
- Economical application of alternative blending components (miscibility)
- Compatibility with lubricants
- Not be toxic
- Easy and safe handling (storage, distribution, retail)
- Low production cost compared to other fuels
- Low noise level
- Odorless burning products in exhaust gas
- Nonpolluting and noncorroding unburned fuel components in piston or any engine part
- Additivation at required levels
- Exhaust gas composition slightly or not at all harmful for living organisms

The corresponding four quality categories as defined for diesel fuels in the WWFC are listed in Tables 2.39 to 2.44 [32]. Summarized in Table 2.44 are the most critical specifications of diesel fuels to be abided by industrially advanced regions.

TABLE 2.40   Diesel fuel—WWFC category 1

Properties	Limit
Cetane number	Min. 48
Cetane index	Min. 48
Density on 15 °C, kg/m^3	820–860
Viscosity on 40 °C, mm^2/s	2.0–4.5
Sulfur content, mg/kg	Max. 2000
T95, °C	Max. 370
Flashpoint, °C	Min 55
Carbon residue, %	Max. 0.30
Water content, mg/kg	Max. 500
FAME content, v/v%	Max. 5
Copper corrosion	Class 1
Ethanol/methanol content, v/v%	Nondetectable
Ash content, %	Max. 0.01
Particulate contamination, mg/kg	Max. 10
Lubricity (HFRR) wear scar diameter at 60 °C, micron	Max. 400

Note: Additives must be compatible with engine oils, and addition of ash-forming components is not allowed. Good industry practice is to reduce contamination (dust, water, other fuels, etc.). Also adequate labeling of pumps must be defined and used.

TABLE 2.41   Diesel fuel—WWFC category 2

Properties	Limit
Cetane number	Min. 51
Cetane index	Min. 51
Density on 15 °C, kg/m^3	820–850
Viscosity on 40 °C, mm^2/s	2.0–4.0
Sulfur content, mg/kg	Max. 300
Total aromatics content, %	Max. 25.0
PAH content, %	5.0
T90, °C	Max. 340
T95, °C	Max. 355
Final boiling point, °C	365
Flashpoint, °C	Min. 55
Carbon residue, %	Max. 0.30
Water content, mg/kg	Max. 200
FAME content, v/v%	Max. 5.0
Copper corrosion	Class 1
Ethanol/methanol content, v/v%	Nondetectable
Ash content, %	Max. 0.01
Particulate contamination, mg/kg	Max. 10
Lubricity (HFRR) wear scar diameter at 60 °C, micron	Max. 400

Note: Additives must be compatible with engine oils, and addition of ash-forming components is not allowed. Good industry practice is reduce contamination (dust, water, other fuels, etc.). Also adequate labeling of pumps must be defined and used.

TABLE 2.42 Diesel fuel—WWFC category 3

Properties	Limit
Cetane number	Min. 53
Cetane index	Min. 53
Density on 15 °C, kg/m³	820–840
Viscosity on 40 °C, mm²/s	2.0–4.0
Sulfur content, mg/kg	Max. 50
Total aromatics content, %	Max. 20.0
PAH content, %	3.0
T90, °C	Max. 320
T95, °C	Max. 340
Final boiling point, °C	Max. 350
Flashpoint, °C	Min. 55
Carbon residue, %	Max. 0.20
Water content, mg/kg	Max. 200
FAME content, v/v%	Max. 5.0
Copper corrosion	Class 1
Ethanol/methanol content, v/v%	Nondetectable
Ash content, %	Max. 0.01
Particulate contamination, mg/kg	Max. 10
Lubricity (HFRR) wear scar diameter at 60 °C, micron	Max. 400

Note: Additives must be compatible with engine oils, and addition of ash-forming components is not allowed. Good industry practice is to reduce contamination (dust, water, other fuels, etc.). Also adequate labeling of pumps must be defined and used.

TABLE 2.43 Diesel fuel—WWFC category 4

Properties	Limit
Cetane number	Min. 55
Cetane index	Min. 55
Density on 15 °C, kg/m³	820–840
Viscosity on 40 °C, mm²/s	2.0–4.0
Sulfur content, mg/kg	Max. 50
Total aromatics content, %	Max. 15.0
PAH content, %	2.0
T90, °C	Max. 320
T95, °C	Max. 340
Final boiling point, °C	Max. 350
Flashpoint, °C	Min. 55
Carbon residue, %	Max. 0.20
Water content, mg/kg	Max. 200
FAME content, v/v%	Max. 5.0
Copper corrosion	Class 1
Ethanol/methanol content, v/v%	Nondetectable
Ash content, %	Max. 0.01
Particulate contamination, mg/kg	Max. 10
Lubricity (HFRR) wear scar diameter at 60 °C, micron	Max. 400

Note: Additives must be compatible with engine oils, and addition of ash-forming components is not allowed. Good industry practice is to reduce contamination (dust, water, other fuels, etc.). Also adequate labeling of pumps must be defined and used.

TABLE 2.44   Changing of the quality specifiations of gasoils

Properties	US and CARB (2008)	European Union EN 590 (2009)	Japan (2007)	WWFC Category 4
Cetane number, minimum	53	51	50	55
Density, kg/m³	—	820–845	< 860	820–840
Total aromatic content, % maximum	10	— 15–20*	—	15
Policyclic aromatic content, % maximum	3.5	8.0 (2.0–4.0)[a]	—	2.0
Sulfur content, mg/kg maximum	15	10	10	10

Note: CARB: California Air Resources Board; WWFC: WorldWide Fuel Charter (2006). The dash means "no requirement."

[a] Requirement expected in the near future.

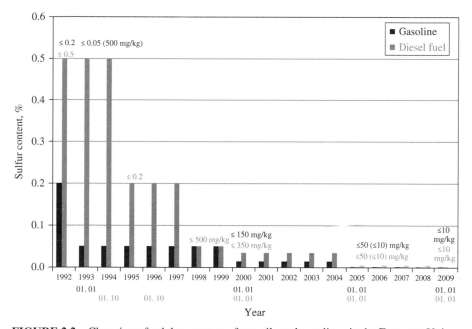

FIGURE 2.2   Changing of sulphur content of gas oils and gasolines in the European Union

The data clearly show that the quality of diesel fuels have been continuously improved over the last two decades. Especially, the maximum allowable sulfur content of diesel fuels has been substantially reduced from the level 2000–5000 mg/kg to the level 10 to 50 mg/kg (see Figure 2.2). The 10 mg/kg maximum allowable sulfur content was not implemented for further reducing the harmful effect of the SO_x emission due to sulfur sensitivity to the exhaust gas treating catalyst. For example,

TABLE 2.45 Generaly applicable requirements and test methods

Property	Unit	Limits Class A		Limits Class B		Test Method*
		Minimum	Maximum	Minimum	Maximum	
Cetane number		70.0	—	51.0	66.0	EN ISO 5165 EN 15195
Density at 15 °C	kg/m³	770.0	800.0	770.0	800.0	EN ISO 3675 EN ISO 12185
Total aromatics content	%	—	1.0	—	1.0	EN 12916 UOP 495 SS 155116
Polycyclic aromatic hydrocarbons content	%	—	0.1	—	0.1	EN 12916 UOP 495 SS 155116
Total olefin content	%	—	0.1	—	0.1	ASTM D1159 ASTM D2710
Sulfur content	mg/kg	—	5.0	—	5.0	EN ISO 20846 EN ISO 20884
Flash point	°C	>55	—	>55	—	EN ISO 2719
Carbon residue (on 10% distillation residue)	%	—	0.30	—	0.30	EN ISO 10370
Ash content	%	—	0.01	—	0.01	EN ISO 6245
Water content	mg/kg	—	200	—	200	EN ISO 12937
Total contamination	mg/kg	—	24	—	24	EN 12662
Copper strip corrosion (3h at 50 °C)	Rating	Class 1		Class 1		EN ISO 2160
Oxidation stability	g/m³	—	25	—	25	EN ISO 12205
Lubricity, corrected wear scar diameter (wsd 1.4) at 60 °C	μm	—	460	—	460	EN ISO 12156-1
Viscosity at 40 °C	mm²/s	2.00	4.50	2.00	4.50	EN ISO 3104
Distillaton 95% (v/v) recovered at	°C	—	360	—	360	EN ISO 3405

the NO$_x$ emission can be reduced indirectly by the activity of the SCR (selective catalytic reduction) system. In regard to the polyaromatic content, the reduction in the particulate matter emission leads to better combustion overall.

Clearly, among the motor gasoline and diesel fuel quality requirements of WWFC category 4, the standards of developed countries, as demanded by engine and vehicle manufacturers in respect to the quality of fuels, are stricter than the specifications for developing regions. European Union has thoroughly standardized the high isoparaffin content diesel gas oils. They are to be produced from renewable sources, from natural triglycerides, biomass waste, and using Fischer–Tropsch synthesis. The diesel gas oils are ranked by three different groups [34]:

- Gas oil product from Fischer–Tropsch synthesis (XTL),
- Hydrotreatment of vegetable oils (HVO),
- Conversion of light olefins to distillates (COD).

Table 2.45 gives the main properties of these products (CWA 15940). From the table, it is clear that these product have high cetane numbers, low temperatures, and high hydrogen contents, enabling cleaner burning.

In the next chapters we will explore ways in which the production of fuels affects internal combustion engines. We will consider both the fossil and the alternative sources. We will introduce the blending components, additives, the theory and the practice of the fuel blending, and the most important analytical and performance properties.

REFERENCES

1. Hancsók, J., Kasza, T. (2011). The importance of isoparaffins at the modern engine fuel production. *Proceedings 8th International Colloquium Fuels* January 19–20, Stuttgart/ Ostfildern.
2. The Auto-Oil II Programme—A Report from the services of the European commission. (2000). Report by the Directorates of EU, Brussels.
3. The Auto-Oil II Programme—Executive Summary. (2000). Report by the Directorates of EU, Brussels.
4. Sanger, R. P. (1995). Motor vehicle emission regulations and fuel specifications in Europe and United States. Concawe, Report No. 5/95, Brussels.
5. UOP (1994). *The Challenge of Reformulated Gasoline; An Update on the Clean Air Act and the Refining Industry*. UOP, Des Plaines, IL.
6. California Air Quality Legislation (2011). *Annual Summary*. Sacramento.
7. Kinugasa, Y. (2006). Role of JCAP II. for fuel quality improvement in Japan. *Proceedings of 4th Asian Petroleum Technology Symposium*, 2006.
8. International Energy Agency (2011). *CO$_2$ Emission from Fuel Combustion*. Highlights, Paris.

9. Shell International B.V. (2011). Shell Energy Scenario to 2050. London.

10. International Energy Agency, 2011. IEA, Paris.

11. BP Statistical Review of World Energy, June 2011. London.

12. Wester, H. J. (2010). New challenges for the world's automotive industry and its consequences of the power train development. 31st International, Vienna Motor Symposium, April 29–30.

13. Xin Q. (2011). *Diesel Engine System Design*. Woodhead Publishing, Oxford.

14. 2011/510/EU Directive: "Setting emission performance standards for new light commercial vehicles as part of the Union's integrated approach to reduce CO_2 emissions from light-duty vehicles." *Official J. E U*, L145, 1–18.

15. 2003/30/EC Directive: "The promotion of the use of biofuels or other renewable fuels for transport." *Official J. E U*, L123, 42–46.

16. 2009/28/EC Directive: "The promotion of the use of energy from renewable sources." *Official J. E U*, L140, 16–62.

17. 2002/51/EC Directive: "On the reduction of the level of pollutant emissions from two- and three-wheel motor vehicles and amending Directive 97/24/EC." *Official J. E U*, L252, 20.

18. 98/69/EC Directive: "Relating to measures to be taken against air pollution by emissions from motor vehicles and amending Council Directive 70/220/EEC." *Official J. E U*, L350, 1.

19. 2007/717/EC Directive: "On type approval of motor vehicles with respect to emissions from light passenger and commercial vehicles (Euro 5 and Euro 6) and on access to vehicle repair and maintenance information." *Official J. E U*, L171, 1–16.

20. US regulations on fuels, http://www.epa.gov/otaq/fuels.htm.

21. Walsh, M. P. (2005). *Global Motor Vehicle Emissions Regulations*. Walsh Car Lines.

22. National auto fuel policy (2003). India.

23. Indian emission regulations (2011). Complied by Automotive research association of India, Pune.

24. Ministry of Environmental Protection (2006). *Vehicle Emission Control in China*. China.

25. Hancsók J. (1997). *Fuels for Engines and JET Engines. Part I: Gasolines*. University of Veszprém, Veszprém.

26. Hancsók J. (1999). *Fuels for Engines and JET Engines. Part II: Diesel Fuels*, University of Veszprém, Veszprém.

27. California Air Pollution Control Laws (2012). *Bluebook*. Sacramento.

28. Commission of the European Communities, COM. (2006). *848: Renewable Energy Road Map—Renewable Energies in the 21st Century: Building a More Sustainable Future*. Brussels.

29. 2009/30/EC directive: "The specification of petrol, diesel and gas-oil and introducing a mechanism to monitor and reduce greenhouse gas emissions and the specification of fuel used by inland waterway vessels." *Offical J. E U*, L140, 88–113.

30. Conservation of Clean Air and Water in Europe (Oil Companies European Organisation for Environment, Health & Safety). www.concawe.be, 2012.05.06.

31. Magyar, Sz., Hancsók, J. (2007). Key factors in the production of modern engine gasolines. In Proceeding of 6th International Colloquium, Fuels 2007. January 10–11, Esslingen, Germany.

32. ACEA, Alliance, EMA, HAME (2006). *Worldwide Fuel Charter.*
33. Nagy, G., Hancsók, J. (2009). Key factors of the production of modern diesel fuels. In Proceedings of 7th International Colloquium Fuels, Mineral Oil Based and Alternative Fuels. January 14–15, Stuttgart/Ostfildern.
34. CWA 15940: CEN Workshop Agreement—Automotive fuels—Paraffinic diesel from synthesis or hydrotreatment—Requirements and test methods (2009).

CHAPTER 3

Fuels from Crude Oil (Petroleum)

3.1 CRUDE OIL

Crude petroleum (crude oil), as it comes out of a well, contains impurities such as water, salts, and dispersed solids (dust and dirt). These impurities have to be removed before the oil is transported. Crude oil is desalted and dewatered, again, prior to processing it in an oil refinery. Dispersed solids are then separated by settling and dissolved salts and water are separated by desalting/dewatering chemicals (usually surface active compounds) with a high electrical potential to agglomerate and separate salts associated with water in the crude oil. The process is called electrostatic desalination, which involves washing the crude with water, mixing in chemicals, and settling the oil mixture under electrostatic potential [1,2].

Desalted crude oil is a natural material that contains hydrocarbons (compounds from carbon and hydrogen), heteroatoms containing compounds (metal, sulfur, nitrogen, oxygen), resinous and asphaltenic compounds. Generally, the elemental composition of conventional crude oil varies in the following ranges [1–3]:

Carbon: 83.0–87.0 wt%; hydrogen: 10.5–13.5 wt%; sulfur: 0.05–6.0 wt%, nitrogen: 0.02–1.7 wt%; oxygen: 0.05–1.8 wt%; metals: 0.00–0.15 wt%.

The types of hydrocarbons occurring in the crude oil are as follows: (strongly depend on the origin of crude oil)

- Straight-chain paraffins or alkanes (40–45 wt%):
  - Normal paraffins or linear alkanes, for example,

$$CH_3\text{--}CH_2\text{--}CH_2\text{--}CH_2\text{--}CH_3 \quad CH_3\text{--}(CH_2)_n\text{--}CH_3$$
n-Pentane

where n = ca. 3–50

Fuels and Fuel-Additives, First Edition. S. P. Srivastava and Jenő Hancsók.
© 2014 John Wiley & Sons, Inc. Published 2014 by John Wiley & Sons, Inc.

○ Branched alkanes or isoparaffins, for example,

$$CH_3-CH-CH_2-CH_3 \qquad -CH_2-(CH_2)_n-CH_2-CH-$$
$$\quad\ \ | \qquad\qquad\qquad\qquad\qquad\qquad\qquad |$$
$$\quad\ \ CH_3 \qquad\qquad\qquad\qquad\qquad\quad -(CH_2)_m-$$

Isopentane

where $(n+m)<60$

C_1–C_4 straight-chain paraffins are in the gaseous state at a temperature of 20 °C and atmospheric pressure. C_5–C_{16} paraffins are in the liquid state, and they occur in the lighter fractions of the crude oil (naphtha, kerosene, gasoil). C_{17} and heavier hydrocarbons under normal conditions (20 °C, 1 bar) are in the solid state, and they occur dissolved in the higher boiling point fractions (gasoils and base oils).

• Cycloparaffins (40–45 wt%), for example:

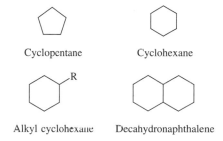

Cyclopentane   Cyclohexane

Alkyl cyclohexane   Decahydronaphthalene

R: alkyl group

Cycloparaffins (or naphthenes) with five- or six-membered ring and their lighter alkyl derivates (C_6–C_8) occur in naphthas, while other non-aromatic cyclic compounds with higher boiling points occur in kerosene, gasoil and base oil fractions.

• Aromatic hydrocarbons:
  ○ Monoaromatics, for example,

Benzene 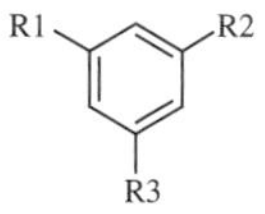   Benzene derivates

R1:<2–50; R2 and R3: 0–20

  ○ Polyaromatics, for example,

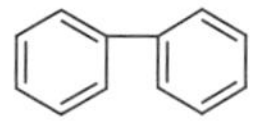

Biphenyl 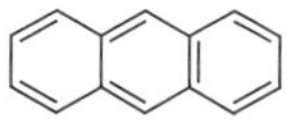   Antracene

Benzene and its homologues with their alkyl sides are in the lighter fraction. Polyring and partially saturated polyring aromatics and their alkyl and dialkyl derivates occur in the heavier fractions (gasoil, base oil).

- Olefins (alkenes):
  - Open-chain unsaturated hydrocarbons rarely (usually <0.1%) occur in crude oil.
  - These compounds are produced in refinery conversion technologies.
- Sulfur compounds (typical sulfur content of conventional crude oil is 0.6–0.8 wt%):
  - H_2S (hydrogen-sulfide)
  - Mercaptanes: R–SH (R: hydrocarbons with a carbon number of 3–30)
  - Mono- and disulfides, R–S–R′ (R, R′: alkyl groups),
    CH_3–S–CH_3, C_2H_5–S–CH_3,
    dimethyl-sulfide   ethyl-methyl-sulfide
    CH_3–S–S–CH_3, C_2H_5–S–S–C_2H_5,
    dimethyl-disulfide   diethyl-disulfide
  - Thiophenes,

    Thiophene   Alkyl-thiophenes   Benzothiophenes   Dibenzothiophenes

    R: alkyl groups

- Nitrogen-containing compounds (typical nitrogen content of conventional crude oils is 0.05–0.1 wt%):
  - Amines,

    R – NH_2,

    where R: <3–30.
  - Nitriles, R–CN, where R: <3–30.
  - Pyridine derivates (pyridine, quinoline, acridine) (basics),

  - Pyrrole derivates (pyrrole, indole, carbazole),

- Oxygen-containing compounds:
  - Aliphatic carboxylic acids, R–COOH

○ Naphthenic (carboxylic) acids (in the highest proportion),

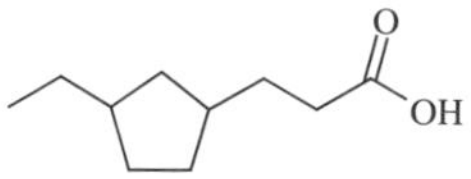

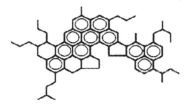

○ Furan derivates,

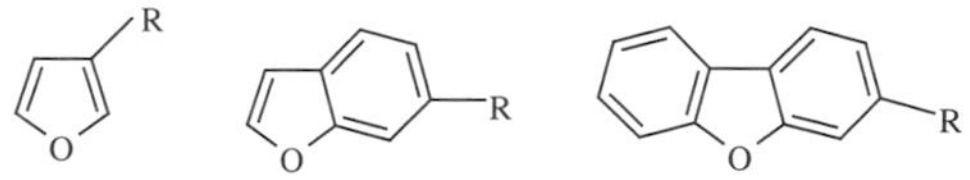

R: alkyl groups

○ Phenols,

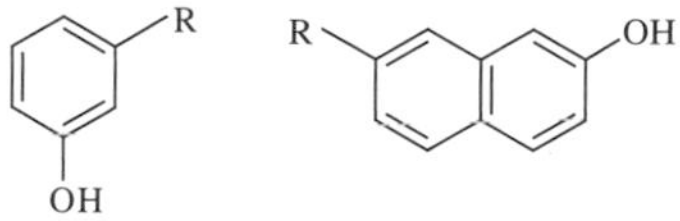

R: alkyl groups

• Resinic and asphaltenic compounds, for example:
  ○ Asphalthenes (M > 3000 g/mol, C/H = 0.79),

  ○ Resins (M = 800 g/mol, C/H = 0.72),

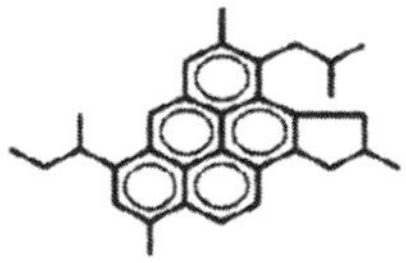

  ○ Other polyaromatic compounds (M = 400–600 g/mol, C/H = 0.54),

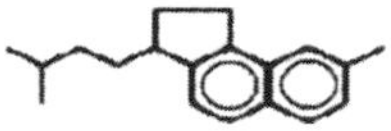

• Metal-containing compounds (especially Ni: Nickel and V: vanadium containing compounds), for example:

Metallic compounds mainly occur in the resinic and asphaltenic parts.

3.2 CRUDE OIL REFINING

Crude oil is a hydrocarbon mixture with a very wide boiling point range. Crude oil is first separated to narrower boiling range fractions, depending on the final product demand, mainly by distillation processes. These fractions are then transported to one or more different conversion and/or separation processes, and after additivation, the required products are obtained. Figure 3.1 shows a basic flow system of a flexible modern refinery that produces both naphtha and gasoil [4].

In advanced crude oil processing (in a modern oil refinery) the processes used to achieve high white product (gasoline, kerosene, and diesel) yields can be divided into two main groups:

- Separation and recovery processes
- Conversion processes
  - Without change of feedstock carbon number
  - With change of feedstock carbon number (increase or decrease)

3.2.1 Separation and Extraction Processes

Separation and extraction processes are carried out based on the difference in physical properties of the hydrocarbons. In the refining industry, during distillation, which may be carried out either at atmospheric pressure or in a vacuum, compounds are separated based on their relative volatility. Liquid–liquid extraction may also be used in refining. And it is widely used in recovery of aromatic hydrocarbons from naphtha and in heavy-base oil production (in extraction of bright-stock). In addition, the following processes are generally used in the petroleum industry:

- Adsorption (e.g., hydrogen/hydrocarbon separation)
- Absorption (e.g., in removing H_2S from H_2S-containing hydrogen gas)

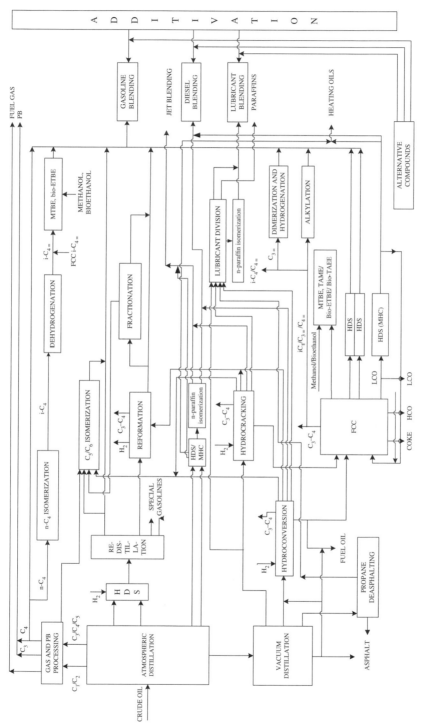

FIGURE 3.1 Basic scheme of modern crude oil refinery system used for motor gasoline and diesel fuel production

- Molecular sieve filtration (depending on size of molecules, e.g., in separation of normal and isoparaffins)
- Crystallization (depending on the difference of freezing points, e.g., for light and medium raffinate)
- Other separation technologies (e.g., in supercritical extraction or membrane separation)

Distillation of the Crude oil to Fractions   Crude oil refining starts with atmospheric and vacuum distillation. During atmospheric distillation, crude oil is separated to various fractions with narrow boiling ranges at atmospheric pressure. The crude oil then interchanges heat with product distillates in a heat exchanger train, where it is partly vaporized in a tube furnace at temperatures of 280 to 340 °C prior to being fed to a distillation column that contains trays or partly structured packing. (Under these conditions, the thermal decomposition/coking of hydrocarbons is negligible; for, such simplified flow schemes, refer to Figures 3.2 and 3.3[5].) The vapor- and liquid-phase components are separated next in the evaporation zone

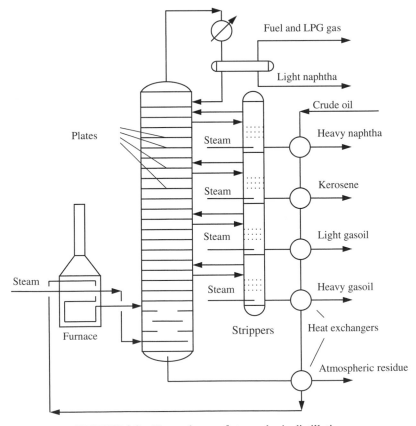

FIGURE 3.2   Flow scheme of atmospheric distillation

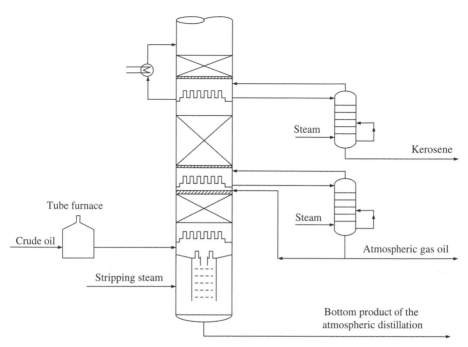

FIGURE 3.3   Atmospheric distillation column with improved distillation efficiency

of the distillation column. The light components in the vapor phase rise upward, while components with higher boiling points condense due to the down flow of the cold liquid (reflux). In the vapor phase the hydrocarbons with the lowest boiling points (below 120–135 °C) exit the column at the top. After condensation, the overhead product is separated into gas (C_1–C_4 gases) and liquid (light naphtha) products. In the order of their increasing boiling points, these fractions are heavy naphtha, light kerosene, and heavy gasoils (Figure 3.4) [6].

These fractional distillates are not separated into very precise chemical compounds. These product fractions consist of mixtures with overlapping boiling ranges, and the lighter fractions are removed by stripping in a later operational unit.

The heavy hydrocarbons that remain in the liquid phase flow down from the evaporation zone. The dissolved lighter compounds are stripped by the steam at the bottom of column. The product that settles at the bottom is called atmospheric residue. This fraction may be used as heating oil. A commonly used alternative is to feed the residue to a vacuum distillation unit to produce further distillates and base oil. It also can be routed to any residue upgrading unit to convert heavy components into valuable light products.

No further separation of the atmospheric residue can be carried out by distillation at atmospheric pressure once the molecules start to break down (referred to as decomposition or cracking) at higher temperatures. However, the residue can still be separated in a vacuum distillation unit by employing low pressure

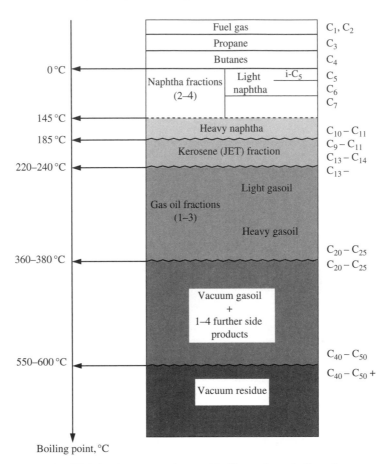

FIGURE 3.4 Carbon numbers and boiling points of oil products

(usually 5–10 kPa). At low pressures, the distillation proceeds without the molecules cracking. The heavy vacuum gas oil is separated at a temperature of approximately 150 °C, and lubricating oil (base oil) cuts can be obtained at a temperature range of 250–350 °C, (usually light, medium, and heavy cuts are drawn off; see Figure 3.5) [6]. The vacuum gasoil can be further processed for to produce light lubricating oils or routed to the conversion units to increase the yields of valuable light hydrocarbon products. The residue collected at the bottom of the vacuum column, called vacuum residue is then used as feedstock for bitumen and heavy-base oil production. At very high pressures, the residue can also be catalytically converted to high-value products.

To produce feedstock for cracking technologies, heavy vacuum gasoil (HVGO) is required. This is produced by dry vacuum distillation, that is, without the use of steam (Figure 3.6) [5]. The vacuum for this process is generated either by a barometric condenser with a steam jet ejector or by a vacuum pump.

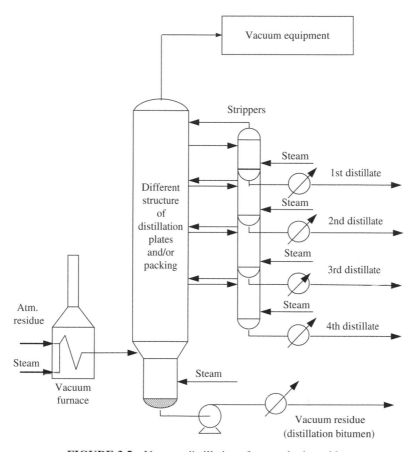

FIGURE 3.5   Vacuum distillation of atmospheric residue

3.2.2   Change of Quality and Yield of Hydrocarbon Fractions

The crude petroleum found in different geographical regions has wide variation in its chemical composition. Therefore the yields of the distillates vary accordingly. The market demands products with constant quality and quantity, and for that reason there are secondary, tertiary, and subsequent, conversion processes that add flexibility to the production process, allowing production to shift in the direction of market-driven products. For example, currently there is an increased demand for engine fuels, namely for gasoline in the United States and for diesel in India and in the European Union. Technologies are consequently employed in modern refineries to meet both types of product quality specifications based on market demand.

Conversion of straight-run distillates results in a higher yield of more valuable products. The options are as follows:

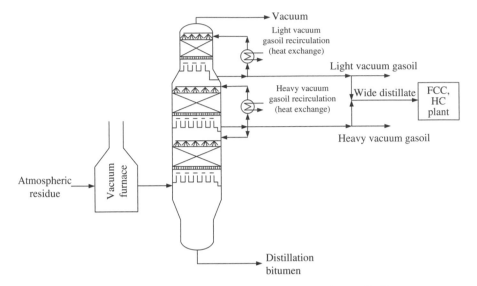

FIGURE 3.6   Packed distillation column producing heavy distillates

- Change of H:C ratio
  - Increase
    - hydrogen intake
    - carbon removal (e.g., coking)
  - Decrease (e.g., catalytic naphtha reforming)

- Change of the molecule weight:
  - Larger molecules from smaller ones (e.g., alkylation, oligomerization, etherification)
  - Smaller molecules from larger ones (e.g., thermal, catalytic and hydrocracking; coking)

Conversion Processes   In refining conversion processes, the chemical structure of the hydrocarbons in the feedstock can be changed in lower or higher degrees. One way to accomplish this is without changing the carbon number of molecules (as during desulfurization, skeletal isomerization, etc.). The other way is to change the carbon number. The carbon number can be changed in two different ways. One way is to increase carbon number, by converting the hydrocarbon with a lower carbon number into a more valuable hydrocarbon with a higher carbon number. Some examples are oligomerization of $C_3^=$–$C_5^=$ olefins to naphtha, kerosene, and gasoil by different boiling ranges of hydrocarbons or alkylation of isobutane with olefins to isoparaffins with naphtha by boiling ranges of isoparaffins. The other way is to change the carbon number by decreasing it; that is, the hydrocarbon with a high carbon number is converted into a more valuable hydrocarbon with lower carbon number. For example, by thermal and catalytic cracking, or hydrocraking, high molecular weight hydrocarbons can be changed to valuable, low molecular weight such as naphtha and gasoil.

Classification of Conversion Processes

Basically there are thermal and catalytic processes.

Thermal processing changes the molecules' size and structure due to the effect of heat, with decomposition or cracking occurring mainly at high temperatures [5,6]. These process can take the form of:

- Thermal cracking
- Visbreaking (possible in the presence of hydrogen as well)
- Coking (delayed coking, fluid coking, flexi-coking)

Catalytic conversion processes (conversion processes in the presence of catalysts) can be divided into two groups based on the presence or absence of hydrogen [5,6] Catalytic conversion processes without applying hydrogen take the form of:

- Catalytic cracking (fixed, moving, or sliding and fluid bed reactors)
- Alkylation of isobutane with olefins
- Oligomerization of light olefins (with carbon numbers of 3–5)
- Synthesis of ethers (by the reaction of isobutylene or isoamylene with methanol or ethanol)

Catalytic conversion processes in a hydrogen atmosphere enable:

- Heteroatom removal by hydrogenation (sulfur, nitrogen, and oxygen removal)
- Hydrocracking
- Hydroisomerization (e.g., affecting the skeletal rearrangement of naphtha and the base oil boiling range of hydrocarbons)
- Hydrogenation paraffin conversion (e.g., causing mainly isomerization and mild cracking of the n-paraffin hydrocarbons that are in the gas oil or base oil boiling range to create target products)
- Reforming of naphtha (to produce gasoline blending components with high octane number and aromatic content and/or to produce feedstock for the extraction of unique aromatic compounds)
- Aromatic saturation with hydrogen (reducing the aromatic content of middle distillates and base oil fractions with catalytic hydrogenation to produce kerosene, gasoil and white oils for the cosmetics and pharmaceutical industry)
- Hydrofinishing (at conventional base oil production, the final refining of the solvent extracted oil with reduced paraffin content under mild conditions beside the low degree of desulfurization and denitrogenation to enhance the stability and quality of the product)

There are further some technologies where both variants of these catalytic processes can be found. Such technology is the indirect alkylation: dimerization of isobutylene with itself or with other C_4-olefins in absence of hydrogen, then the catalytic hydrogenation of the generated C_8-olefins to isooctanes. The production of middle distillates by oligomerization of the light olefins and the subsequent hydrogenation of the produced isoolefins to their end products consists of these steps.

In this book, however, we are concerned only with engine fuels and their additives, and thus the refinery processes that are suitable for producing the two main engine fuel types, namely gasoline and diesel, which we categorize and discuss accordingly.

Production of Crude Oil Based Motor Gasoline-Blending Components

Gasoline is a motor fuel with high energy and good antiknock properties (due to a high octane number). MoGas is a blend of conventional (crude oil based) and alternative blending components, with additives used to ensure that product quality standards are achieved [7].

Refinery technologies have been developed to produce gasoline-blending components that are essential to raising the compliance of motor gasolines with the quality specifications and the projected quantity demand. The main quality requirements are as follows [7]:

- High research and motor octane number (RON: min. 95–98; MON: min. 85–95)
- Low sulfur content (≤ 10–15 mg/kg)
- Low benzene content (≤ 0.5–1.0 v/v%)
- Reduced aromatic content (≤ 25–35 v/v%)
- Reduced olefinic content (≤ 10–18 v/v%)
- Unleaded (free of metals)
- Free of halogens
- Can be blended with alternative components
- Appropriate volatility (vapor pressure, distillation curve)
- Additives in appropriate quality and quantity
- Nontoxic
- Ease of handling
- The pollution of exhaust products on the environment as little as possible
- Low production costs
- Consistent quality

In the oil industry, the main processes used to increase the octane number and improve the quality of the gasoline-blending components are the following [6]:

- Desulfurization
- Dimerization

- Alkylation (direct and indirect)
- Ether synthesis
- Hydroisomerization of n-paraffins
- Catalytic naphtha reforming
- Catalytic cracking

Some other processes used in producing gasoline-blending components that requires further improved quality are as follows [6]:

- Hydrocracking of distillates and residues
- Thermal cracking
- Coking (delayed, fluid and flexi-coking)
- Visbreaking
- Medium and heavy distillates, as well as catalytic hydrodesulfurization of residues
- Catalytic processes of base oil production

Feedstocks of processes producing gasoline-blending components are straight-run naphthas, vacuum distillates, and various residues. Straight-run naphthas with an octane number of 80 or above can only be blended into the gasoline pool. Otherwise, the octane number is improved by conversion of the straight-run naphthas [6].

Gasolines mainly consist of hydrocarbons in the approximate boiling ranges of 20–185 °C (max. 210 °C), and they have carbon numbers of 5 to 11 (see Figure 3.4). The main characteristics of the hydrocarbon groups in motor gasoline are listed in Table 3.1 [6].

Gasoline-blending components with high octane numbers have been proved to be beneficial. These gasoline-blending components are produced by:

- Upgrading naphthas with low octane numbers (e.g., isomerization, reforming)
- Separating heavier molecular weight hydrocarbons (e.g., C_7–C_9 isoparaffins by alkylation) from lighter molecules (e.g., C_1–C_4)
- Cracking long-chain hydrocarbons (>C_{20}–C_{25}) into shorter molecules with carbon numbers of 5 to 11 (e.g., catalytic cracking).

Naphtha desulfurization technologies are discussed in the following section.

Desulfurization of Naphtha Fractions   Naphtha desulfurization must comply with product quality standards [7]:

- to prevent catalyst poisoning in downstream processes
- to prevent corrosion (refinery equipment and vehicles)
- to protect catalytic converters of vehicles
- to meet general and specific environment protection regulations

TABLE 3.1 Main characteristics of hydrocarbon groups as gasoline-blending compounds

| Properties | Hydrocarbons Found in Crude Oil | | | | Hydrocarbons Obtained from Conversion (e.g., Cracking) |
	n-Paraffins	Isoparaffins	Naphthenes	Aromatics	Olefins
Density of liquid	Low	Low	Medium	High	Low
Research octane number (RON)	Very low (−19)	High enough (89) / High (100)	Low (80)	Very high (>100)	High enough (90)
Motor octane number (MON)	Very low (−19)	High enough (86) / High (100)	Low (77)	High (>95)	Low (75)
Sensibility (RON-MON)	Very low (0)	Low (<+3)	Medium (+3)	High (>+12)	Very high (+15)

There are options for desulfurization of naphtha that involve different boiling ranges (yielding light, middle, heavy, and full-range naphthas). In the modern refinery, the full-range naphtha that is produced via atmospheric distillation is directly desulfurized for the economic reason of saving energy. Then it is routed to redistillation to produce naphtha fractions with narrower boiling ranges. The quality of these fractions is subsequently improved in downstream catalytic process units (see Figure 3.1). In less modern refineries, each naphtha feedstock to the catalytic conversion units is desulfurized separately.

Various sulfur compounds are present in the different naphtha fractions. These compounds can be mercaptans (e.g., ethyl-mercaptan, n-nonyl-mercaptan), sulfides (e.g., dimethyl-sulfide, n-butyl-sulfide), disulfides (e.g., dimethyl-disulfide, ethyl-disulfide) and/or thiophenes (e.g., thiophene, benzothiophene). Generally, there are two warp to desulfurize naphtha fractions (boiling point range: 20–70 °C). For light naphtha, the malodorous and very corrosive mercaptans (thiols) are extracted by an alkali wash (a solution of NaOH, KOH + additives to improve solvency). The alkali wash is followed by steam flush-out. The sulfur content of the end product is <2–10 mg/kg [1–3].

The remaining mercaptanes are oxidized by exposing to air the disulfides that are less corrosive. Any remaining thiols that cannot be dissolved in the excess alkali solution, are oxidized by a cobalt compound (e.g., cobalt phthalocyanine) catalyst that is added to the base solution [1–3].

The second, more economical way is to desulfurize the medium, heavy, and full-range naphtha fractions (typical boiling point ranges of 65–120 °C; 110–210 °C; 30–220 °C, respectively) by catalytic hydrogenation. For example, with the $CoMo/Al_2O_3$, $NiMo/Al_2O_3$ catalyst at 280–320 °C temperature, 20–35 bar pressure, 3.0–5.0-h^{-1} liquid hour space velocity (LHSV: flow of feedstock in m^3 through 1-m^3 catalyst during 1 hour catalyst at normal condition of 20 °C and 101.3 kPa), and a 100–250 Nm^3/m^3 hydrogen/hydrocarbon ratio, the sulfur is removed from each sulfur compound as hydrogen-sulfide simultaneously by saturating the unsaturated hydrocarbons (see Figure 3.7). These products have a sulfur content of <1–10 mg/kg depending on the desired end product.

For the desulfurization of the naphtha fractions produced by thermal and catalytic cracking processes (e.g., FCC, coking), there are a number of options, (namely adsorption, oxidation, hydrogenation, or alkylation) [7–14]. The most economical, and therefore industrially widely used, is selective hydrogenation with a special $CoMo/Al_2O_3$ [7,15] (or even NiMo, Pt, Pd/USY [7,15]) catalyst. This catalyst is developed especially for desulfurization. The unfavorable olefin saturation reactions cause significant octane number loss (the octane numbers of the olefins are higher by 10 to 30 units than the octane numbers of the n-paraffins with the same carbon number), and this happens at a low rate. For this reason the process is called selective hydrogenation.

The feedstock is processed in the cracked naphtha at an initial boiling point of 70–90 °C, and it contains as high as 20–100 mg/kg sulfur. The resulting sulfur content of the stabilized product is less than 2–10 mg/kg and the octane number loss is as low as 1.0–2.5 units at process conditions of 180–260 °C, 20–30 bar, LHSV of 1.0–3.0 h^{-1}, and 150–250 Nm^3/m^3 hydrogen/naphtha volume ratios.

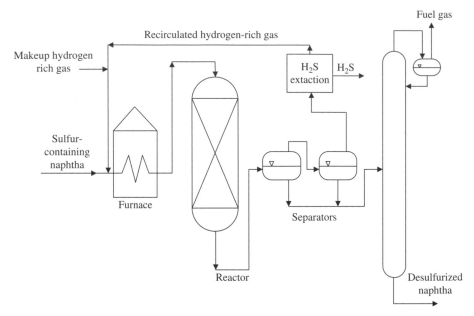

FIGURE 3.7   Basic scheme of heteroatom removal of naphthas by catalytic hydrogenation

During hydrodesulfurization of naphthas, other heteroatoms, namely nitrogen and oxygen, may be converted and removed as ammonium and water, and the extent of this depends on the composition of the catalyst used and the process parameters applied [9].

Processes of Producing Gasoline-Blending Components with High Octane Numbers

DIMERIZATION AND OLIGOMERIZATION OF LIGHT OLEFINS   Unsaturated dimers, trimers, tetramers, and even higher molecular weight liquid products may form during the catalytic oligomerization (a low-degree polymerization of 2–6 molecules) of the olefins with carbon numbers of 3 to 5 that are obtained from the cracking technologies (Figure 3.8) [16]. The dimers can be a blend of gasoline components, while the trimers and higher molecular weight compounds can be a blend of middle distillate components (Jet, gasoil), after saturation (hydrogenation) of the olefinic double bonds. The catalysts used in oligomerization are conventionally phosphoric acids at temperatures of 150–250 °C and pressures of 15–70 bar. [3]. In more recent times, the catalyst support has been either ion exchange resin operating at 75–90 °C and 15–25 bar [17,18] or zeolites operating at 200–280 °C and 10–50 bar [19] and ionic liquids [3,20,22]. The yield of C_6–C_{10} and C_{11+} compounds is controlled by the reaction conditions. The research octane number of the oligomer gasolines produced from propene and butane is about 95, but its motor octane number is only about 83.

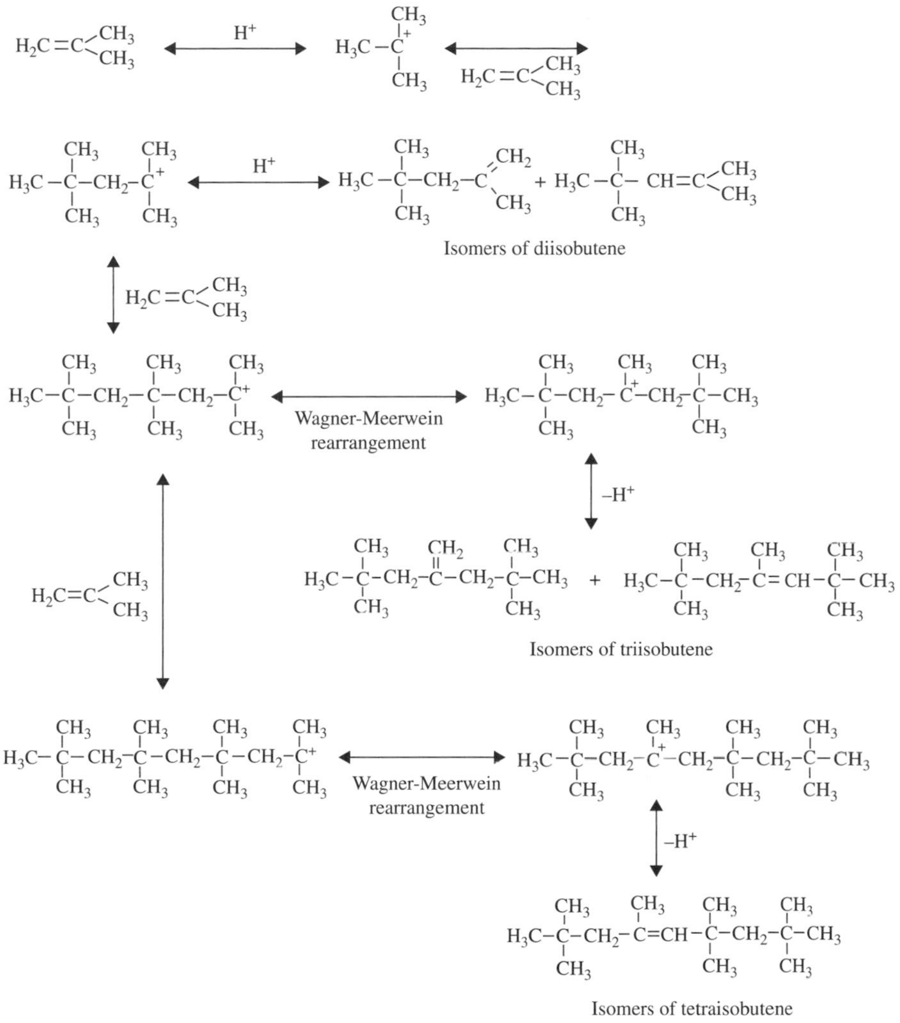

FIGURE 3.8 Basic mechanism and reactions of the oligomerization of isobutene

ALKYLATION In the crude oil industry, the term of "alkylation" means the alkylation of isobutene with C_3–C_5 olefins (direct alkylation) [1–3,7,16] or the dimerization of isobutylene mostly with C_4 olefins (including isobutylene), and the hydrogenation of the produced iso–$C_8^=$ product mixture to isoparaffins having carbon number 8 (indirect alkylation) [7,16,23].

In *direct alkylation, mainly* C_3–C_5 olefins and isobutylene-rich fractions from FCC technology are used as feedstocks, with strong acid catalysts (sulfuric acid [1–3,16,24,25], hydrogen fluoride [1–3,25], and lately solid super acids [26–29]) being employed. Reactions take place through tertiary butyl carbenium ions [1–3,16].

Reactions   The initial reaction of proton to isobutene, for example, is as follows:

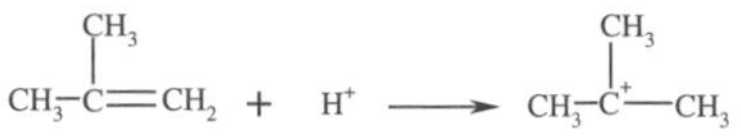

$$CH_3-C(CH_3)=CH_2 + H^+ \longrightarrow CH_3-\overset{+}{C}(CH_3)-CH_3$$

With the addition of a strong acid, the reaction is

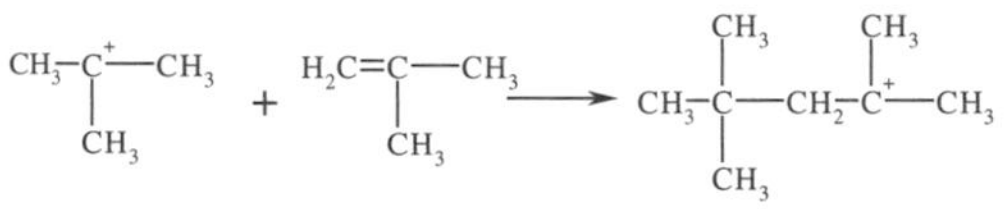

$$CH_3-\overset{+}{C}(CH_3)-CH_3 + H_2C=C(CH_3)-CH_3 \longrightarrow CH_3-C(CH_3)_2-CH_2-\overset{+}{C}(CH_3)-CH_3$$

Chain propagation

$$CH_3-C(CH_3)_2-CH_2-\overset{+}{C}(CH_3)-CH_3 + CH_3-CH-CH_3(CH_3) \longrightarrow CH_3-C(CH_3)_2-CH_2-CH(CH_3)-CH_3 + CH_3-\overset{+}{C}(CH_3)-CH_3$$

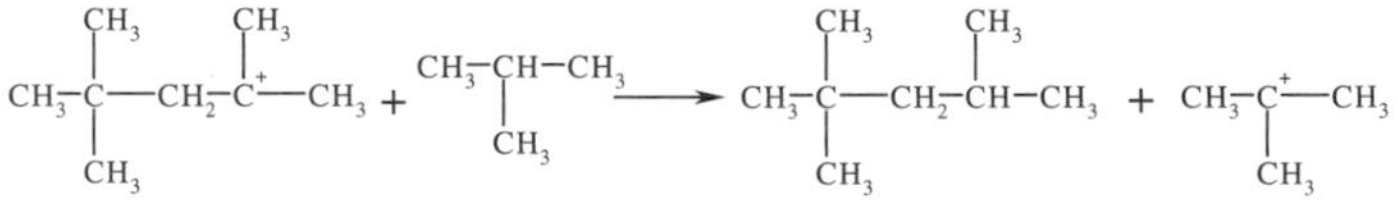

2,2,4-TMP

Chain termination

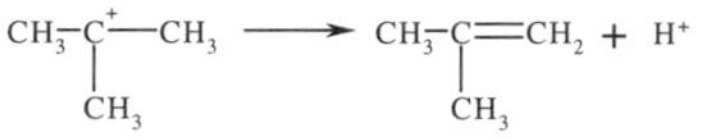

$$CH_3-\overset{+}{C}(CH_3)-CH_3 \longrightarrow CH_3-C(CH_3)=CH_2 + H^+$$

Besides these main reactions, hydrogen transfer, isomerization, polymerization, cracking and especially, in case of liquid catalysts, complexes of the catalysts may from as well. Exothermic reactions occur at a temperature of about 5–10 °C in the case of the sulfuric acid catalyst and 25–40 °C in case of the HF catalyst, at pressures of about 20 bar and 15 bar, respectively. The isobutylene/olefin ratio is between 10 and 20 per volume [1–3,24,25]. The carbon number of the end products is obtained by summing the carbon numbers of isobutylene and the olefin reactants. The general composition and octane numbers of the alkylates produced with the different catalysts using the same feedstock are [30]

Catalyst	Reaction Product $C_5/C_6-C_7/C_8/C_{9+}$, %	RON/MON
H2SO4	7/76/10/7	95.6/93.6
HF	2/91/4/3	95.7/94.2
Solid acid	5/71/11.5/12.5	97.0/93.2

When the carbon number of the olefins is 3 or 5, the octane number of the alkylate is 3 to 4 points lower than that of the alkylate produced from the C_4 olefins.

In *indirect alkylation*, the dimerization of the olefins results from the ion exchange with the resin or zeolite. One component of the feedstock is isobutylene (a C_4 fraction

TABLE 3.2 Octane numbers of isooctanes obtained with oligomerization of isobutene and different butenes and then hydrogenation of isooctenes

Isobutene Reaction with	Isomer of Isooctane	RON	MON
1-Butene	2,2-Dimethyl-hexane	72.1	77.6
1-Butene	2,3-Dimethyl-hexane	71.5	78.8
2-Butene	2,3,4-Trimethyl-pentane	102.5	95.8
2-Butene	2,3,3-Trimethyl-pentane	106.0	99.6
2-Butene	2,2,3-Trimethyl-pentane	109.8	99.8
Isobutene	2,2,4-Trimethyl-pentane	100.0	100.0

from either FCC or steam cracking technologies, or produced by dehydrogenation of isobutane), and generally the other component is an olefin with also a carbon number of 4 [23,25,31,32]. The octane numbers of iso-C_8-olefins and isooctanes produced by hydrogenation of iso-C_8-olefins are listed in Table 3.2.

ISOMERIZATION OF LIGHT NAPHTHA   In the crude oil industry, the definition of "isomerization" relates to the skeletal realignment of low-value hydrocarbons with less favorable properties to isoparaffins with desirable properties. The process takes place without change of carbon number; that is, the number of carbons in the feedstock is equal to the number of carbons in the product.

With the isomerization of n-butane, isobutane is produced, and this is basically the feedstock of the alkylate production [1–3,7]. Furthermore isobutylene can be produced by dehydrogenation of isobutene, which can be the feedstock of the synthesis of alkylates in indirect way or ethers (e.g., methyl-tertiary-butyl-ether: MTBE; (bio) ethyl-tertiary-butyl-ether: (bio) ETBE) [23].

The isomerization of n-butane is carried out in fixed bed reactors connected in a series on platinum (0.3–0.4%)/alumina/chlorine (8–10%) catalyst at conditions of 130–200 °C, 20–35 bar, 1.5–2.5 m^3/m^3h liquid hourly space velocity, and a hydrogen–hydrocarbon mol ratio of 2–3. The isomerization product has a yield of at least 98.5% on a volume basis and contains about 50–60 v/v% isobutane. The isobutane fraction from the product is obtained in a distillation column as an overhead product. The isobutane content is about 96 v/v%. An n-butane-rich fraction is drawn off as a side product, leaving a bottom mixture of unreacted C_{5+} hydrocarbons. The side product is blended with the feedstock mixture and is then recycled into the isomerizer reactor [33].

With the isomerization of the pentane/hexane fractions, C_5 and/or C_6-fractions rich in n-paraffins and having low octane numbers (RON: 30–65) are transformed into a gasoline-blending component rich in isoparaffins and with high octane numbers (RON: 80–91). The carbon number, namely the number of carbons in the feedstock and product, is unchanged during the process. Isomerization reactions are reversible, the same as they are in thermodynamics (Figure 3.9) [6]. Lower temperatures enable iso-compounds, especially those of multi-branched compounds with higher octane numbers (e.g., 2,2- and 2,3-dimethyl-butane: DMB), to form [6,34,39].

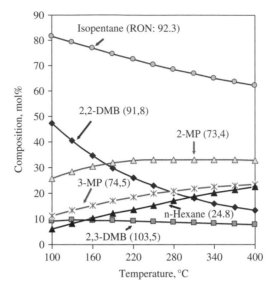

FIGURE 3.9   Equilibrium concentration of isopentane and isohexane isomers as a function of temperature

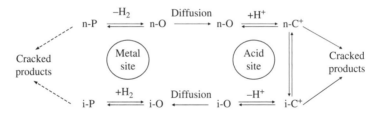

n-P: n-paraffin; n-O: n-olefin; n-C+:n-carbenium ion;
i-C+: iso-carbenium ion; i-O: iso-olefin; i-P: iso-paraffin

FIGURE 3.10   Isomerization of n-C$_5$/n-C$_6$-paraffins on bifunctional catalysts

The general (conventional) mechanism of isomerization of n-C$_5$/n-C$_6$ paraffins on the bifunctional catalyst is shown in Figure 3.10 [6]. These bifunctional catalysts may be as Pt/Al$_2$O$_3$/chlorine (6–12%) [34,39–41], Pt/sulphated metal-oxides [42–44], metal-oxide mixtures [45,46], or Pt/H-mordenite and other zeolites [36,38,47–50]. The first three catalysts work at low temperature (<200 °C), and the zeolites work at medium temperature (200–<300 °C) (Figure 3.11).

The processes operate at conditions of 25 to 35 bar, 1.0 to 2.5 h^{-1} LHSV and H$_2$/ hydrocarbon molar ratio of 1.0–2.0. During isomerization of a feedstock with a research octane number of approximately 70 will result in a mixture of isomerates with research octane number of 84 on the first (at 110–140 °C), 80–82 on second and third (at 180–210 °C), and 79 on the fourth catalyst types, respectively, in once-through operation mode [1–3,6,35,41].

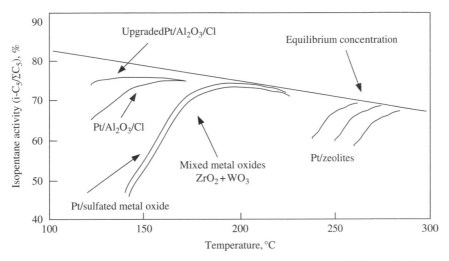

FIGURE 3.11   Comparison of the activity of commercial catalysts for isomerization

Isomerates with higher octane numbers can be achieved using the same catalysts when the C_5 and C_6 components with the low octane numbers (e.g. methyl-pentanes with RON of 73–74) are separated from the reactor effluents and then recycled. The isomerates produced on the first, the second and third, and the fourth catalyst types have research octane numbers of 91–93,88–89, and 86–88, respectively, resulting in overall octane number increments of 16–23 compared to the feedstock. The products have excellent sensibility: around +1. The yield of the liquid products is 99.5–99.8% on Pt/Al_2O_3/chlorine, while it is 95–96% on other catalysts [1–4,6,34,36,39,51].

With a slight change in composition and technology, the catalysts mentioned before, excluding the mixed metal-oxides, are suitable for benzene saturation of light gasoline fractions with a benzene content of 0.5–4% and for isomerization of n-paraffins in one step. The benzene content of the products is <0.01 v/v%, and they are also sulfur free (<1–5 mg/kg) [41,52–57].

SYNTHESIS OF ETHERS   The synthesis of ethers (e.g., MTBE: methyl-tertiary-butyl-ether, ETBE: ethyl-tertiary butyl-ether, TAME: tertiary-amyl-methyl-ether) occurs on acidic ion exchange resins by the reacting methanol or ethanol and the light cracked gasoline having a high olefin content (the isobutylenes are obtained by dehydrogenating isobutane from the C_4 fraction of the steam cracker or by isomerization of n-butylenes) or C_4–C_5 olefins with high activity [58–61]. The synthesis of MTBE can be expressed by the equilibrium reaction equation

In conventional processes, the catalyst is installed in a shell-and-tube and a fixed-bed reactor. The separation of the product mixture is performed by distillation [62]. Modern processes also use reactive (catalytic) distillation to aid the reaction, consequently a higher conversion can be obtained in one step. (In some process variants of the selective hydrogenation of diolefins, and in TAME production, the isomerization of 3-methyl-1-butylene to the reactive 2-methyl-2-butylene is carried out in this unit, beside the separation of the ether from the other compounds [63]). In conventional MTBE, ETBE, and TAME syntheses, the olefin conversion is 95–96%, >85%, and 67–70%, respectively, and in catalytic distillation, it is 99%, >95%, and 90%, respectively [61,64,65].

The temperatures required for the synthesis of the ethers increase with the increasing carbon numbers of the feedstock (55–110 °C). In the commonly applied pressures of 5–20 bar with a 1.1–1.3:1 alcohol/isobutylene molar ratio and LHSV value of 3.0–6.0 h^{-1}, the conversion to ether decreases from 99% obtained in the case of isobutylene to 30–40% in the case of C_7 and C_8 olefins. The research octane number of the C_5–C_6 ethers is about 110–118 and their sensibility is 13–18 units [65–67].

The ethers contain oxygen atoms; hence they are classed among the oxygenate type of engine gasoline-blending components.

CATALYTIC NAPHTHA REFORMING   The catalytic naphtha-reforming produces fractions rich in aromatic hydrocarbons (>70–75 wt%). With this conversion process the component composition of the fraction is changed, and the target product has a lower hydrogen content than the feedstock, which is a rare objective in the refinery industry [6]. The product is used partly as a high octane number blending component (98–102), and partly to produce individual aromatic compounds (benzene, toluene, xylenes: BTX). Moreover, because of the high amount of produced hydrogen, this technology is an important hydrogen source for refineries [1–3,6,68].

The boiling range of the reforming feedstock fractions (e.g., desulfurized straight-run naphtha, hydrocracked naphtha, hydrogenated naphthas from thermal conversion processes, which do not yet contain olefins) is between 70 °C and 180 °C; it is advantageous if the naphthene content is high. The compounds with sulfur, nitrogen, and metal content can poison the catalyst, so the amounts of these compounds have to be below 1 mg/kg [1–3,6,69].

The processes differ in catalyst composition, in regeneration and activation methods and parameters, in the reactor system, in the process parameters, and so forth [1–3,6,68,70,71]. To enhance the expected reactions, a bifunctional (dehydrogenating and skeletal rearranging) catalyst is needed. The catalysts usually contain different metals mainly with an alumina-oxide support [72–74]. The most applied combinations are 0.2–0.4% platinum+0.15–0.5% rhenium, 0.15–0.4% platinum+0.1–0.3% rhenium and/or 0.15–0.5% tin or iridium, and they usually contain 0.6–1.2% halogen, mainly chlorine [1–3,6,72,74–80].

Of the previously mentioned catalysts, the following main reactions take place (according to the process parameters): dehydrogenation of naphthenes and dehydrocyclization of paraffins to aromatics, isomerization of n-paraffins and

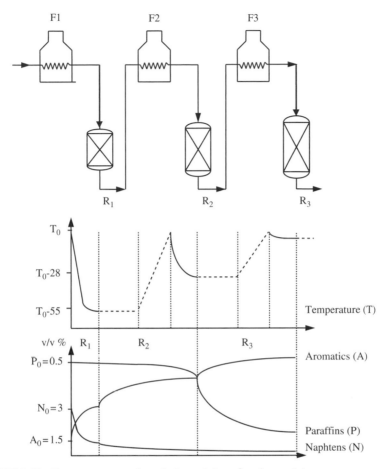

FIGURE 3.12   Reactor system of catalytic naphtha reforming and the temperature profile of each reactors

naphthenes (to isoparaffins and furthermore that of the cyclopentane derivates to compounds containing the cyclohexane ring), hydrocracking of paraffins, dealkylation of alkyl-aromatics (e.g., from toluene to benzene), alkylation, disproportionation, and coke formation [1–3,6,16,69,72,74,81–85]. All these reactions are endothermic, so they need high temperatures to take place.

The generally applied process parameter combinations are the following: temperature 480–540 °C, pressure 4–25 bar, liquid hourly space velocity 1.5–2.5 m³/m³h, and hydrogen to hydrocarbon molar ratio 4:1–8:1 [1–3,6,70,72,86–88]. The heat of the feedstock and the recirculated gas does not provide enough heat for the endothermic reactions, so the catalyst bed cools down vertically. To prevent the extreme cooling, reactors are applied in series with increasing volumes (e.g., 1:2:4), and after leaving each reactor, the reaction mixture is heated in a fire heater before moving to the next rector (see Figure 3.12) [6].

TABLE 3.3 Yields and quality of the products of naphtha reforming

Products	Yield (wt%)	Properties
Hydrogen-rich gas	7–10	Hydrogen concentration: 60–80 v/v%
Fuel gas (C_1–C_2)	1–3	
Propane	3–5	
Butanes	5–8	ca. 50% isobutane
Reformate	74–84	RON: 97–101 MON: 86–88 End boiling point (EBP): with 20–30 °C higher then the EBP of the feedstock Aromatic content >70% Density: 0.760–0.790 g/cm³

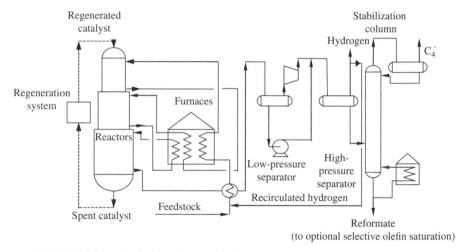

FIGURE 3.13 Catalytic naphtha reforming with continuous catalyst regeneration

By decreasing the pressure and the recirculation rate of the hydrogen, the reformate having the high octane number (about 100) can be produced with more than 70 wt% yield (see Table 3.3) [6], but in this case the catalyst deactivation will be faster. The main cause of the faster deactivation is the coke formation on the catalytic active sites.

If the activity of the catalyst decreases significantly, it has to be regenerated and activated periodically (e.g., 10–16 months; Figure 3.12) or continuously (Figure 3.13) at oxidative conditions [1–3,6,89–97]. In the older technologies with high pressure and hydrogen recirculation, a "semi-regenerative" solution [84,97] is applied. In technologies with lower pressure and hydrogen recirculation, the continuous operation is done in reactors above each other with a platinum/tin/alumina catalyst and by moving (sliding) the catalyst bed (Figure 3.13). With this solution, it is possible to regenerate and activate the catalyst continuously [1–3,6,86,92,98].

In reforming naphtha in continuous operation, the saturation (hydrogenation) of the olefins is carried out in small batches (1–4%) if the reformate is to be used for the production of aromatics [99].

FLUID CATALYTIC CRACKING   Catalytic cracking is a conversion process that changes the fractional composition of the crude oil, in that cleavage of the C–C bonds of the heavier molecular weight hydrocarbons also takes place; therefore lower value hydrocarbons with higher carbon numbers can be transformed to more valuable products with lower carbon numbers. The relatively higher hydrogen content of the produced lower molecular weight components is covered by the transformation of a part of the feedstock to heavier molecular weight liquid products with less hydrogen content and to coke. The yield of light products from crude oil can significantly be increased by catalytic cracking [1–3,6,86,100–106]. Catalytic cracking technologies are distinguished in terms of feedstock, distillate and residue [1–3,6]. The distillate feedstock for cracking can be hydrogenated or nonhydrogenated [1,107–109]. The feedstock of residue cracking may be heavy distillates (gasoils, oil distillates), bitumen free oils, or heavy products of mild and severe hydrocrackers.

The efficiency of the cracking is unfavorably affected by hydrocarbons with heteroatoms and by different metal compounds. The basic nitrogen compounds, sodium and vanadium, can neutralize the acid centrums of the catalyst, or at least decrease its acid strength, and change the pore structure of the catalyst. Nickel catalyzes undesirable dehydrogenation reactions. Accordingly, only adequately treated distillates (e.g., middle and heavy distillates of mild hydrocracking) and residues (e.g., oil produced with extraction from heavy vacuum gasoil) are suitable as feedstocks for this process [86,107–109].

Nowadays, catalysts for cracking are made of synthetic zeolites (e.g., Y-type zeolites) with or without rare earth metals (e.g., lantane, cerium). These are applied in dispersed form in amorphous alumina or alumina/silica matrix because of their very high activity and low mechanical strength [1–3,6,86,110–112]. Additives that promote the formation of light olefins and raise the octane number, reduce sulfur emission, promote carbon-monoxide burning, trap metal or deactivate, reduce residue, improve catalyst strength, are employed and blended into the catalyst [1,86, 113–125]. The catalyst particles must have sizes of 50–80 micrometers to keep the catalyst bed in a fluidized state [1,6,86].

The transformation of hydrocarbons takes place through carbonium ions on the strongly acidic catalyst. The electron-deficient carbonium ion containing positively charged carbon atom forms either via subtraction of a hydride from a saturated hydrocarbon or via addition of a proton to an olefin or an aromatic hydrocarbons. (The catalyst must be capable of electron uptake or proton release.) Carbonium ions are able to act in various reactions due to their electron-deficient structure. These reactions are the following: isomerization with hydride and methyl group migration, resulting in a high amount of isocompounds in the product; beta-positioned chain split, which results in a significant amount of olefins produced with carbon number of 3 and more; and hydrogen transfer, during which hydride ions are subtracted from the paraffin hydrocarbons or proton transfer takes place to olefins. Olefinic hydrocarbons that are in the gasoline boiling range are stabilized after the primary reactions. In the final step of the reaction, the carbonium ions form molecules through either hydride-ion subtraction or proton addition to the acidic sites of the catalyst. The main reactions taking place during fluid catalytic cracking are the following [1–3,6,86]:

Cracking	Examples
Cracking of paraffins to olefins and paraffins of lower molecular weight	$C_{10}H_{22} \rightarrow C_4H_{10} + C_6H_{12}$
Cracking of olefins to olefins of lower molecular weight	$C_9H_{18} \rightarrow C_4H_8 + C_5H_{10}$
Splitting of side chains of aromatics	$Ar–C_{10}H_{21} \rightarrow Ar–C_5H_9 + C_5H_{12}$
Cracking of naphthenes to olefins and low molecular weight cyclic compounds	$Cyclo–C_{10}H_{20} \rightarrow C_6H_{12} + C_4H_8$
Isomerization	
Double bond shift	$1-C_4H_8 \rightarrow trans-2-C_4H_8$
Skeletal isomerization of normal olefins	$n-C_5H_{10} \rightarrow iso-C_5H_{10}$
Skeletal isomerization of normal paraffins	$n-C_5H_{12} \rightarrow iso-C_5H_{12}$
Isomerization of cyclohexane to methyl cyclopentane	$C_6H_{12} \rightarrow C_5H_9CH_3$
Hydrogen transfer	Naphthenes + Olefins $\rightarrow$ Aromatics + Paraffins
	$C_6H_{12} + 3C_5H_{10} \rightarrow C_6H_6 + 3C_5H_{12}$
Alkyl-group shift	$C_6H_4(CH_3)_2 + C_6H_6 \rightarrow 2C_6H_5CH_3$
Cyclization of olefins	$C_7H_{14} + C_6H_6 \rightarrow 2C_6H_5CH_3$
Dehydrocyclization	$n-C_8H_{18} \rightarrow 2C_6H_5CH_3$
Dealkylation	$iso-C_3H_7–C_6H_5 \rightarrow C_6H_6 + C_3H_6$
Condensation	$Ar–CH=CH_2 + R_1CH=CHR_2 \rightarrow Ar–Ar + H_2$

Coke-like deposits develop on the surface of the catalyst during the reactions, but these substances mostly originate from the nonvolatile polycyclic (likely heteroatom-containing) hydrocarbons in the feedstock and are not the product of catalytic cracking reactions. These deposits strongly adhere to the hot surface of the catalyst and become involved in further condensation reactions with other nonvolatile hydrocarbons. The amount of coke produced mainly depends on the temperature and the residence time [1–3,6,86].

Nowadays, the catalytic cracking takes place in fluidized bed reactor (Figure 3.14) [6]. This part of the reactor is the riser ("long rising pipe"). The process is called fluidized catalytic cracking (FCC). The chemical reactions take place in the catalyst riser at 480–530 °C, 2–3.5 bar pressure, and with a catalyst-to-feedstock ratio of 6–7 at the mass base. The mixture of fresh feed and hot catalyst, that contains the fresh catalyst, and the so-called equilibrium catalyst returning from the regenerator, is continuously moved through the reactor. The mixture of the feedstock and the enormous amount of catalyst is kept moving partly by the kinetic energy of the feedstock but mostly by the upward movement of the fast evaporated lighter components forming on the hot catalyst particles.

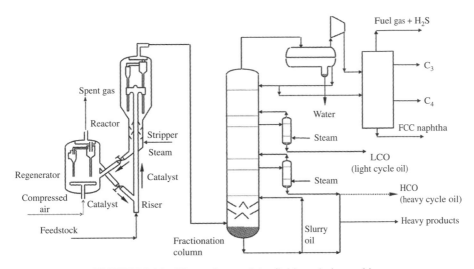

FIGURE 3.14 Flow scheme of the fluid catalytic cracking

Above the riser, the hydrocarbons are separated from the catalyst and routed to fractionation and gas concentration units. The catalyst goes to the steam-stripping section where any hydrocarbons that adhered to the surface are removed; it is then routed to the regenerator where the deposited coke on its surface is burned off in the presence of air. The small particles heat up, and the catalyst drops and flows to the feedstock injection point via a slide valve. Hot flue gas that was produced during the burning-off of coke goes to a heat recovery system through a cyclone system. The temperature in the regenerator is 680–710 °C, and the pressure is 2–2.5 bar [1–3,6,86,126–128]. (In the cracking of distillates or residue, the catalyst regeneration is carried out in one step or two steps, respectively [129].)

The main products of the catalytic cracking are the following: 3–5 wt% C_1–C_2 fraction, which is suitable for alkylation of benzene to produce ethyl-benzene and styrene due to its ethylene content; 7–20 wt% C_3–C_4 fraction, which is an important feedstock for alkylation, oligomerization, and synthesis of ethers due to its high unsaturated content; 30–60 wt% cracking naphtha with octane numbers of 91–94 and with a sensibility of only 11–14 units; 10–20 wt% high aromatic and olefin content gasoil fraction (light cycle oil [LCO] cetane number <35), which is suitable for gas-oil blending followed by hydrotreating; and 10–15 wt% cracking residue (HCO), which is suitable for heating oil or, when routed, for coking. The coke deposited on the surface of the catalyst, of quantity 3–6 wt%, is burned off in the regenerator unit [1–3,6,86,126].

The FCC units primarily run in a so-called gasoline-producing mode (40–60 wt% of the gasoline pool consists of cracked gasoline) [1–3,6,86,126], but the olefin–producing mode (the yield of olefins is 40–45 wt%) has been widely applied for years [106,130–132]. Nowadays a gasoil mode is also used (the yield of cracked gasoil is about 45 wt%) [133–135]. Figure 3.15 gives the yields of products in a volume basis.

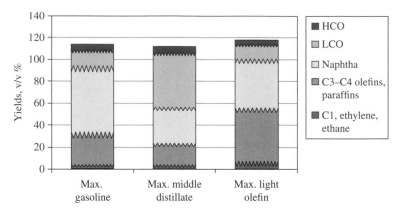

FIGURE 3.15   Olefin, naphtha, and gasoil flexibility of a FCC plant

For the past ten years, the FCC process has been widely used as well for the conversion of residual fractions. In these conversions there is a need for more rigorous reaction parameters, a higher catalyst-to-feedstock ratio, and a higher temperature in the two-step regenerator unit [1,86,136].

Other Gasoline-Blending Components Derived from Crude Oil   Some of the production processes of naphthas are suitable for gasoline blending after one or more of their properties are improved.

Hydrocracking of Distillates   Hydrocracking is a process used for changing the composition of crude oil hydrocarbons. During hydrocracking the bigger molecules of distillates and residues are transformed into smaller molecules in the presence of atmospheric hydrogen, both with and without use of a catalyst (as is typical of residue processing) [1–3,6,86,137]. In addition, the hydrogen content of products substantially increases compared to the amount of hydrogen in the feedstock.

The light and heavy naphthas produced during hydrocracking have low octane numbers of 70–75 and 60–65, respectively [1–3,6,86,126]. Therefore these naphthas cannot be directly used for gasoline-blending despite their sulfur and nitrogen content being lower than 1–5 mg/kg and their aromatic content being below 15% (typically 2.0–5.0%). In order to increase octane number of light naphtha, the naphta has to be routed to the isomerization process (after separation of isoparaffins); furthermore heavy naphtha has to be sent for reforming [1–3,6,86,126]. Hydrocracking naphtha without any further quality improvement produces excellent feedstock for steam cracking (light olefin producing) units [6].

The chemistry, catalysts, process parameters, and industrial hydrocracking of distillates are discussed in a later section on the blending components of diesel gasoils.

Naphthas of Residue Processing   The thermal conversion of residues [138] to more valuable lighter products is realized by different variants of thermal cracking [1–3,86,139]. The thermal stability of hydrocarbons in ascending order is as follows: n-paraffins, isoparaffins, cycloparaffins (naphthenes), aromatics,

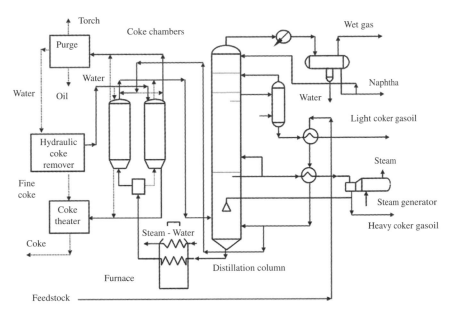

FIGURE 3.16 Flow scheme of delayed coking process

naphthenic-aromatics, polycyclic naphthenes, and polycyclic aromatics. Thermal cracking at temperatures of 500 °C and above and at pressures of 5–70 bar takes place through a free radical mechanism. The free radicals formed by the cleavage of C–C and C–H bonds of hydrocarbon molecules may break down further. The dehydrogenation process needs more severe conditions.

The velocity of cracking is basically determined by the applied temperature while the distribution of the products is determined by temperature, pressure, and mainly by residence time. Gas formation is favored by higher temperature (>550 °C) and lower pressure (1–10 bar), while formation of naphtha and middle distillates is promoted by changing parameters the other way round. Further severe process conditions increases coke formation.

Visbreaking is a thermal cracking process operating at mild conditions, which increases the yield of distillates and reduces the viscosity of fuel oil. The process is carried out in one- or two-tube furnaces, and in an additional reaction chamber in some cases, at pressures of 5–20 bar and at temperatures of 450–500 °C. Besides the lower viscosity residue product, fuel gas (1%), naphtha (5–7%), and gasoil (10–13%, which can be 35% when using two-tube furnaces) are produced [3,140,141]. Visbreaking distillates have a high sulfur content, so they need to be reprocessed, or desulfurized, if intended for use as fuel.

Coking is a thermal cracking process carried out at relatively severe conditions, and used for mainly gasoil production. A variant of coking is *delayed coking* (Figure 3.16) [6]. In this process, the feedstock is heated up to temperatures of 480–520 °C in a tube furnace and then routed to a coke drum (chamber), operating at pressures of 2–8 bar and applying long reaction times. Coke (20–40% of one feedstock) is removed periodically from the chambers (by alternating between two or more chambers); all the while, the coke drum overhead vapor is continuously passed to a distillation column and separated to produce fuel gas (7–12%), naphtha (14–18%), and gasoil (30–50%). The heavier part of the gasoil is usually recirculated for further cracking. With heavier feedstock, a higher recirculation ratio, and higher pressure in coke drums, the result is an increased yield of coke and lower liquid production [86,142].

The other variant of coking of residues is *fluid coking*, which takes place in a fluidized bed. The feedstock is injected into a fluid bed of coke particles (reactor) where the coke formation takes place. Coke mainly forms by a stacking of the coke particles, but fine coke particles also form. Depending on the feed, 75–85% of the gross coke production is drawn off as a product while the rest of coke is burned with air to satisfy process heat requirements. Hot coke is recycled to the reactor to maintain the temperatures required for sustaining cracking reactions. The quantity of drawn-off coke is about 20–26 wt% of the feedstock. The process operates at high temperatures and with short reaction time, which results in low yields of gases and coke but high yields of liquid products.

The reactor temperatures and pressures are 510–520 °C, 1.5–2 bar, respectively; the burner operates at a temperature range of 610–630 °C and at slightly higher pressures than the reactor. The reactor overhead vapors are routed to a distillation column and separated into gases (about 6–10%), naphtha (kb. 16–17%), and gasoils (45–70%). These products have to be desulfurized; moreover, the naphtha fraction with a boiling range of 80–180 °C has to be routed for reforming to increase octane number [3,86,143].

Steam cracking is a thermal process by which hydrocarbons (from ethane to gasoils) are cracked at close to atmospheric pressure and at temperatures of 700 °C and above in presence of steam. Products of the process are predominantly light olefins (ethylene, propylene). If naphtha or heavier hydrocarbon mixtures are used as feedstock, a light liquid product called *pyrolysis naphtha* is also produced. [144–147]. Following the selective hydrogenation of dienes and acetylenes (saturated to olefins) in the naphtha product, a gasoline-blending component with a high octane number (RON $\geq$ 100), but with high aromatic and olefin content, can be obtained. Since significant quality improvement is required to comply with the fuel standards, naphtha from the steam cracker is rarely blended to the gasoline pool.

During the catalytic *heteroatom removal* of middle and heavy distillates as well as residual fractions, the sulfur, nitrogen, and oxygen atoms in the feedstock are removed as hydrogen sulfide, ammonia, and water, respectively. Usually alumina (in some cases zeolite) supported catalysts such as sulphided cobalt/molybdenum and nickel/molybdenum are employed. Depending on the composition of feedstock, the operating conditions are as follows: temperatures are 330-380 °C, pressures are 35–100 bar, the LHSV is 1.0–2.5 m^3/m^3h, while the hydrogen–hydrocarbon ratio is 300–800 Nm^3/m^3. In addition to the desulfurized main product, there is a gasoline fraction produced with a yield of approximately 3–6%, which can be used as a blending component after its quality improvement. The heteroatom removal of middle distillates will be discussed in a later section of on the production of diesel-blending components.

Production (and Quality Improvement) of Diesel-Blending Components Diesel fuels, like motor gasolines, are mixtures of conventional (crude oil derived) and alternative blending components having high energy content and high cetane numbers (i.e., ease of ignition). Moreover additives help the diesel satisfy the product specifications.

The development of refinery technologies for producing diesel-blending components has chiefly been driven by the increasingly stringent compliance requirements, as well as by the growing market demand for diesel fuel. The main quality requirements are as follows [4,5,148]:

- Distillation properties (e.g., 95 vol% of diesel recovered at 360 °C),
- Easy ignition and good combustion (cetane number: 50–58),
- Low sulfur content (≤10–500 mg/kg)
- Low aromatic content (≤5–20)
- Low polyaromatic content (<0.5–8%)
- Appropriate density (810–845 kg/m³)
- Good cold flow, namely a cold filter plugging point (CFPP) suitable to various climatic conditions (e.g., from +5 °C to −35 °C), in terms of pour point and viscosity properties
- Good storage stability
- Low content of olefins and other compounds susceptible to resinification
- Ease of blending with alternative components
- Optimized additive content
- Nontoxicity
- Ease of handling
- Exhaust products that pollute as little as possible
- Low production costs
- Consistent quality

Some of the product requirements can be met by use of additives (e.g., by detergent-dispersant and antifoaming effects). However, the most relevant specifications, (relating to the boiling point range, the sulfur and aromatic contents, and density, can only be ensured by employing certain refinery technologies, and some other properties (e.g., cold flow) need a combination of additives and complex processing to meet regulatory standars.

Diesel-blending components are produced in complex, Processing systems with interdependent units. The complexities of the processes and interconnections are apparent in Figure 3.1. The gas oil producing part of the processing can be divided into two groups [4,5,148]:

- Yield-enhancing processes
  - Atmospheric distillation (previously discussed)
  - Cracking of distillates and residues (thermal cracking, visbreaking, coking, fluid catalytic cracking, and hydrocracking)

- Oligomerization of light olefins ($C_2^=$–$C_5^=$) and hydrogenation of gas oil boiling range olefins
- Other, nondiesel producing gas oil boiling point range fractions, for example, catalytic base oil production, hydrotreating and mild hydrocracking processes used for pretreating feedstocks of fluid catalytic cracking and hydrocracking
- Quality-improving processes
  - Heteroatom removal (sulfur, nitrogen, and oxygen removal)
  - Catalytic dewaxing (enhancing cold flow properties by selective hydrocracking and skeletal isomerization of n-paraffins)
  - Hydrogenation of aromatics

Gas oils produced in yield-enhancing processes may directly be blended into the diesel pool, but these processes are usually combined with quality-improving processes to produce advanced gas oil blending components. For example straight-run, cracking, and coker gas oils—as explained later—belong to the first group, and products of the diesel hydrotreating process belong to the latter group.

We will discuss the importance, objectives, and chemistry of each process before we advance to the fundamentals of their input to gas oil production.

The gas oil fractions distilled from diverse crude oils can meet neither the quantitative nor the qualitative demand of advanced diesel fuels. These straight-run distillates have a yield of 20–30 vol% from crude oil, while their sulfur content is in the range of 0.1–1.5 wt%, their cetane number is approximately 40–54 and pour point is between −2 °C and −25 °C. Therefore both quantity and quality enhancements are essential. The yield improvement is usually done by cracking the heavy hydrocarbons or residues. Basically quality enhancement means lots of desulfurization. The cetane number and cold flow properties are determined by the distribution of the different hydrocarbon groups (i.e., paraffin, naphthene, aromatics, and olefin) and with different structures (e.g., normal and isoparaffins) [148–151]. Table 3.4 [5] lists some properties of gas oil boiling point range hydrocarbon groups; Figure 3.17 [5] and Table 3.5 [5] show their cetane numbers, Table 3.6 [5] and Figure 3.18 [4] show their freezing points in function of the carbon numbers of their compounds. It can be seen that the highest cetane number and the lowest freezing point (pour point, cold filter plugging point in the case of mixtures) cannot be reached simultaneously in these hydrocarbons, so a compromise is needed. With consideration of the hydrocarbon structures, environmental and human biological effects, and other aspects, the most reasonable solution is to produce gas oil blending components with relatively high paraffin (mainly isoparaffin) contents. This conclusion closely relates to the principle of how diesel engines work. The diesel fuel, which is injected into the hot air compressed in the combustion chamber, be capable of easily igniting and then continuously to burn during atomization. These are requirements fulfilled mostly by paraffins.

The distillation and thermal processes during gasoil production were previously discussed. Therefore only heteroatom removal, aromatic saturation, catalytic n-paraffin conversion, and hydrocracking processes are described in the following discussion.

TABLE 3.4 Characteristics of some gasoil boiling point range hydrocarbon compounds

Properties	Hydrocarbons Found in Crude Oil				Hydrocarbons Obtained from Conversion (e.g., Cracking)
	n-Paraffins	Isoparaffins	Naphthenes	Aromatics	Olefins
Density of liquid	Low	Low	Medium	High	Low
Cetane number	Very high (85–105)	High–medium	Medium	Low–very low	Low
Freezing point	High (> +10 °C)	Medium–low	Medium–low	Very low–low	Medium–low

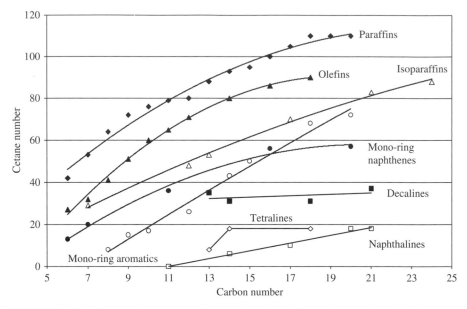

FIGURE 3.17   Change of cetane number as a function of the carbon number for hydrocarbon groups occurring in straight-run products

Desulfurization of Gasoil Fractions (Heteroatom Removal)   The unfavorable effects of heteroatoms (sulfur, nitrogen, and oxygen) on gasoil and their products of combustion on the storage stability of motor gasoline have already been discussed. Here we will analyze these effects in more detail.

Heteroatom removal from gasoil fractions refers to the removal of sulfur-, nitrogen-, and oxygen-containing compounds by the transformation of the heteroatoms to compounds that can easily be removed or retrieved from the end products [1–3,5,150].

Both the heteroatom content and the types of the heteroatom-containing compounds of gasoils, in particular, of crude oils, vary depending on the oil deposits. The sulfur compounds in gasoils are mercaptans, sulfides (straight-chain and saturated cyclic sulfides), disulfides, and thiophenes (thiophene ring containing heterocyclic sulfur compounds with a varying number of aromatic and naphthenic rings, and with a different number and length of alkyl-chains). In addition to sulfur compounds, nitrogen, and oxygen-containing compounds occur in gasoils. It is important to know that the presence of these compounds in gasoils inhibits the desulfurization reactions. For example, during hydrodesulfurization, the nitrogen and oxygen removals take place on the same catalytic active sites as the sulfur removal. In addition, the nitrogen-containing heterocyclic compounds in gas oils reduce the storage stability of the products (visible as solid segregations and color decay), during the improvement of the gasoil properties (e.g., catalytic paraffin removal) the basic nitrogen-containing compounds are catalytic poisons. The nitrogen content of gasoils is generally in a

TABLE 3.5 Cetane numbers of some typical compounds in diesel gasoils

Component	Type of Compound	Formula	Cetane Number
n-Decane	n-Paraffin	$C_{10}H_{22}$	76
n-Tridecane	n-Paraffin	$C_{13}H_{28}$	88
n-Pentadecane	n-Paraffin	$C_{15}H_{32}$	95
n-Hexadecane	n-Paraffin	$C_{16}H_{34}$	100
n-Heptadecane	n-Paraffin	$C_{17}H_{36}$	105
n-Octadecane	n-Paraffin	$C_{18}H_{38}$	110
n-Eicosane	n-Paraffin	$C_{20}H_{42}$	110
3-Ethyl-decane	Isoparaffin	$C_{12}H_{26}$	48
4,5-Diethyl-octane	Isoparaffin	$C_{12}H_{26}$	20
2,7-Dimethyl-4,5-diethyloctane	Isoparaffin	$C_{14}H_{28}$	39
Hepta-methyl-nonane	Isoparaffin	$C_{16}H_{34}$	15
7,8-Dimethyl-tetradecane	Isoparaffin	$C_{16}H_{34}$	40
8-Propyl-pentadecane	Isoparaffin	$C_{18}H_{38}$	48
7,8-Diethyl-tetradecane	Isoparaffin	$C_{18}H_{38}$	67
9-Methylheptadecane	Isoparaffin	$C_{18}H_{38}$	66
9,10-Dimethyl-octadecane	Isoparaffin	$C_{20}H_{42}$	59
Decalin	Naphthene	$C_{10}H_{18}$	48
3-Cyclohexyl-hexane	Naphthene	$C_{12}H_{24}$	36
n-Butyldecalin	Naphthene	$C_{14}H_{26}$	31
2-Methyl-3-cyclohexyl-nonane	Naphthene	$C_{16}H_{32}$	70
n-Octyldecalin	Naphthene	$C_{18}H_{34}$	31
2-Cyclohexyl-tetradecane	Naphthene	$C_{20}H_{40}$	57
1-Methyl-naphthalene	Aromatic	$C_{11}H_{10}$	0
n-Pentyl-benzene	Aromatic	$C_{11}H_{16}$	8
Biphenyl	Aromatic	$C_{12}H_{10}$	21
1-Butyl-naphthalene	Aromatic	$C_{14}H_{16}$	6
2-Phenyloctane	Aromatic	$C_{14}H_{22}$	33
n-Nonyl-benzene	Aromatic	$C_{15}H_{24}$	50
n-Octylxylene	Aromatic	$C_{16}H_{26}$	20
2-Octyl-naphthalene	Aromatic	$C_{18}H_{21}$	18
n-Dodecylbenzene	Aromatic	$C_{18}H_{30}$	68
n-Tetradecyl-benzene	Aromatic	$C_{20}H_{34}$	72

range of 40 to 300 mg/kg, which can be found in aliphatic amines, amine derivatives, and heterocyclic nitrogen compounds (pyridine and pyrrole derivatives). The oxygen content of gasoils is generally in a range of 5 to 30 mg/kg, mainly in form in organic acids, but phenol and furan derivatives are also present [1–3,86,152].

Some alternative gas oil desulfurization (heteroatom removal) processes are the following:

• Extraction, by partly reactive extraction [3,5]
• Selective adsorption/chemisorption of sulfur compounds [152–154]

TABLE 3.6 Boiling and freezing points of some typical compounds found in diesel gasoils

Component	Formula	Type of Hydrocarbon	Boiling Point, C°	Freezing Point, C°
Naphthalene	$C_{10}H_8$	Aromatic	217.7	78
Tetralin	$C_{10}H_{12}$	Aromatic	207	−35
cis-Decalin	$C_{10}H_{18}$	Naphthene	193	−43
1,2-Diethyl-benzene	$C_{10}H_{14}$	Aromatic	183	−31
n-Butyl-cyclohexane	$C_{10}H_{20}$	Naphthene	179	−78
n-Decane	$C_{10}H_{22}$	n-Paraffin	174	−30
Biphenyl	$C_{12}H_{10}$	Aromatic	255	71.5
Dodecane	$C_{12}H_{24}$	n-Paraffin	216	−9.4
1,4-Dimethylnaphthalene	$C_{12}H_{12}$	Aromatic	263	−18.0
2-Methil-1-undecene	$C_{12}H_{24}$	Isoparaffin	212	−41
Anthracene	$C_{14}H_{10}$	Aromatic	341.1	216
Dibenzyl	$C_{14}H_{14}$	Aromatic	284	51
Tetradecane	$C_{14}H_{30}$	n-Paraffin	253	5.5
2,6-Diethylnaphtalene	$C_{14}H_{16}$	Aromatic	110	50
1-Pentyl-naphthalene	$C_{15}H_{18}$	Aromatic	306.1	−23.9
n-Nonyl-cyclohexane	$C_{15}H_{30}$	Naphthene	282.1	−10
n-Decyl-cyclopentane	$C_{15}H_{30}$	Naphthene	178.9	−22.2
n-Pentadecane	$C_{15}H_{32}$	n-Paraffin	271.1	9.9
2-Methyl-tetradecane	$C_{15}H_{32}$	Isoparaffin	265	−7.8
1-Octyl-naphthalene	$C_{18}H_{10}$	Naphthene	354.8	−2,0
Dodecyl-benzene	$C_{18}H_{30}$	Aromatic	288–300	2,8
n-Octadecane	$C_{18}H_{38}$	n-Paraffin	317	28,7
5-Methyl-heptadecane	$C_{18}H_{38}$	Isoparaffin	253,3	−19,8
1-Decyl-naphthalene	$C_{20}H_{28}$	Aromatic	378.9	15
n-Tetradecyl-benzene	$C_{20}H_{34}$	Aromatic	353.9	16.1
n-Tetradecyl-Cyclohexane	$C_{20}H_{40}$	Naphthene	353.9	25
n-Penta-decyl-Cyclopentane	$C_{20}H_{40}$	Naphthene	352.8	17.2
n-Eicosane	$C_{20}H_{42}$	n-Paraffin	343.9	36.1
2-Methyl-nonadecane	$C_{20}H_{42}$	Isoparaffin	338.9	17.8

- Selective oxidation [152,155,156]
- Biocatalytic desulfurization [157–159]
- Catalytic hydrodesulfurization (heteroatom removal by catalytic hydrogenation) [86,160–163].

Nowadays, it has been proved that the most economical and popular solution is catalytic hydrodesulfurization by which nitrogen and oxygen removal takes place as well. During the catalytic hydrotreatment, the formation of the corresponding hydrocarbons—such as hydrogen-sulfide, ammonia, and water in the hydrogen atmosphere on the catalyst—cleaves the carbon–sulfur, carbon–nitrogen, and carbon–oxygen bonds of the sulfur-, nitrogen-, and oxygen-containing compounds (hydrogenation: bond cleavage on the metal component

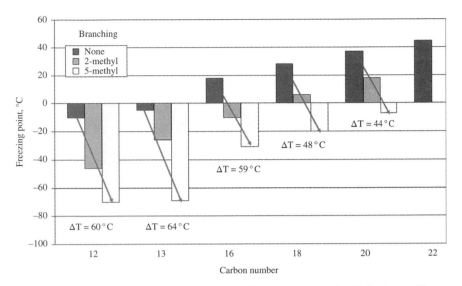

FIGURE 3.18   Freezing point as a function of carbon number for aliphatic paraffins

of the catalyst). Although same 50 to 60 years ago desulfurization technology was discussed in the literature as hydrogenation and/or catalytic desulfurization, now catalytic desulfurization is used not only for distillated gasoil fractions, but for other gasoil boiling range products as well (e.g., as gas oils of coker or FCC plants). Some examples of the main reactions involved are the following [1–3,5,86,152,161–163]:

Desulfurization
- Mercaptans: $R-S-H+H_2 \rightarrow R-H+H_2S \quad R \geq C_{10}$
- Disulfides: $R_1-S-S-R_2+3\,H_2 \rightarrow R_1-H+R_2-H+2H_2S$
- Thiophenes:

Alkyl-thiophenes:

$$R-CH=CH \text{ (ring with S)} +4\,H_2 \rightarrow R-CH-CH_2-CH_3 \text{ (}CH_3\text{)} + H_2S$$

Benzothiophenes:

$$\text{benzothiophene} + 3H_2 \rightarrow \text{aromatic}-CH_2-CH_3 + H_2S$$

R: alkyl-group

Dibenzothiophenes:

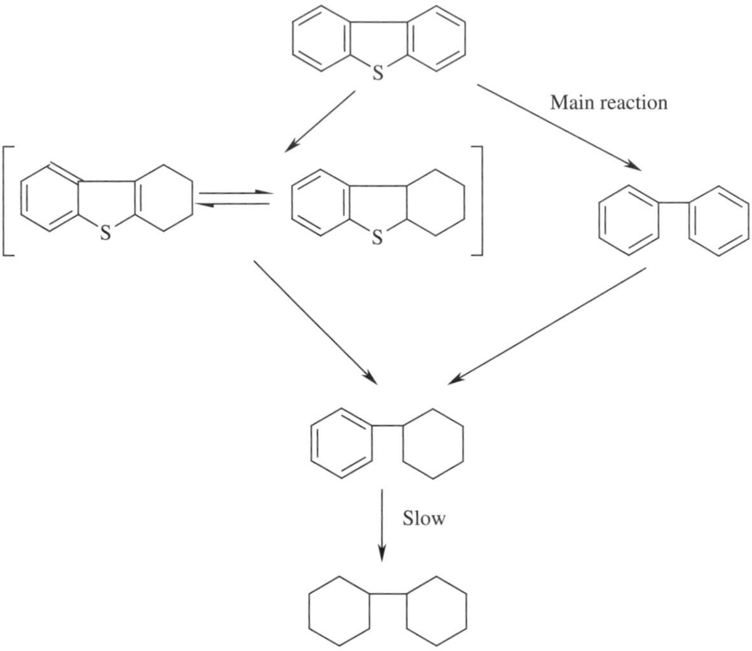

Remember, the relative desulfurization reactivity of sulfur-containing compounds in gas oils decreases in the following order: thiophenes (1)>benzothiophenes (ca. 0.6) >>dibenzothiophenes (ca. 0.04) >>4, and/or 6 methyl-dibenzothiophene (ca. 0.004) [164,165].

Nitrogen removal [161]

- Amines: $R–NH_2 + H_2 \rightarrow R–H + NH_3$
- Pyrolline derivatives:

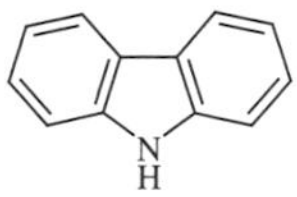

Carbazole

$+ \, xH_2 \longrightarrow$ hydrocarbon $+ NH_3$

- Pyridine derivatives:

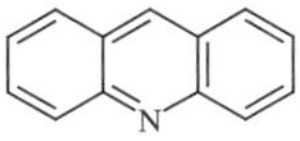

$+ \, yH_2 \longrightarrow$ hydrocarbon $+ NH_3$

Oxygen Removal [163]

- Organic acids:

$$R – COOH + 3H_2 \rightarrow R – CH_3 + 2H_2O,$$

where R: alkyl group, naphthene, or aromatic ring.

- Phenols:

+ zH$_2$ $\longrightarrow$ hydrocarbon + H$_2$O

Alkylphenols

- Furan derivatives:

+ wH$_2$ $\longrightarrow$ hydrocarbon + H$_2$O

Benzofurans

In practice, during heteroatom removal, side reactions also take place. These are hydrogenation (saturation of aromatics and olefins), hydrocracking, and condensation reactions. Moreover, any metal compounds present may also be transformed. Application of suitable catalysts and reaction conditions can facilitate saturation reactions (e.g., hydrogenation of aromatics and olefins if there are any). This is an important points, since the hydrogenation of polycyclic aromatics can take place simultaneously with the heteroatom removal to a certain degree (approximately in 40–60%) and improve the cetane number [1–3,86,162,166,167]. The saturation of olefins also increases oxidation and the thermal stabilities (ease of storability) of the product.

The above-listed reactions are exothermic. The heat released in the process is directly proportional to the hydrogen consumption. The reaction heat is much lower at the hydrogenation of heteroatom-containing compounds than at unsaturated compounds. The difference in reaction heats is important to consider especially if the sulfur content is high (above 0.5–1.0 wt%) and if the concentration of unsaturated components is high (e.g., in coker gas oil) because the temperature will suddenly increase in the reactor. To avoid the sudden spike in temperature (called "temperature excursion" or "temperature runaway"), The reactor temperature must be controlled by use of relatively cold hydrogen-rich gas. This is mixed with the liquid flowing down from the preceding catalyst bed right before the mixture passes into the next bed. Recall that reactors are usually connected in series with intermediate quenching, or a divided bed reactor will have a side quench attached to it [5,86,160,168–173].

For hydrodesulfurization, there are various types of catalysts employed. Most are different combinations of oxides and sulfides of cobalt, molybdenum, nickel, iron, and wolfram on a γ-alumina or an alumina/silica/zeolite support, or on their mixture (e.g., CoMo/Al$_2$O$_3$ [174,173–176], NiMo/Al$_2$O$_3$ [177–181], CoMoNi/Al$_2$O$_3$ [169,174,182,183]).

In earlier years, the presulfidation of catalysts was carried out only in situ (in the reactor) [1,5,86,174,184,185]. Over the last 15 to 20 years, catalysts have become available for purchase from catalyst suppliers in a presulfided form, that means an exsitu (outside the reactor) catalyst pretreatment [1,5,174,184].

Catalysts can now even be regenerated in the reactor (in situ [1–3,5,86,174]) or out of reactor (ex situ [1–3,5,86,185,186]) at the manufacturer of the catalyst.

Technology of Hydrodesulfurization   A hydrodesulfurization plant has three basic parts [1–3,5,86]:

- A catalytic hydrotreater
- A stabilizer
- A hydrogen sulfide removal unit

The desulfurization processes of licensors differ only slightly from each other. The main differences are usually the following [148,169–172,187,188]:

- Number of the reactors
- Configuration of multiple reactors (serial or parallel arrangement)
- Reactor internals, and numbers and types of catalyst beds (e.g., divided bed, feedstock-distribution)
- Numbers and types of inlet(s) of feedstock and hydrogen-rich gas
- Number and composition of catalysts (e.g., number of catalyst beds)
- Process parameters (temperature, pressure, liquid hourly space velocity, and hydrogen/hydrocarbon volume ratio)

A typical flow scheme of a hydrogenation heteroatom removal process is shown in Figure 3.19. The mixture of the preheated (in a tube furnace) gas oil and the

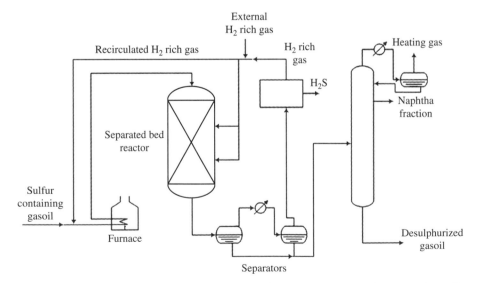

FIGURE 3.19   Flow scheme of catalytic hydrogenation heteroatom removal process of gas oils

hydrogen-rich gas flows through a fixed bed of catalysts, and the reactor effluent is then separated into liquid and vapor phases in two steps. From the hydrogen-rich gas, the hydrogen sulfide is recovered by absorption and then recycled to the front of the process. The liquid products mixture is separated in a fractionation column into tails gas (C_{4-}), naphtha, and desulfurized gas oil. The first two products are formed by side reactions; their amounts are 0.5–2% and 3–5% on the feedstock base. The gas products can be used as fuel gas, and the naphtha fraction can be used as a gasoline-blending component. The sulfur content of the target product is less than 10–50 mg/kg, which depends on the specification of the sulfur content, the sulfur content of the feedstock, the applied catalyst, and the operating conditions.

Process parameters of hydrodesulfurization are determined by the composition of the feedstock, the product specifications, the available amount of hydrogen, and economics. Furthermore the applied catalyst and the inside structure of reactor have important roles.

The generally applied process parameters are:

- Temperature: 320–380 °C [1–3,5,86,189]
- Pressure: 35–80 bar [1–3,5,86,187,190]
- Liquid hour space velocity: 1.5–2.5 h^{-1} [1–3,5,86,169–171,189]
- H_2/hydrocarbon ratio: 200–350 Nm^3/m^3 [1–3,5,86,187,188,191]

It should be noted that the aromatic content of the products of (deep) hydrodesulfurization having low sulfur content (< 10–50 mg/kg) is significantly lower than that of the feedstock. As we mentioned earlier, the partial saturation of aromatics (mono-, di-, and polyaromatics) takes place in degree, depending on the applied process parameters and the catalysts. For a supported sulfided transition metal catalyst, the total aromatic content of the products decreases only slightly; in the case of CoMo metals, it is because only the ratio of the aromatic hydrocarbon groups changes in the products (polyaromatics are converted to monoaromatics in the consecutive ring saturation reactions). For a sulfided $NiMo/Al_2O_3$ catalyst that has a higher hydrogenation activity, a hydrogenation of monoaromatics also takes place beside the saturation of polyaromatics. Consequently the total aromatic content of the products is significantly lower (10–30%) than that of the feedstock [166].

The hydrogen sulfide generated during the hydrodesulfurization of gas oils is absorbed from the purge gas in ethanolamines (e.g., monoethanolamine, diethanolamine, and nowadays mainly methyldiethanolamine) and then separated by stripping, pressure swings, or other processes that break the easily dissociable bonds between hydrogen-sulfide and the ethanolamines [1–3,5,86]. From the desorbed hydrogen sulfide, elementary sulfur can be produced by partial oxidation—along with the production of water—with 99.95% efficiency [1–3,5,86].

Decreasing Aromatic Content of Gasoils   Similarly, not only heteroatom concentration is decreased during hydrodesulfurization but some aromatic compounds, mainly polycyclic aromatic hydrocarbons, are also hydrogenated

[1–3,86,174,192] in degree, depending on the composition of the catalyst and the process parameters. However, the total aromatics content can be decreased only by 10%–30% because the process parameters favorable for deep desulfurization do not match the conditions, especially temperature, required for an extensive aromatic saturation. This is partly because the deep desulfurization at a relatively high temperature range (350–390 °C) limits the hydrogenation of aromatics, since the process is highly exothermic, and partly because the heteroatom removal reactions do not completely cover the catalyst sites that catalyze the aromatic hydrogenation. Nevertheless, as we noted in the previous chapter, the cetane number of the aromatic hydrocarbons is relatively low and thus their autoignition is also low. Therefore, imperfect combustion of gasoils with high aromatic content, and low cetane number, results in increased pollutants, especially particulate matter emissions [193,194]. Consequently, in some developed countries (e.g., Finland, Sweden, and the state of California), the strict specifications of polycyclic aromatics (0.02–1.4 v/v%) are accompanied by a limit on the total aromatic content in the highest quality diesels (5–20 v/v%) [4,148,195]. In the future, the aromatic content of the diesel fuels is expected to be significantly reduced (total aromatic content: ≤ 5–20%, polycyclic aromatic content <1–2%) [148,195,196]. As a result of such actions, the emissions of the diesel engines and the loads of particle filters will be significantly reduced. For the same reasons, in the future, the quality of gasoils deriving from fluid catalytic cracking and residue processing technologies will improved by the reduction of aromatics in all cases, and even in straight-run gasoils.

For the reduction of their aromatic content, middle distillates are divided into two candidate process groups [5]:

- Noncatalytic
- Catalytic

Noncatalytic Processes   Some examples are as follows:

- Aromatic sulfonation
- Physical separation:
  ○ Adsorption
  ○ Extractive distillation
  ○ Solvent extraction
  ○ Liquid membrane permeation

Aromatic sulfonation is no longer competitive in terms of its economical and environmental (e.g., acidic-resin destruction) aspects. The most important reason for its physical separation is still by solvent extraction, but this is not widely done because of the high cost, relatively low yield (65–80%) of the target product (raffinate), the high amount of extract (20–35%), the limited marketability of aromatic-rich mixtures, and the environmental pollution problems mentioned above.

TABLE 3.7 Some typical data of gasoils of different refinery processes

Blending Component	Boiling Point Range, T_{10}–T_{90} (°C)	Sulfur Content (%)	Polycyclic Aromatic Content (%)	Total Aromatic Content (%)	Density (kg/m³)	Cetane Index
SR LGO	240–320	0.2–1.5	10–15	15–25	820–860	42–54
SR HGO	260–360	0.5–2.5	15–20	20–30	850–890	45–55
SR LVGO	350–400	0.05–1.5	15–25	20–30	840–880	45–55
LGO from coking	210–340	0.5–3.0	20–45	30–50	830–870	36–46
FCC LCO	250–350	0.5–3.0	40–85	60–90	900–940	15–30
LGO from hydrocracking	220–325	0.0005–0.005	0,5–5	1–30	820–860	50–65
LGO of FCC feedstock prehydrogenation	200–340	0.01–0.05	10–20	30–40	840–880	35–45

Note: SR—straight run; LGO—light gasoil; HGO—heavy gasoil; LVGO—light vacuum gasoil; LCO—light cycle oil.

Also hydrocatalytic processes have been able to reduce the aromatic content of middle distillates to some degree.

Catalytic Processes The aromatic content of middle distillates in a heterogeneous catalytic can be reduced by partial or complete hydrogenation (the primary aim is the saturation of the aromatic hydrocarbons to naphthenes) or by different intensities of hydrocracking (the aim is to open the saturated rings in the required ratio). Here we will only discuss the thermodynamics, catalysts, and major processes of hydrogenation and reserve our discussion of the hydrocracking processes for a later chapter.

Chemistry of Hydrogenation of Aromatics Essential to an understanding of the chemistry of the hydrogenation of aromatic compounds is knowledge of the types of aromatic compounds in the middle distillates as well as their quantitative distributions. Table 3.7 lists the main properties of gasoils obtained from different refinery processes and their mono- and polycyclic aromatic content [5]. In the gasoil boiling point range, the monocyclic aromatic compounds, can be alkyl benzenes, benzo-cycloparaffins, and benzo-bicycloparaffins, and the bicyclic aromatic compounds, can be naphthalenes and alkyl naphthalenes, biphenyls, indenes, and dibenzo-cycloparaffins; the three-ring aromatic compounds can be anthracenes, phenanthrenes, and fluorenes [5,196–199].

The hydrogenation rate of the different aromatic hydrocarbons is determined by the following main factors [5,198,199]:

- Structure of the compound
- Number, character, and position of the substituents
- Catalyst(s)
- Process parameters.

Generally, in relation with the chemistry of the aromatic hydrogenation, the following characteristics are considered in optimizing the process parameters [196,200–208]:

- Among aromatic hydrocarbons, the hydrogenation rate of the benzene derivatives is the lowest, given identical substituents.
- In the first step of the hydrogenation reaction, the equilibrium constant of the polycyclic aromatics is lower than that of benzene.
- The more rings there are in aromatic hydrocarbons, the faster the reaction is in the first step of hydrogenation of polycyclic compounds.
- If there are unsaturated bonds inside the rings or in the side chain(s), the rate of hydrogenation increases significantly.
- The more saturated the molecules are, the harder it is for hydrogenation to take place; that means that the first step of hydrogenation is the most favorable from thermodynamics and the kinetics point of view.
- The rate of hydrogenation of aromatic compounds decreases with the increase of the number of side chains.
- In polycyclic condensed and noncondensed aromatic compounds (e.g., biphenyl), the point of equilibrium is complicated because their hydrogenation takes several successive steps, with each step achieving equilibrium.

The steps of hydrogenation of aromatic hydrocarbons and the change of cetane number are illustrated by the following example [209]:

| Naphthalene | Tetralin | Decalin | n-Decane |
| Cetane number: 1 | Cetane number: 10 | Cetane number: 36 | Cetane number: 77 |

While, side reactions do occur, hydrocracking mainly produces lighter molecular weight hydrocarbons.

Figure 3.20 shows the kinetic and thermodynamic effects on the hydrogenation of aromatic hydrocarbons [5]. The curve of the aromatic content of the product shows saturation to be a function of temperature. Consequently, the efficiency of the aromatic saturation changes from a minimum at a very low temperature, and the total aromatic content of the product decreases relatively slowly with the temperature increases because the catalyst is not active enough at the low temperature, namely the efficiency of aromatic saturation increases relatively slowly as well. This is the range of so-called kinetic inhibition. With increasing temperatures, a point is reached where aromatic content of the product is minimal, while the efficiency of aromatic hydrogenation is maximal. With even further temperature increases, the efficiency of aromatic hydrogenation decreases. This is the range of thermodynamic inhibition, the principle behind the reactions. Consequently the saturation of aromatics has an optimal temperature range where the catalyst activity is maximal.

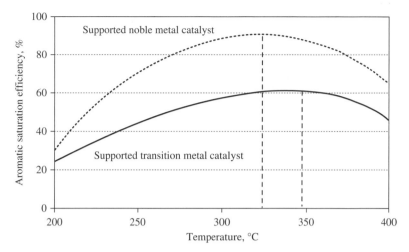

FIGURE 3.20   Kinetic and thermodynamic effects on the hydrogenation of aromatic hydrocarbons

From the figure, it is clear that the aromatic saturation reactions occur at the lowest temperature at which the reaction rate is still adequate at the industrial scale. The increase of temperature enhances the reaction rate, but as thermodynamic equilibrium is attained, it heads in the reverse direction. These two opposite effects can be explained to some extent, by the equilibrium of the reactions having shifted due to the volume reduction of the products with increasing pressure (based on Le Chatelier's principle). But, in practice, the increase of pressure is limited for economical reasons. Consequently the general aim is to develop an aromatic hydrogenation catalyst that is active enough to increase the reaction rate at relatively low pressures and temperatures.

The hydrogenation of aromatic hydrocarbons of the middle distillates is carried out on supported metal or metal sulfide catalysts, depending on the sulfur and nitrogen content of the feedstock. In feedstocks having high (over 200 mg/kg) sulfur and nitrogen content, in situ or ex situ sulphided Co–Mo, Ni–Mo, Ni–W [210–216], CoNiMo [217,218], and NiMoW catalysts supported on γ-alumina are applied [219,220]. The supported metal catalysts (Pt/support or Ni/support) are applied when feedstocks are practically sulfur and nitrogen free (less than 5 mg/kg combined) feedstocks [221,222], or Pt–Pd/zeolite type catalysts are applied when feedstocks have about 200–250 mg/kg sulfur and 20–80 mg/kg nitrogen content [223–226]. For example, the noble metals on Y-type zeolite are relatively sulfur tolerant (up to about 200 mg/kg) [227–231].

On platinum and palladium catalysts, the hydrogenation takes place faster, even at mild conditions, than on non-noble metal-sulfide catalysts. But, during the hydrogenation reactions, the metals of the VIII group are deactivated by the adsorption of sulfur-containing molecule fragments in the industrial feedstocks. The alumina-supported catalysts lose significant activity in the presence of even a small amount of sulfur, but these metals on zeolite supports resist sulfur better. On these catalysts–noble metal (e.g., Pt and/or Pd) on Y or other zeolite types of support—the metals exist as particles with high dispersity or as clusters in the zeolite pores. The acidic zeolite modifies the electron configuration of the metal particles (Pt or Pd) with an electron withdrawing effect that decreases the strength of the metal–sulfur bond [228,231].

For the reduction of the aromatic content of the middle distillates, many processes have been developed that differ in the following ways [5,199,232]:

- Sequence of the heteroatom removal and the aromatic saturation (at the same time or consecutively)
- Number of the applied reactors
- Applied catalyst(s)
- Configuration of the catalysts (e.g., divided bed)
- Distribution method of the streams to the catalyst bed
- Method of quenching
- Applied process parameters

Industrial aromatic saturation processes can be divided into two groups, such as one-step and two-step processes [5,233,234]. The *one-step processes* include those technologies that carry out the hydrogenation of the high sulfur-containing feedstocks (above 200–1000 mg/kg and up to 1.5%) on some metal-sulfide catalyst at severe conditions. These processes work at 310–340 °C, 70–100 bar total pressure, 1.0–2.0 m^3/m^{3}h liquid hourly space velocity, and 600–800 m^3/m^3 hydrogen/hydrocarbon volume ratio. At these conditions the saturation of the aromatic compounds is limited (up to 50% of aromatics in feedstock may saturate), but the decrease of the heteroatom content is significant. As a result, the increase of the cetane number is only 3–8 units. To achieve higher aromatic reduction, the partial pressure must be at least 60 bar. The hydrogen sulfide, the ammonia, the water, and the gas phased hydrocarbons are separated from the product at the last step of the technologies, even in the case of divided catalyst beds or reactors connected in series [1–3,5,86].

In the first step of the two-step technologies, the same process is carried out at relatively mild conditions (330–360 °C, 40–60 bar). The high quantity of hydrogen sulfide formed is separated, and less severe hydrogenation is carried out (160–310 °C, 35–50 bar 0.5–1.5 m^3/m^{3}h liquid hourly space velocity and 800–1000 m^3/m^3 hydrogen/hydrocarbon volume ratio) on a noble metal containing a catalyst that is more active for aromatic saturation but not (below 2–5 mg/kg) or partially (about 20–250 mg/kg) tolerant of sulfur and nitrogen. Therefore hydrogenation of the aromatics is significant (80–85%) resulting in an increased cetane number of 10–20 units [5,228].

As the one-step processes have a better efficiency if the sulfur content of the feedstock is low, these processes are usually run with prehydrogenated distillates. Based on these considerations the one-step and two-step technologies cannot be sharply distinguished.

TABLE 3.8 Comparison of the aromatic hydrogenation processes of gas oils in one or two steps

Catalyst System	NiMo	NiMo+NiW		CoMo+ Pt/Al$_2$O$_3$	NiMo+Relatively Sulfur-Tolerant Catalyst	
Number of Steps	One	Two		Two	Two	
Aromatic content of feedstock, %	Base	Base	Base	Base	Base	Base-5
Partial pressure of hydrogen, bar	90	90	60	60	45	45
Average reactor temperature, °C	Base	Base	Base	Base	Base	Base-20
Aromatic content of product, %	Base/2	Base/2	Base/2	Base-10	Base-8	Base-15

Aromatic saturation reactions are exothermic, so lower temperatures are preferable for these reactions. Consequently, in all industry technologies, inter-stage cooling is applied either by feeding hydrogen-rich gas with the appropriate temperature to the catalyst beds (mostly widespread) or to the reactors connecting in series, or by partial vaporization of the feedstock [235–239]. A comparison of a typical one-step and two-step aromatic saturation processes is presented through a real example in Table 3.8.

In the case of aromatic reduction of the gasoil distillates, the most attractive is the high efficiency of the two-step processes because, after the first desulfurization step (e.g., to 200 mg/kg sulfur content, or to 10–15 mg/kg in the developed countries), the aromatic saturation catalyst results in a further decrease of sulfur and nitrogen content. So the two-step processes are important for the production of environmentally sound gasoils.

Catalytic "Dewaxing"   The cold flow properties of the gasoil distillates (e.g., pour point, cold filter plugging point: CFPP) do not comply with the product standards, especially of the distillates obtained from paraffinic and intermediate crude oil. The reason is that gasoil distillates contain high quantities of n-paraffin hydrocarbons having higher carbon numbers than 12, and thus high melting points (see Tables 3.9 and 3.6). In addition, during the desulfurization and aromatic saturation processes, the paraffinic nature of the gasoils becomes dominant, so the CFPP drops in the value. However, the compounds formed during the hydrogenation of naphthalene and anthracene have lower freezing points (see Table 3.6).

In order to improve these properties, the n-paraffins have to be extracted from the gasoil distillates, or be transformed to hydrocarbons that have suitable cold flow properties. For this purpose, any of the following technologies is used [5,86,240]:

- Solvent dewaxing
- Extractive crystallization (with carbamide)
- Dewaxing with adsorption

- Microbiological dewaxing
- Catalytic dewaxing (selective hydrocracking and hydroisomerization).

A disadvantage of *solvent dewaxing* is that a significant decrease of CFPP requires a very low temperature to chill out the normal paraffins, and the high cost of the technology limits its application for the quality improvement of gas oil distillates.

The molecular sieve deactivates relatively fast during *adsorptive dewaxing*, so impurities (e.g., sulfur) cause serious problems as they adsorb to the n-paraffin hydrocarbons. For this reason it is necessary to apply a final treatment to these products. Both the adsorptive and *extractive dewaxing* technologies require careful and costly processing. These two technologies are only suitable for n-paraffin extraction, not for the quality improvement of gasoil distillates.

Microbiological dewaxing of gasoil distillates yet there are still some plants in Japan and in the United State is not an economical solution considering the using crude oil prices and worldwide oil consumption that use microorganisms to transform the heavy molecular weight n-paraffin hydrocarbons to low value proteins that are only suitable for forage.

From this critical review, it is understandable that *catalytic dewaxing* has come into widespread use. This practice to improve the cold flow properties of the gasoils involves the use of additives, but only in gasoils having a certain CFPP value. Economical estimates are done to determine where, when, and what additive will be applied. Generally, the catalyst is applied first, and the additives to improve the properties further to the desired quality.

As we mentioned earlier, cold flow properties depend on the open-chain paraffin content of gasoils, in particular, on the ratio of normal to isoparaffin content (Table 3.6). The freezing point of the higher carbon number n-paraffins is high, so there is high clouding and a high pour point in these hydrocarbon mixtures. Moreover, the high freezing point can cause a high cold filter plugging point. So it is necessary to transform the higher carbon number n-paraffins to hydrocarbons that have lower freezing points and are easy to blend. These hydrocarbons are the isoparaffins, the lighter molecular weight n-paraffins, and the olefins. The latter components are undesirable in gasoils because they have weak thermal and oxidation stability. These hydrocarbons are produced by the isomerization and selective cracking of heavy molecular weight paraffins. Of course, the severity of the hydrocacking can only be to the extent that ensures that the desired products are obtained at the diesel gasoil boiling point range. The processes with this aim are called catalytic "dewaxing" in the crude oil industry. (It should be noted that "dewaxing" does not describe the process exactly because it is about transformation of n-paraffins instead of an almost full extraction of them.) Since the transformation of the n-paraffins is carried out in hydrogen atmosphere, these technologies are also called *selective hydrocracking*. Depending on the type of reactions—either cracking or isomerization reactions—which ever is more dominant, the technologies are differentiated as conventional catalytic dewaxing [1–5,86,240–246] and isomerization catalytic "dewaxing" [86,247–250] respectively. (The conventional technologies were developed in the 1970s, and the isomerization technologies were developed in the 1990s and have been continually modernized since that time.)

Conventional catalytic dewaxing is carried out on a metal-support catalyst. The metal is mostly nickel or platinum, while the support is mainly ZSM-5 [241] or H-mordenite [243,244] type zeolite. The ranges of the applied process parameters are temperature 310–400 °C, pressure 35–45 bar, liquid hourly space velocity 1.0–3.0 m³/m³h, hydrogen/hydrocarbon ratio 400–600 m³/m³ [1–3,5,241–246].

Isomerization catalytic dewaxing is carried out on a Pt-Pd/zeolite (e.g., β-zeolite, SAPO-11, HZSM-22, and ferrierite zeolite) catalyst at the following process parameters: temperature 320–380 °C, pressure 35–50 bar, liquid hourly space velocity 1.5–2.5 m³/m³h, hydrogen/ hydrocarbon volume ratio 350–500 m³/m³ [1–5,86,247–250].

In both variations of catalytic dewaxing, the products with the desired diesel gas oil boiling point ranges are produced from feedstocks having higher final boiling points. The yield of products is 85–95% with isomerization catalytic dewaxing, and 55–70% by the conventional process; the quantity of the gas products is 1–3% or 6–8%, the yield of the naphtha fraction is 4–14% or 24–37%, respectively, provided that there is an equal decrease of the cold filter plugging point ($\Delta T = 15$–30 °C) [1–3,5,86, 241–250]. With catalytic dewaxing, the decrease of the cold filter plugging point depends on the quality of the feedstock, the applied catalyst, and the process parameters. The catalysts of isomerization catalytic dewaxing are sensitive to the sulfur content of feedstocks. For this reason the gasoils must be desulfurized [4,247–250].

Both conventional and catalytic dewaxing processes can be integrated into the crude oil refining scheme in various ways. They are rarely installed as stand-alone plants. More often a catalytic dewaxing reactor or an isomerization catalytic reactor is applied upstream or downstream of hydrodesulfurization reactor(s), respectively.

A further advantage of isomerization catalytic dewaxing—beyond the nearly 20–30 absolute % higher gasoil yield—is the high isoparaffin-containing naphtha, which is a more valuable final product than the naphtha produced by conventional catalytic dewaxing [4,247–249].

Hydrocracking of the Distillates   Hydrocracking is a process that converts large and high boiling point molecules of the crude oil fractions and distillate, including vacuum distillates, into to smaller and lower boiling point molecules, by the cleavage and hydrogenation of the carbon–carbon bond. This conversion process can be with or without a catalyst, the latter case is typical only for residue processing. As a result the hydrogen content of the product will be substantially larger than that of the feedstock. Hydrocracking is used in the petroleum industry to produce sufficient quantity, high-quality motor fuels and to make products from less valuable refinery streams suitable for further processing.

The catalytic hydrocracking processes, according to the feedstock, are divided into distillates (heavy naphthas, middle and heavy distillates, raffinates, and heavy oil solvents) and residue hydrocracking [5].

The feedstocks of the hydrocracking used to produce diesel-blending components are heavy gasoils, vacuum gasoils, heavy vacuum oil distillates, FCC cycle oil (fluid catalytic cracking), coker gas oil, visbreaking gasoil, and Deasphalted oils [1–3,5,86,254–256]. Hydrocracking (isomerization hydrocracking) of the heavy (paraffinic) parts of Fischer–Tropsch products obtained from different sources of synthesis gas is also becoming widely used to produce diesel-blending components [257–261].

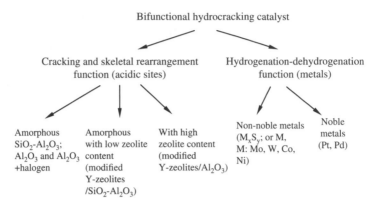

FIGURE 3.21   Composition of bifunctional hydrocracking catalysts

In most hydrocracking technologies—except those already mentioned thermal processes that operate in a hydrogen atmosphere—bifunctional (cracking–skeletal rearrangement and hydrogenation–dehydrogenation function) catalysts are used (Figure 3.21) [5].

Because of their acidic nature, the zeolite-containing hydrocracking catalysts are especially sensitive to the basic character of the compounds (e.g., nitrogen-containing compounds).

When selecting the hydrocracking catalyst(s), the purposes of the implemented technology, the product(s), and the feedstock quality properties must be considered. For example, during the hydrocracking of the distillates, the amorphous compounds and/or lower acidity zeolite-containing catalysts have high diesel selectivity. The hydrocracking activity of the diesel selective catalysts is lower than that of the gasoline selective catalysts. The amorphous catalysts result in a high middle distillate (C_{10}–C_{24}) yield (sometimes over 87%), but they are very sensitive to the organic nitrogen-containing compounds, while the zeolite-containing catalysts are less sensitive. The middle distillate selectivity of the acid-treated and partially dealuminated Y-zeolites is high. The favorable SiO_2/Al_2O_3 ratio is 8.5–11.5. To achieve high diesel yield, a mixture of amorphous and crystalline (zeolite) components are often used [1–3,5,86].

In hydrocracking, the yield and the quality of the products are also greatly affected by the ratio of cracking to hydrogenating activity of the catalyst. In general, if the hydrogenating activity of the catalyst is superior, then because of the higher amount of the hydrogenated product, the yield of overall liquid product is greater. Then the octane number of the lighter gasoline fraction decreases because a significant amount of the aromatics are saturated, and the iso/normal-paraffin ratio decreases, as well. The increase in the acidity of the catalyst compared to its hydrogenation activity has a contrary effect. To sum up, a high middle distillate yield can only be obtained by applying a catalyst with high hydrogenation activity and low acidity.

As a result of the targeted development of catalysts, several zeolite-containing catalysts have become suitable for obtaining maximum gasoline and gasoil yields depending on the applied process parameters. Consequently the flexibility of hydrocracking is decisively affected by the selection of the catalyst based on its composition.

The hydrocracking takes place under three conditions:

- During hydrogen absorption, splitting the carbon–carbon bond on the bifunctional catalyst. The cracking occurs especially on the acidic support, and saturation only at the metal centers. This is typical for the hydrocracking processes applied in refineries.
- During hydrogen absorption on the monofunctional catalyst, splitting the carbon–carbon bond. The catalyst is a metal (Pt, Pd, Ni) or a metal oxide, such as metal sulfide (so this process is hydrogenolysis).
- Through hydrocarbon radicals, during hydrogen absorption (hydropyrolysis), splitting of the carbon–carbon bond under heat.

During hydrocracking the following major reactions take place [1–3,5,86]:

Heteroatom Removal

Hydrodesulfurization

$$R\text{—}CH_2\text{—}SH + H_2 \xrightarrow{\text{Catalyst}} R\text{–}CH_3 + H_2S$$

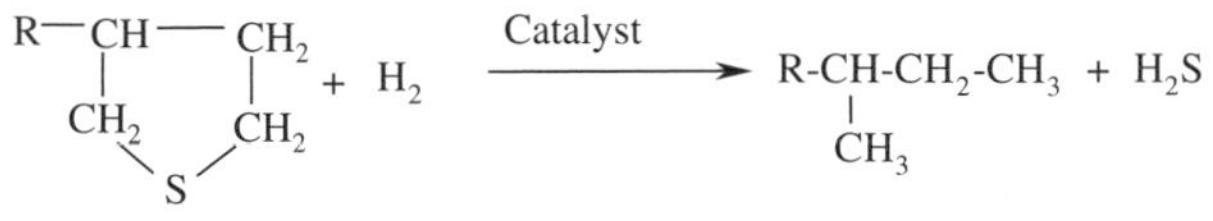

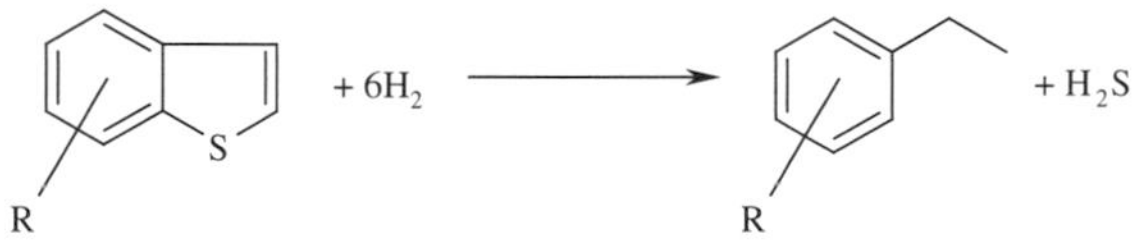

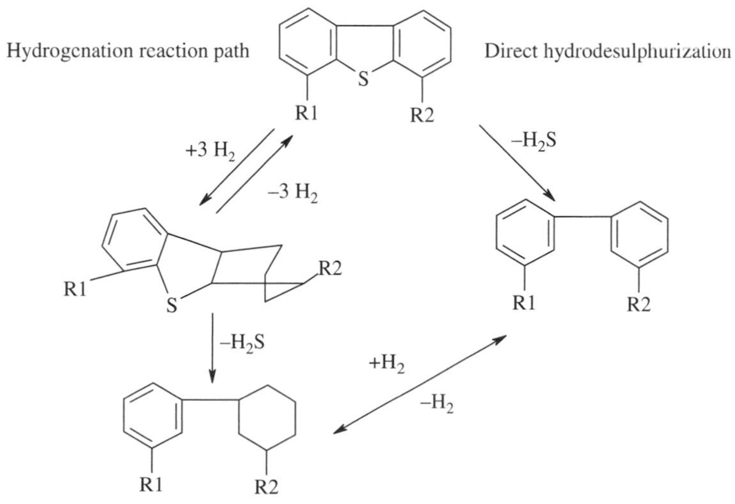

Hydrodenitrogenation

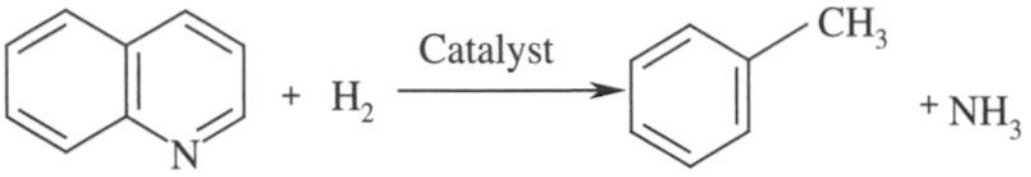

Hydrodeoxygenation

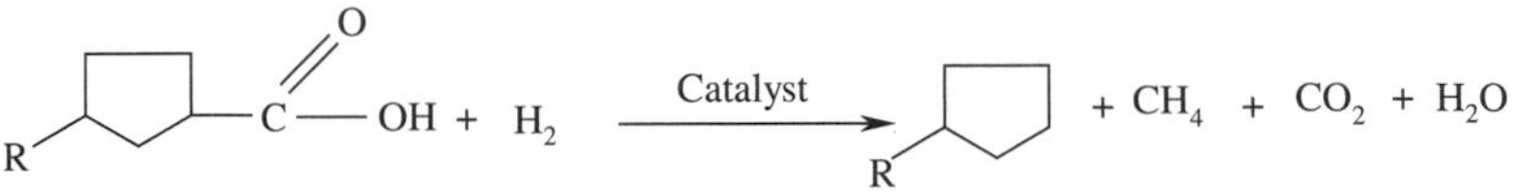

Olefin saturation

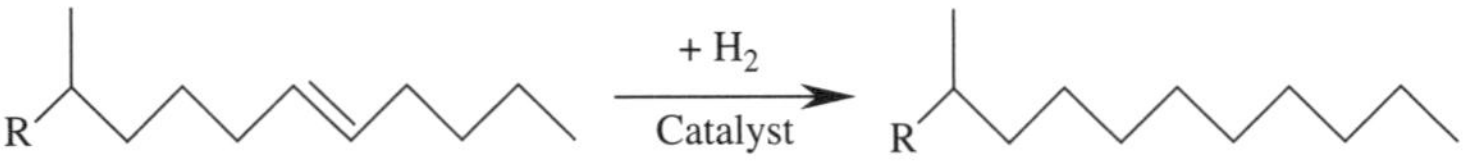

Saturation, Dealkylation

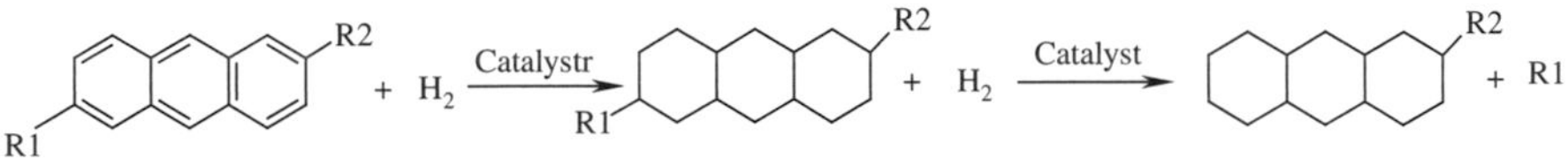

Cracking

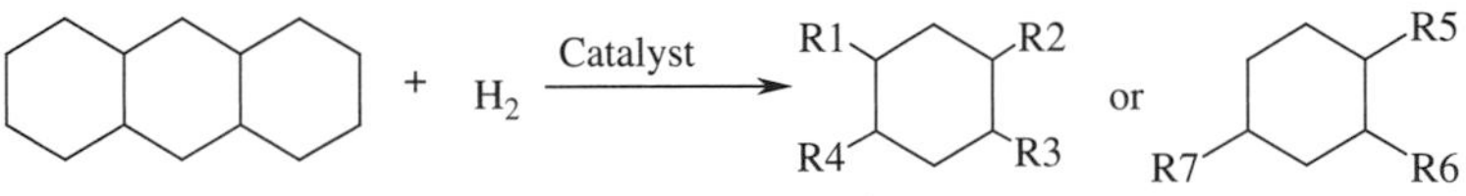

$$C_nH_{2n+2} + H_2 \rightarrow C_aH_{2a+2} + C_bH_{2b+2} \quad (a+b=n)$$

Paraffin Isomerization

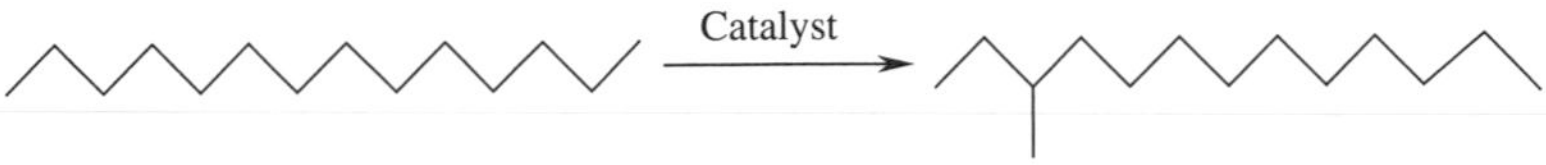

Hydrodemetalization

Hydrocracking of residues and especially heavy vacuum residues mainly results in

Coke Formation

$$\text{Polyaromatics} \xrightarrow{olefin} \text{Alkylation} \xrightarrow{-H_2} \text{Cyclization} \xrightarrow{-H_2} \text{Coke precompounds}$$

In general, the feedstock of hydrocracking is pre-hydrogenated on different CoMo/Al$_2$O$_3$ and/or NiMo/Al$_2$O$_3$ catalysts. In the course of this activity, heteroatom-removing reactions occur to various degrees, almost completely saturating the ole-fins (if they are in the feedstock) and partially saturating the aromatic compounds. A hydrocracking catalyst (e.g., NiW/Al$_2$O$_3$, NiW/zeolite, NiW//Al$_2$O$_3$–SiO$_2$, Pt,Pd/zeo-lite and Pt,Pd/Al$_2$O$_3$–SiO$_2$) is used to fully saturate the aromatic compounds further (hydrogenation), in order to force ring-opening (hydrodecyclization) of the formed naphthenes. Then the lateral chains of the alkyl cycloparaffins break away (hydrode-alkylation), and the long-chained paraffins get cracked and isomerized. During these processes, coke may form as a side reaction on the surface of the catalyst.

With the complete saturation of the naphthenes and aromatic compounds, the side chains of the aromatic hydrocarbons are easily removed, but the aromatic ring needs to be hydrogenated prior to cracking. The cracking of the n-paraffins is relatively easy. Their isomerization results in branched molecules, while in the naphthenes, cracking causes changes in the ring sizes.

Hydrocracking reactions are strongly exothermic. Hence the application of cooling is necessary between the divided catalyst beds or between the reactors. Whether one or two stage the distillate hydrocracking technologies, can operate with or without (Figures 3.22 and 3.23) recirculation at the bottom fraction of the second reactor [5]. Besides these two variants, some major licensors use hydrocracking

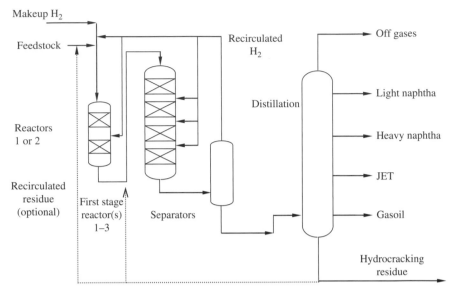

FIGURE 3.22   Flow scheme of single-stage hydrocracking, with (lines shown in dotted) and without recirculation of residue

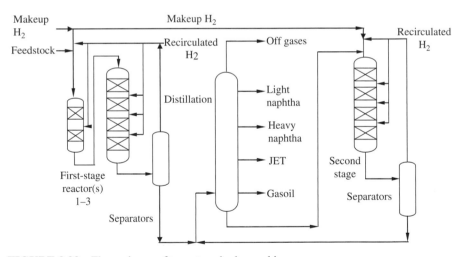

FIGURE 3.23   Flow scheme of two-stage hydrocracking

technologies that satisfy certain requirements. Generally, these are modified versions of earlier technologies. In the case of two-stage technologies, the hydrogen-rich gas is separated from the product mixture after the first stage, and the product mixture is blended with hydrogen sulfide- and ammonia-free fresh hydrogen-rich gas and then fed to the reactor of the second stage. More reactors may be used in both stages but mostly in the second stage.

TABLE 3.9 Typical process parameters of hydrocracking

Process Parameters	General Range	Distillate Hydrocracking (gasoil Mode)
Temperature, °C	300–460	350–420
Pressure, bar	70–300	120–200
Liquid hourly space velocity, m^3/m^3h	0.3–3.0	0.5–2.5
Hydrogen/hydrocarbon ratio, m^3/m^3	800–2000	1000–1500
Makeup hydrogen consumption, Nm^3 H_2/m^3 feedstock	100–500	200–350

The first step does the actual hydrotreating (heteroatom removal and partial aromatic saturation); the second step does the total aromatic saturation and mainly hydrocracking (hydro conversion), as well as hydroisomerization. The one-stage process is used in relatively light feedstocks. Of course, the reactor has divided beds. In the two-stage process, the operating conditions may be milder in the second stage. Moreover the advantage of this process is that heavier and very different quality feedstocks (FCC and coker gasoils, high nitrogen content, etc.) can be processed; it also ensures higher production flexibility (e.g., winter and summer grades, flexibility in naphtha and gas oil cuts). However, the disadvantages are higher capital and operation expenditures.

The typical process parameters of hydrocracking and distillate hydrocracking are summarized in Table 3.9 [1–3,5,86,271–282]. Generally, distillate hydrocracking technologies are classified according to the applied pressure into mild (up to about 100 bar) [280–282], medium (from 100 to 140 bar) [272–274] and high (from 140 to 200 bar) [5] pressure technologies.

Hydrocracking is always the most economical if the lowest temperatures and pressures are employed. In the use of heavier feedstocks and a less active catalyst, higher pressure must be applied. Hydrocracking is very sensitive to changing temperatures, since a rise of temperature results in higher light fraction production. In the use of feedstocks with higher nitrogen content, higher temperatures must be employed. The decrease of catalyst activity is compensated by an increase of temperature. In addition, this is frequently combined with a decrease of liquid hourly space velocity, but only if a capacity surplus is available. The process conditions of distillate hydrocracking must be as extreme as possible to ensure that the required product quality is isolated from the feedstock properties.

The major properties of the feedstock and the products of a distillate hydrocracker operating with total conversion and high middle distillate selectivity are presented in Table 3.10.

The hydrocracking of distillates is accomplished not only with high or total conversion and with severe process parameters, but also works with conversion lower than 50% and with mild reaction parameters (temperature: 370–420 °C, pressure ≤100 bar, liquid hourly space velocity 0.5–2.5 m^3/m^3h, hydrogen/hydrocarbon volume ratio 300–600 m^3/m^3). The main product is generally low-sulfur and

TABLE 3.10 Characteristics of liquid products gained from middle distillate selective hydrocracking of vacuum distillate with total conversion

Product Characteristics	Value
Light naphtha (heavy naphtha)	
Total yield, %	18.5
Density, g/cm^3	0.665
Research octane number	76.5
Paraffin/naphthene/aromatic content, %	75/20/5
Sulfur and nitrogen content, mg/kg	5
Jet	
Yield, %	34.0
Density, g/cm^3	0.805
Smoking point, mm	29
Freezing point, °C	−62
Sulfur content, mg/kg	<10
Nitrogen content, mg/kg	<1
Gas oil	
Yield,%	42.5
Density, g/cm^3	0.815
Cetane number	68
sulfur content, mg/kg	<50
Nitrogen content, mg/kg	<2
Pour point, °C	−10

Note: Boiling point range: 340–530 °C; density: 0.900 g/cm^3; sulfur content: 2.0%; nitrogen content: 600 mg/kg.

aromatic-containing FCC feedstock, but diesel gasoil boiling point range fraction can also be produced. The cetane number of the gas oil from the mild advanced hydrocracking processes is 40–45; the sulfur content is lower than 100 mg/kg [280–282].

The other distillate hydrocracking process that is important in diesel gasoil production is selective hydrocracking, or catalytic dewaxing, as was discussed in the previous chapter. In a yet another variation of hydrocracking, namely in lube (or base oil producing) hydrocracking, a good quality gasoil is produced [257,282,283]. In residue hydrocracking, the gasoil produced at the vacuum distillate boiling range cannot be directly blended into the diesel poolly for the applicable gasoil blending. It must be transformed to motor fuel via fluid catalytic cracking or distillate hydrocracking [284–289].

Other Refinery Processes for Production of Gasoil Boiling Range Hydrocarbon Mixtures   The gasoil (light cycle oil: LCO) yield can be as high as 45% in a fluid catalytic cracking unit run in the gasoil mode [130–132]. Recall the use of distillate FCC technology that was presented earlier in the chapter in the discussion of motor gasoline. The gasoils discussed were produced in FCC for the purpose of olefin and gasoline production (see Table 3.10). Similarly, the quality of LCO has to be improved by desulfurization and aromatic saturation or hydro-cracking. Depending on the sulfur content of the feedstock, a gasoil fraction with a high sulfur content (0.2–0.9%) and low cetane number (<25–35) is produced and used as a gasoil-blending component after hydrogenation.

Recall that the product quality of the FCC unit can significantly be improved by removal of heteroatoms from the feedstock prior to processing, for example, by mild hydrocracking. This way the sulfur content of the FCC gasoil fraction can be decreased below 500 mg/kg, the aromatic content even slightly lower, but the cetane number raised is moderately higher.

The FCC process has been widely used since the mid-1990s for the conversion of residues as well. In this process, it is necessary to use other catalyst compounds or catalyst additives, more catalysts as a function of the feedstock, more severe reaction parameters, and higher temperatures in the regeneration unit. Generally, the feed-stock and the catalyst are passed separately into the riser, where a gas lift of light hydrocarbons develops [290,291]. The light gasoil fraction of the residue fluid catalytic cracking can only be used after hydrotreating, whereas the heavier products must be hydrocracked before their use as blending components [86,129,136].

As we mentioned previously, the production of motor gasoline-blending compo-nents can include other goals than producing gasoil blending components, with gas oil boiling range products forming the by-products. Such processes consist of differ-ent types of the thermal cracking, namely visbreaking (12–35% gasoil), delayed cok-ing (20–30% gasoil) and fluid coking (45–65% gasoil). Gasoil streams from the abovementioned sources must be desulfurized, and their aromatic content may need to be reduced via catalytic hydrogenation. The gasoil boiling range hydrocarbons produced during hydrodesulfurization of residues are converted to good quality gasoil-blending components via mild or full conversion distillate hydrocracking.

REFERENCES

1. Lucas, A. G., ed. (2000). *Modern Petroleum Technology*. Wiley Chichester.
2. Speight, J. G., ed. (2006). *Petroleum Chemistry and Refining*. Taylor & Francis, Laramie.
3. Hobson, G. D., ed. (1984). *Modern Petroleum Technology*. Wiley Chichester.
4. Hancsók, J., Kasza, T. 2011. The importance of isoparaffins at the modern engine fuel pro-duction. *Proceedings of 8th International Colloquium Fuels 2011*, January 19–20, Stuttgart/Ostfildern.
5. Hancsók, J. (1999). *Fuels for Engines and JET Engines. Part II: Diesel Fuels.* Veszprém University Press, Veszprém.
6. Hancsók, J. (1997). *Fuels for Engines and JET Engines. Part I: Gasolines.* University of Veszprém, Veszprém.
7. Magyar, Sz., Hancsók, J., Holló, A. (2007). Key factors in the production of modern engine gasolines, *Proceedings of 6th International Colloquim, Fuels 2007*, January 10–11, Stuttgart/Ostfildern.
8. Song, C. (2003). An overview of new approaches to deep desulfurization for ultra-clean gasoline, diesel fuel and jet fuel. *Catal. Today* 86, 211–263.

9. Magyar, Sz., Hancsók, J., Kalló, D. (2005). Hydrodesulfurization and hydroconversion of heavy FCC gasoline on PtPd/H-USY zeolite. *Fuel Proc. Technol.*, 86, 1151–1164.

10. Magyar, Sz., Hancsók, J. (2005). Comparison of catalysts for FCC gasoline desulphurtization. *Chem. Eng. Trans.*, 7, 285–290.

11. Magyar, Sz., Hancsók, J. (2005). Investigation of the desulfurization of FCC gasoline and pyrolysis gasoline/FCC gasoline mixture over precious metals/support catalyst, *Prepr. Pap.—Amer. Chem. Soc., Div. Pet. Chem.*, 50(4), 395–398.

12. Hancsók, J., Magyar, Sz., Kalló, D. (2005). Investigation of the desulfurization of FCC gasoline and pyrolysis gasoline/FCC gasoline mixture over non-precious metals/support catalyst. *Prepr. Pap.—Amer. Chem. Soc., Div. Pet. Chem.*, 50(3), 328–331.

13. Hancsók, J., Magyar, Sz., Juhász, K., Kalló, D. (2007). HDS, HDN and HDO of FCC gasoline spiked with benzothiophene over PtPd/H-USY. *Topics Catal.*, 45(1–4), 207–212.

14. Magyar, Sz., Hancsók, J., Kalló, D. (2008). Reactivity of several olefins in the HDS of full boiling point range FCC gasoline over PtPd/USY. *Fuel Proc. Technol.*, 89, 736–739.

15. Debuisschert, Q., Nocca, J. -L., Jean-Paul C. (2002). *Prime-G + ™: The Key to FCC Gasoline Desulfurization*, Proceedings of the Interfaces 2002 Conference, September 19–20. 2002. Budapest, Hungary.

16. Olah, G. A., Molnár, A., eds. (2003). *Hydrocarbon Chemistry*. 2nd edition, Wiley Hoboken, NJ.

17. Yoon, J. W., Chang, J. S., Lee, H. D., Kim, T. J., Jhung, S. H. (2006). Trimerization of isobutene over cation exchange resins: effect of physical properties of the resins and reaction conditions. *J. Molecul. Catal. A: Chem.*, 260, 181–186.

18. Krivián, E., Marsi, G., Hancsók, J. (2010). Investigation of the oligomerization of light olefins on ion exchange resin catalyst. *Hungarian J. Indust. Chem.*, 38(1), 53–57.

19. Yoon, J. W. Chang J. S., Lee, H. D., Kim, T. J., Jhung, S. H. (2007). Trimerization of isobutene over a zeolite beta catalyst. *J. Catal.*, 245, 253–256.

20. Gu, Y., Shi, F., Deng, Y. (2003). SO3H-functionalized ionic liquid as efficient, green and reusable acidic catalyst system for oligomerization of olefins. *Catal. Commun.*, 4, 597–601.

21. Stenzel, O., Brüll, R., Wahner, U. M., Sanderson, R. D., Raubenheimer H. G. (2003). Oligomerization of olefin sin a chloroaluminate ionic liquid. *J. Molecul. Catal. A. Chem.*, 192, 217–222.

22. Fehér, Cs., Krivián, E., Hancsók, J., Skoda-Földes, R. (2012). Oligomerization of isobutene with silica supported ionic liquid catalysts. *Green Chem.*, 14, 403–409.

23. Trotta, R., Marchionna, M., Di Girolamo, M., Pescarollo, E. (1998). Provide superior alkylate through the use of the isoether process. *Oil Gas Eur. Mag.*, 24(3), 32–36.

24. Ackerman, S., Chintnis, G. K., McCaffrey Jr., D. S. (2002). Advances in sulfuric acid alkylation process improve gasoline blending economics. *World Refin.*, 12(7), 26–32.

25. Meister, J. M., Muldoon, B. S., Wei, D. H. (2000). Making premium alkylate with your MTBE unit. Hart's World Fuels Conference, March 28–30. San Antonio, USA.

26. Hollerbach, S. D., Broekhoven, E. H., Nat., P. J., Nousiainen, H., Jakkula, J. (2002). Alkyclean™ the new solid acid catalyst alkylation process technology. ERTC 7th Annual Meeting, November 18–20., Paris.

27. Mukherjee, M., Nehlsen, J. (2007). Reduce alkylate costs with solid-acid catalysts. *Hydrocarb. Proc.*, 86 (10), 110–114.

28. van Broekhoven, E. H., van Rooijen, E. (2008). Alkylation with solid acid catalyst. *Petrol. Technol. Quart.*, (4), 87–93.

29. Mukherjee, M., Nehlsen, J., Ezeribe, A., Ibe, J. (2009). Revamp of HF alkylation unit employs solid-acid catalyst. *OilGas J.*, 107(42), 46–53.

30. Meister, J. M. (2000). Optimize alkylate production for clean fuels. *Hydrocarb. Proc.*, 79(5), 63–75.

31. Sahay, N., Rock, K., Marchionna, M., Tagliabue, L. (2003). Low cost conversion of MTBE units to alternative gasoline blending componensts production. Hart's World Fuels Conference, March 25–37. San Antonio.

32. Trotta, R., Marchionna, M., Di Girolamo, M., Pescarollo, E., Hyland, M. (1998). How to make alkylate without an acid alkylation unit Isoether-100—an extension of isoether dimerization/etherification process. NPRA, Annual Meeting, AM 98–51, March 15–17., San Francisco.

33. Graeme, S., van der Laan, M. (2003). Butane and light naphta isomerisation. *Petrol. Technol. Quart. Catal.*, 47–49.

34. Asselin, G. F., Bloch, H. S., Donaldson, G. R., Haensel, V., Pollitzer, E. L. (1972). Isomerization of paraffins. *ACS Div. Petr. Chem.*, 17(2), B4–B18.

35. Kuchar, P. J. (1993). Paraffin isomerization innovations. *Fuel Proc. Technol.*, 35(1–2), 183–200.

36. Domergue, B., Watripont, L. (2005). Paraffins isomerization options. *Petrol. Technol. Quart.*, (2), 21–25.

37. Hancsók, J., Magyar, S., Holló, A. (2007). Importance of isoparaffins in the crude oil refining industry. *Chem. Eng. Trans.*, 11, 41–47.

38. Hancsók, J., Magyar, Sz., Nguyen, K.V.S., Keresztúry, L., Valkai, I. (2003). Investigation of the production of gasoline blending component free of sulfur. *Petrol. Coal*, 45(3–4), 99–104.

39. Holló, A, Hancsók, J., Kalló, D. (2002). Kinetics of hydroisomerization of C_5–C_7 alkanes and their mixtures over platinum containing mordenite. *Appl. Catal. A: Gen.*, 229(1–2), 93–102.

40. Anderson, G. C., Rosin, R., Stine, M. A., Hunter, M. J. (2004). New solutions for light paraffin isomerization. In *Proceedings of 2004 NPRA Annual Meeting*, March 21–23, San Antonio.

41. Hancsók, J., Magyar, Sz., Szoboszlai, Zs., Kalló, D. (2007). Investigation of energy and feedstock saving production of gasoline blending components free of benzene. *Fuel Proc. Technol.*, 88(4), 393–399.

42. Furuta, S. (2003). The effect of electric type of platinum complex ion on the isomerization activity of Pt-loaded sulfated zirconia-alumina. *Appl. Catal. A*, 251, 285–293.

43. Paál, Z., Wild, U., Muhler, M., Manoli, J. M., Potvin, C., Buchholz, T., Sprenger, S., Resofszki, G. (1999). The possible reasons of irreversible deactivation of Pt/sulfated zirconia catalysts: structural and surface analysis. *Appl. Catal. A*, 188, 257–266.

44. Corma, A., Serra, J. M., Chica, A. (2003). Discovery of a new paraffin isomerization catalysts based on SO_4^{2-}/ZrO_2 and WO_x/ZrO_2 applying combinatorial techniques. *Catal. Today*, 81, 495–506.

45. Hernandez, M. L., Montoya, J. A., Hernandez, I., Viniegra, M., Llanos, M. E., Garibay, V., Del Angel, P. (2006). Effect of the surfactant on the nanostructure of mesoporous Pt/Mn/

WO$_x$/ZrO$_2$ catalysts and their catalytic activity in the hydroisomerization of n-hexane. *Microporous Mesoporous Materials*, 89, 186–195.

46. Barrera, J.A., Montoya, M., Viniegra, J., Navarette, G., Espinosa, A., Vargas, P., del Angel, G. (2005). Isomerization of n-hexane over mono- and bimetallic Pd-Pt catalysts supported on ZrO$_2$–Al$_2$O$_3$–WO$_x$ prepared by sol–gel, *Appl. Catal. A*, 290, 97–109.

47. Lee, J.-K., Rhee, H.-K. (1998). Sulfur tolerance of zeolite beta-supported Pd–Pt catalysts for the isomerization of n-hexane. *J. Catal.*, 177(2), 208–216.

48. Weyda, H., Köhler, E. (2003). Modern refining concepts—an update on naphtha-isomerization to modern gasoline manufacture. *Catal. Today*, 81, 51–55.

49. Hancsók, J., Holló, A., Perger, J., Gergely, J. (1998). Environmentally friendly possibilities to compensate octane deficiency resulting from benzene content reduction of motor gasolines. *Petrol. Coal*, 40 (1), 33–38.

50. Hancsók, J., Holló, A., Valkai, I., Szauer, Gy., Kalló, D. (2002). Simultaneous desulfurization and isomerization of sulfur containing n-pentane fractions over Pt/H–mordenite catalyst. *Studies in Surface Science and Catalysis—Porous Materials in Environmentally Friendly Processes*. Elsevier, Amsterdam.

51. Domerque, B., Waltripont, L. (2000). Advanced recycle paraffin isomerization technology. *World Refin.*, 10(4), 26–32.

52. Aitani, M., Hamid, S. H. (1995). Benzene reduction in motor gasoline—a review of options. *Oil Gas—Eur. Mag.*, 21(2), 28–32.

53. Gladman, P., Tobin, G. (1998). Benzene reduction: options and opportunities. *Hydrocarb. Eng.*, 3(2), 68–72.

54. Emmrich, G., Ranke, U. (1999). Enforcing benzene removal. *Hydrocarb. Eng.*, 4(6), 50–55.

55. Hancsók, J., Holló, A. (1997). Production of benzene free isohexane fractions. *Petrol. Coal*, 39(2), 6–9.

56. Hancsók, J., Magyar, Sz., Nguyen, K.V.S., Szoboszlai, Zs., Kalló, D., Holló, A., Szauer, Gy. (2005). Simultaneous desulfurization, isomerization and benzene saturation of n-hexane fraction on Pt-H/MOR. *Studies in Surface Science and Catalysis.—Molecular Sieves: From Basic Research to Industrial Applications*, 158, 1717–1724, Elsevier, Amsterdam.

57. Hancsók, J., Holló, A., Debreceni, É., Perger., J., Kalló, D. (1999). Benzene saturating isomerization. *Studies in Surface Science and Catalysis—Porous Materials in Environmentally Friendly Processes*. 125, 417–424, Elsevier, Amsterdam.

58. Unzelmann, G. H. (1989). Ethers have good gasoline-blending attributes. *OilGas J.*, 87(15), 33–37.

59. Palkhe, G., Leonhard, H. Tappe, M. (2000). Mögliche Umweltelastungen durch die Nutzung von MTBE als Kraftstoffzusatz in Deutschland und Westeuropa. *Erdöl Erdgas Kohle*, 116(10), 498–504.

60. Gold, R. B., Lichtblau, J. H., Goldstein, L. (2002). MTBE vs. ethanol: sorting through the oxigenate issues. *Oil Gas J.*, 100(2), 18–32.

61. Hancsók, J., Magyar Sz. (2003). *The Methyl-Tertier-Buthyl-Ether and the Environment*, BME OMIKK, Budapest.

62. Pecci, G., Floris, T. (1977). Ethers ups antiknock of gasoline. *Hydrocarb., Proc.*, 56(12), 98–102.

63. Rock, K., Gildert, G. R., McGuirk, T. (1997). Catalytic distillation. *Chem. Eng.*, (July), 78–84.

64. Singer, R. M. (1993). Review the basics of MTBE catalysis. *Fuel Reformulation*, 3(11–12), 46–48.

65. Unzelman, G. (1992). Ethers. *Fuel Reform.*, 2(6), 24–29.

66. Koskinen, M. (1997). NExETHERS—New way of producing ethers. *Oil Gas Eur. Mag.*, 23(3), 31–34.

67. Dekker, E. (2010). When MTBE outscores ETBE for bioenergy content. *Petrol. Technol. Quart.*, (2), 87–93.

68. Antos, G. J., Aintani, A. M., eds.- (2004). *Catalytic Naphtha Reforming*. Dekker, New York.

69. Antos, G. J., Aitani, A.M., Parera, J. M., eds. (1995). *Catalytic Naphta Reforming Science and Technology*. Dekker, New York.

70. Sertic-Bionda, K. (1997). Effect of process severity on the quality of reformate. *Oil Gas—Eur. Mag.*, 23(3), 35–37.

71. Zhou, T., Baars, F. (2010). Catalytic reforming options and practices. *Petrol. Technol. Quart.*, (2), 21–25.

72. Gates, B. C., Katzer, J. R., Schuit, G. C. A. (1999). *Chemistry of Catalytic Processes*. McGraw-Hill, New York.

73. Stiles, A. B., ed. (1987). *Catalyst Supports and Supported Catalysts*. Butterworth, Stoneham, MA.

74. Le Goff, P.-Y. (2009). Consider advanced multi-promoted catalysts to optimize reformers, *Hydrocarb. Proc.*, 88(10), 81–85.

75. Absi-Halabi, M., Stanislaus, A., Qabazard, H. (1997). Trends in catalysis research to meet future refining needs. *Hydrocarb. Proc.*, 76(2), 45–55.

76. Baronetti, G. T., de Miguel, R., Scelza, O. A., Fritzler, M. A., Castro, A. A. (1985). Pt–Sn/Al_2O_3 catalysts: studies of the impregnation step. *Appl. Catal.*, 19(1), 77–85.

77. Foger, K., Jaeger, H. (1981). Oxidation of silica-supported Pt, Ir and Pt/Ir catalysts. *J. Catal.*, 70(1), 53–71.

78. Wijngaarden, R. J., Kronberg, A., Westerterp, K. R. (1998). *Industrial Catalysis*. Wiley, Weinheim.

79. Sivasankar, S., Ratnasamy, P. (1995). Reforming for gasoline and aromatics: recent developments. In *Catalytic Naphta Reforming* (ed. G. J. Antos), Dekker, New York, 483.

80. Clause, O., Mauk, L., Martino, G. (1998). Trends in catalytic reforming and paraffin isomerization. *Proceedings of the 15th World Petroleum Congress*, Wiley, New York.

81. Paál, Z. (1995). Basic reactions of reforming on metal catalysts. In *Catalytic Naphta Reforming* (ed. G. J. Antos), Dekker, New York, 35–75.

82. Satterfield, C. N., ed. (1991). *Heterogeneous Catalysis. in Industrial Practice*. McGraw-Hill, New York.

83. Bragin, O. V., Karpinski, Z., Matusek, K., Paál, Z., Tétényi, P. (1979). Comparative study of various platinum catalysts in skeletal reactions of C6-hydrocarbons. *J. Catal.*, 56(2), 219–228.

84. Bond, G. C. (2000). Relativistic effects in coordination, chemisorption and catalysis. *J. Molecul. Catal. A: Chem.*, 156(1–2), 1–20.

85. Sachtler, W. M. H., Zhang, Z. (1993). Zeolite-supported transition metal catalysts. *Adv. Catal.*, 39, 129–220.

86. Meyers, R. A., ed. (2007). *Handbook of Petroleum Refining Processes*. McGraw-Hill, New York.

87. Antos, G., Moser, M., Lapinski, M. (2004). The new generation of commercial catalytic naphtha-reforming catalysts. In *Catalytic Naphtha Reforming* (eds. G. Antos, A. Aitani). Dekker, New York, 335.

88. Aitani, A. (2006). Catalytic naphta reforming, In *Encyclopedia of Chemical Processing* (ed. S. Lee). Taylor & Francis, New York.

89. Beltramini, J. N., Wessel, T. J., Datta, R. (1991). Deactivation of the metal and acid functions of Pt/Al$_2$O$_3$-Cl reforming catalyst by coke formation. *Studies Surf. Sci. Catal.*, 68, 119–128.

90. Barbier, J., Churin, E., Marecot, P. (1990). Coking of Pt–Ir/Al$_2$O$_3$ and Pt–Re/Al$_2$O$_3$ catalysts in different pressures of cyclopentane. *J. Catal.*, 126(1), 228–234.

91. Al-Saggaf, R. M., Padmanabhan, S., Abuduraihem, H. Z., Bhattacharya, N. (2008). Innovative approach solves CCR regenerator pinning problem. *Hydrocarb. Proc.*, 87(10), 93–96.

92. Carvalho, L. S., Pieck, C. L., do Carmo Rangel, M., Fígoli, N. S., Parera, J. M. (2004). Sulfur poisoning of bi- and trimetallic γ-Al$_2$O$_3$–supported Pt, Re, and Sn catalysts. *Ind. Eng. Chem. Res.*, 43, 1222–1226.

93. Gjervan, T., Prestvik, R., Holmen, A. C. eds. (2004). *Catalytic Reforming.* Springer, München.

94. Zhang, D.-Q., Sun, Z.-L. (2000). Studies on sulfate pollution of Pt–Re reforming catalysts. 2nd International Symposium on Deactivation and Testing Catalysts, March 26–31, San Francisco.

95. Macleod, N., Fryer, J. R., Stirling, D., Webb, G. (1998). Deactivation of bi- and multimetallic reforming catalysts: influence of alloy formation on catalyst activity. *Catal. Today*, 46, 37–54.

96. Marr, G. A. (2003). New catalyst formulations for semi-regenerative reformers. *Petrol. Technol. Quart.*, (2), 50.

97. Coats, R. (1978). Semi-regenerative reforming process providing continous hydrogen production. US Patent 4,208,397, filed Jun. 12, 1978 and Jun. 17, 1980.

98. Lapinski, M. P., Rosin, R. R., Anderle, C. J. (2004). Innovating for increased reforming capacity. *Hydrocarb. Eng.*, 9(9), 29–31.

99. Xuhong, M., Enze, M., Baoning, Z., Mingyuan, H. (2001)., Selective hydrogenation of olefins in reformate using amorphous nickel alloy catalyst in magnetically stabilized bed reactor. *China Petrol. PetroChem. Technol.*, 2001(2), 37–43.

100. Tian, H.-P., He, M.-P., Long, J. (2005). Upgrade catalytic cracking operations. *Hydrocarb. Proc.*, 84(2), 85–88.

101. Dries, H., Muller, F., Willbourne, P., Fum, M., Williams, C. P. (2005)., Consider using new technology to improve FCC unit reliability. *Hydrocarb. Proc.*, 84(2), 69–74.

102. Letzsch, W. (2005). Improve catalytic cracking to produce clean fuels. *Hydrocarb. Proc.*, 84(2), 77–81.

103. Korpelshoek, M., Podrebarac, G., Rock, K. Samarth, R. (2010). Increasing diesel production from the FCCU. *Petrol. Technol. Quart.*, (1), 75–79.

104. McLean, J. (2010). Maximising FCC distillate production. *Petrol. Technol. Quart.*, (1), 37–44.

105. Aguilar, J. M., Leon Gil, J. M., Sanchez Arandilla, F. (2009). Increasing olefins. *Petrol. Technol. Quart.*, (1), 19–20.

106. Jain, D., Ganesh, G. S. (2007). Increasing FCC unit performance. *Petrol. Technol. Quart.*, (4), 67–73.

107. Shorey, S. W., Lomas, D. A., Keesom, W. H. (1999). Exploiting synergy between FCC pretreating units to improve refinery margins and produce low-sulphur fuels. NPRA Annual Meeting, March 21–23., San Antonio.

108. Tyas, A. R. (2008). Diesel hydrotreating and mild hydrocracking. ERTC 13th Annual Meeting, Vienna.

109. Patel, R., Moore, H., Duff, H., Hamari, B. (2004). Fluidised catalytic cracker hydrotreater revamp. *Petrol. Technol. Quart.*, (2), 115–121.

110. Fletcher, R. (2004). New method describes FCC catalyst selection for diffusion limited units. *Oil Gas J.*, 102, 54.

111. Maholland, M. K. (2006). Improving FCC catalyst performance. *Petrol. Technol. Quart. Catal.*, 41–42.

112. Yung, K.Y., O'Connor, P., Yanik, S. J., Bruno, K. (2001). Catalytic challenges. *Hydrocarb. Eng.*, 1, 43–48.

113. Bruno, K. (2006). *Maximising octane. Petrol. Technol. Quart.*, 11(2), 54–55.

114. Polato, C. M. S., Rodrigues, A. C. C., Monteiro, J. L. F., Henriques, C. A., (2010). High surface area Mn,Mg,Al–Spinels as catalyst for SOX abatement in fluid catalytic cracking units. *Ind. Eng. Chem. Res.* 49, 1252–1258.

115. Myrstad, T., Engan, H., Seljestokken, B., Rytter, E., (1999). Sulphur reduction of fluid catalytic cracking (FCC) naphtha by an in situ Zn/Mg(Al)O FCC additive. *Appl. Catal. A: General*, 187 (2), 207–212.

116. Lappas, A. A., Iatridis, D. K., Vasalos, I. A., Rhemann, H., Schwarz, G., Lonka, S., Heinonen, P., Spyridaki, G., Psichogios, Y. (2005). Reducing FCCU NO$_x$ emissions. *Petrol. Technol. Quart.*, (1), 59–67.

117. McGuirk, T. (2010). Co-catalysts provide refiners with FCC operational Flexibility. In *Proceedings of 2010 NPRA Annual Meeting*, March 21–23, Phoenix.

118. McLean, J. B., (2003). Distributed matrix structures—a technol. platform for advanced FCC catalytic solutions. In *Proceedings of 2003 NPRA Annual Meeting*, March 23–25, San Antonio.

119. Fletcher, R., Evans, M. (2009). Additives for real-time FCC catalyst optimization. *Petrol. Technol. Quart.*, (3), 90.

120. Voolapalli, R. K., Ravichander, N., Murali, C., Gokak, D. T., Choudary, N. V., Siddiqui, M. A. (2007). Fine-tune catalyst addition rates. *Hydrocarb. Proc.*, 86(8), 109–112.

121. Chester, A. W. (2006). CO combustion promoters: past and present. *Prepr. Pap—Amer. Chem. Soc., Div. Pet. Chem.* 51(2), 389–392.

122. Sexton, J., Fisher, R., (2009). Marathon refineries employ new FCCU CO combustion promoter. *OilGas J.*, 107, 51–52.

123. Stonecipher, D. L., (1997). Passivate vanadium on FCC catalysts for improved refinery profitability In *Proceedings of 1997 NPRA Annual Meeting*, March 16–18, San Antonio.

124. Baillie, C., Kieffer, R., Sargenti, L., (2010). Reduction sulphur oxide emissons from the FCC unit. *Petrol. Technol. Quart.*, 31–35.

125. Reagan, B., Evans, M., (2006). FCC additives: tried and tested. *Hydrocarb. Eng.*, 11(7), 65–69.

126. Sadeghbeigi, R., ed. (2000). *Fluid Catalytic Cracking Handbook*. Gulf Publishing, Houston.

127. Letzsch, W. S., Santner, C., Tragesser, S. (2008). FCC reactor design: part I. *Petrol. Technol. Quart.*, (4), 63–68.

128. Letzsch, W. S., Santner, C. (2008). FCC regenerator design: part I. *Petrol. Technol. Quart.*, (1), 8.

129. Kraus, M., Fu, Q., Kiser, N. Thornton, O. (2010). Resid FCC catalyst technology for maximum distillates yield. *Petrol. Technol. Quart.*, (3), 93–101.

130. Imhof, P., Rautiainen, E. P. H., Hakuli-Pieterse, A. (2005). Boosting FCC propylene with minimal conversion loss. *Petrol. Technol. Quart.*, 38–39.

131. Kapur, S., Vaidyanathan, S., Rajguru, A., Menegaz, D. (2009). Why integrate refineries and petrochemical plants? *Hydrocarb. Proc.*, 88(2), 29–40.

132. Dharia, D. (2004). Increase light olefins production. *Hydrocarb. Proc.*, 83, 61.

133. Fletcher, R. (2009). Opportunities for on-demand LCO maximization. *Proceedings of 2009 NPRA Annual Meeting*, March 22–24, San Antonio.

134. Hunt, D., Hu, R., Ma, H., Langan, L., Cheng, W.-C. (2009). Strategies for increasing production of light cycle oil. *Petrol. Technol. Quart.*, 14(5), 83–85.

135. Yung, K. Y., Pouwels, A. C. (2008). Fluid catalytic cracking—a diesel producing machine. *Hydrocarb. Proc.*, 87(2), 79–83.

136. Guojing, Z., Fuchuan, Y., Liu, G., Zhongchen, B., Meng, X., Butterfield, R. (2007). Upgrading Daqing. *Hydrocarb. Eng.*, 12(6), 49–54.

137. Scherzer, J., Gruia, A. J. eds. (1996). *Hydrocracking science and Technololgy*, Dekker, New York.

138. Wang, Z., Que, G., Liang, W. (1998). Distribution of major forms of sulphur in typical Chinese sour vacuum residues and VRDS residue. *Proceedings of Symposium on Chemical Analysis of Crude Oils for Optimizing Refinery Yields and Economics*, 215th National Meeting, American Chemical Society, March 29 – April 3, Dallas.

139. Gardner, A. (1996). Refining details. *Thermal Cracking in Refining.*, October.

140. Di Carlo, S., Janis, B., Migliorati, P. (1995). Tendency of petroleum residues to be processed in visbreaking: a prediction model. *Proceedings of Symposium on Petroleum Chemistry and Processing*, 210th National Meeting, American Chemical Society, August 20–25, Chicago.

141. Kataria, K. L., Kulkarni, R. P., Pandit, A. B., Joshi, J. B., Kumar, M. (2004). Kinetic studies of low severity Visbreaking. *Ind. Eng. Chem. Res.*, 43, 1373–1387.

142. Pieper, C. J., Stewart, C. W. (2001). Drums designed for longer life. *Hydrocarb. Eng.*, 6(2), 32–33.

143. McCaffrey, D. S., Hammond, D. G., Patel, V. R. (1998). Fluidised bed coking—utilising bottom of the barrel. *Petrol. Technol. Quart.*, (4), 37–43.

144. Albright, L. F., Crynes, B. L., Corcoran, W. H., eds. (1983). *Pyrolysis Theory and Industrial Practice*. Academic Press, New York.

145. Yue, Z., Zheng, C. W. (2006). Fine-tune cracking effeciencies for larger olefins cracker. *Hydrocarb. Proc.*, 85(4), 63–65.

146. Bhirud, V. L. (2007). Improve naphtha quality for olefins cracking. *Hydrocarb. Proc.*, 86(4), 69–74.

147. Buffenoir, M. H., (2008). Consider an innovative method for olefin unit construction. *Hydrocarb. Proc.*, 87(12), 49–56.

148. Nagy, G., Hancsók, J. (2009). Key factors of the production of modern diesel fuels, *Proceedings of 7th International Colloquium Fuels, Mineral Oil Based and Alternative Fuels*, January 14–15., Stuttgart/Ostfildern.

149. Murphy, M. J., Taylor, J. D., McCormick, R. L. (2004). Compendium of experimental cetane number data. Subcontractor Report 540–36805. Golden, CO, National Renewable Energy Laboratory.

150. Hancsók, J., Magyar, S., Holló, A. (2007). Importance of isoparaffins in the crude oil refining industry. *Chem. Eng. Trans.*, 11, 41–47.

151. Yaws, C. L., ed. (2008). *Thermophysical Properties of Chemicals and Hydrocarbons*. William Andrew, Beaumont, CA.

152. Gary, J. H., Handwerk, G. E., eds. (2004). *Petroleum. Refining—Technology and Economics*. Dekker, New York.

153. Babich, I.V., Moulijn, J. A. (2003). Science and technology of novel processes for deep desulfurization of oil refinery streams a review. *Fuel*, 82(6), 607–631.

154. Johnson, B. G. (2001). Application of Phillip's zorb process to distillates—meeting the challenge. *Proceedings of 2001 NPRA Annual Meeting*, March 18–20., New Orleans.

155. Kane, L., Romanow, S. (2001). Oxidative route "cleand up" diesel. *Hydrocarb. Proc.*, 80(5), 29–33.

156. Otsuki, S., Nonaka, T., Takashima, N., Qian, W., Ishihara, A., Imai, T., Kabe, T. (2000). Oxidative desulfurization of light gas oil and vacuum gas oil by oxidation and solvent extraction. *Energy Fuels*, 14, 1232–1239.

157. Arena, B. J., Janssen, A., Kijlstra, S. (2001). Biological desulfurization. *Hydrocarb. Eng.*, 6(3), 68–71.

158. Le Borgne, S., Quintero, R. (2003). Biotechnological processes for the refining of petroleum. *Fuel Proc. Technol.*, 81, 155–169.

159. Greenwood, G. J. (2000). Next generation sulfur removal technology. *Proceedings of 2000 NPRA Annual Meeting*, March 26–28, San Antonio.

160. Furimsky, E. (1998). Selection of catalysts and reactors for hydroprocessing. *Appl. Catal. A: General*, 171, 177–206.

161. Kabe, T., Qian, W., Hirai, Y., Li, L., Ishihara, A. (2000). Hydrodesulfurization and hydrogenation reactions on noble metal catalysts: I. Elucidation of the behavior of sulfur on alumina-supported platinum and palladium using the ^{35}S radioisotope tracer method. *J. Catal.*, 190, 191–198.

162. Furimsky, E. (2005). Hydrodenitrogenation of petroleum. *Catal. Rev.*, 47, 297–489.

163. Varga, Z., Hancsók, J. (2003). Deep hydrodesulphurisation of gas oils. *Petrol. Coal*, 45(3–4), 147–153.

164. Furimsky, E. (2000). Catalytic hydrodeoxygenation. *Appl. Catal. A: General*, 199(2), 147–190.

165. Topsoe, H. (2005). ULSD with BRIM catalyst technology. *Proceedings of 2005 NPRA Annual Meeting*, March 13–15, San Francisco.

166. Nag, N. K., Sapre, A. V., Broderick, D. H., Gates, B. C. (1979). Hydrodesulfurization of polycyclic aromatics catalyzed by sulfided $CoO–MoO_3/\gamma–Al_2O_3$: the relative reactivities. *J. Catal.*, 57(3), 509–512.

167. Cooper, B. H., Donnis, B. B. L. (1996). Aromatic saturation of distillates: an overview. *Appl. Catal. A*, 137(2), 203–223.

168. Varga, Z., Hancsók, J., Tolvaj, G. (2001). Investigation of the deep hydrodesulphurization of gas oil fraction. *Proceedings of 40th International Petroleum Conference*, September 17–19, Bratislava, Slovakia.

169. Ferguson, D., Flenniken, M. (2006). Invsetigating hydrotreater performance. *Petrol. Technol. Quart. Catal.*, 29–33.

170. Lee, S. L., Olivier, S., Plantenga, F. L., Leliveld, R. G. (2006). Maximizing diesel volumes and quality with albemarle hydroProcessing Catalysts. *Proceedings of 2006 European Refining Technology Conference*, 11th Annual Meeting, November 13–15, Paris.

171. Plumail, J. C. (2004). Advanced catalytic Engineering produces dual-activity catalysts. *Proceedings of 5th EMEA Catalyst Technology Conference*, March 3–4, Cannes, France.

172. Low, G., Townsend, J., Shooter, T. (2002). Systematic approach for the revamp of a low pressure hydrotreater to produce 10 ppm "Sulphur Free" diesel at BP Coryton refinery. *Proceedings of ERTC 2002 Conference*, November 18–20, Paris.

173. Hamilton, G. L., van der Linde, B., di Camillo, D. (2000). Revamp options for improved diesel. *Petrol. Technol. Quart.*, 5(4), 33–35.

174. Topsøe, H., Clausen, B. S., Massoth, F. E. eds. (1996). *Hydrotreating Catalysis, Catalysis Science and Technology*. Springer, Berlin.

175. Albemarle Introduces new catalyst to improve ultra-low-sulfur diesel production. Albemarle Company website http://www.albemarle.com/Home-3.html (Jan. 15. 2011).

176. Fujikawa, T. (2007). Development of new CoMo HDS catalyst for ultra-low sulfur diesel fuel production, *J. Jap. Petrol. Inst.*, 50(5), 249–261.

177. Carlson, K. (2010). Active site developments for improved productivity. *Petrol. Quart. Catal.*, (1), 21.

178. Vázquez, P., Pizzio, L., Blanco, M., Cáceres, C., Thomas, H., Arriagada, R., Bendezú, S., Cid, R., García, R. (1999). NiMo(W)-based hydrotreatment catalysts supported on peach stones activated carbon. *Appl. Catal. A: General*, 184, 303–313.

179. Pappal, D. A. (2003). Stellar improvements in hydroprocessing. catalyst activity. 2003 NPRA Annual Meeting, March 23–25, San Antonio.

180. Grange, P., Vanhaeren X. (1997). Hydrotreating catalysts, an old story with new challenges. *Catal. Today*, 36, 375–391.

181. Skyum, L., Zeuthen, P., Cooper, B. (2008). Next generation BRIM™ catalyst technology. *Proceedings of 2008 European Refining Technology Conference 13th Annual Meeting*, November 17–18, Vienna.

182. Sanghavi, K., Torrisi, S. (2007). Converting a DHT to ULSD service. *Petrol. Technol. Quart.*, (1), 45–54.

183. Benyamna, A., Bennouna, C., Leglise J., van Gestel, J., Duyme, M., Duchet, J.C., Dutartre, R., Moreau, C. (1998). Effect of the association of Co and Ni on the catalytic properties of sulfided CoNiMo/Al$_2$O$_3$. *Prepr. Pap—Am. Chem. Soc. Div. Petr. Chem.*, 43(1), 39–42.

184. Alvarez, L., Espino, J., Ornelas, C., Rico, J. L., Cortez, M. T., Berhault, G., Alonso, G. (2004). Comparative study of MoS$_2$ and Co/MoS$_2$ catalysts prepared by ex situ/in situ activation of ammonium and tetraalkylammonium thiomolybdates. *J. Molecul. Catal. A: Chem.*, 210, 105–117.

185. Marklund, S. (2006). Sulfiding solutions. *Hydrocarb. Eng.*, (March), 11(3), 69.

186. Labruyere, F., Dufresne, P., Lacroix, M., Breysse, M. (1998). Ex situ sulfidation by alkylpolysulfides: a route for the preparation of highly dispersed supported sulfides. *Catal. Today*, 43, 111–116.

187. Ohmes R., Sayles, S. (2006). ULSD best practices. *Hydrocarb. Eng.*, 11(9), 45–49.

188. Knudsen, K. G., Cooper, B. H., Topsøe, H. (1999). Catalyst and process technologies for ultra low sulfur diesel. *Appl. Catal. A: General*, 189, 205–215.

189. Marion, P. (2002). Axens clean fuels technologies. *Proceedings of 4th International Conference on Petroleum Refining Technology and Economics in Russia, the CIS and Baltics*, October 21–23, Barcelona.

190. Kabe, T., Aoyama, Y., Wang, D., Ishihara, A., Qian, W., Hosoya, M., Zhang, Q. (2001). Effects of H_2S on hydrodesulfurization of dibenzothiophene and 4,6-dimethyldibenzothiophene on alumina-supported NiMo and NiW catalysts. *Appl. Catal. A: Gen.*, 209 (1–2), 237–247.

191. Sie, S. T. (1999). Reaction order and role of hydrogen sulfide in deep hydrodesulfurization of gas oils: consequences for industrial reactor configuration. *Fuel Proc. Technol.*, 61(1–2), 149–171.

192. Lamourelle, A. P., McKnight, J., Nelson, D. E. (2001). Clean fuels: route to low sulfur low aromatic diesel. *Proceedings of 2001 NPRA Annual Meeting*, March 18–20, New Orleans.

193. Knudsen, K. G. (2000). What does it take to produce ultra-low-sulfur diesel? *World Refin.*, 10(9), 14–18.

194. Durbin, T. D., Zhu, X., Norbeck, J. M. (2003). The effects of diesel particulate filters and a low-aromatic low-sulfur diesel fuel on emissions for medium-duty diesel trucks. *Atmos. Environ.* 37, 2105–2116.

195. Szalkowska, U. (2009). Fuel quality—global overview. *Proceedings of 7th International Colloquium Fuels*, January 14–15, Ostfildern, Germany.

196. Lee, C. K., McGovern, S. (2001). Comparison of clean diesel production technologies. *Petrol. Technol. Quart.*, 6(4), 35–39.

197. Giardino, R., Calemma, V., Carati, A., Ferrari, M. (2007). Hydrogen saving in LCO upgrading. *Prepr. Pap—Amer. Chem. Soc., Div. Petr. Chem.*, 52(2), 3–5.

198. Girgis, M. J., Gates, B. C. (1991). Reactivities, reaction network and kinetics in high-pressure catalytic hydroprocessing. *Ind. Eng. Chem. Res.* 30(9), 2021–2058.

199. Stanislaus, A., Cooper, B. H. (1994). Aromatic hydrogenation catalysis a review. *Catal. Rev.*, 36(1), 75–123.

200. Le Page, J. F., ed. (1987). *Applied Heterogeneous Catalysis*. Editions Technip, Paris.

201. McVicker, G., Daage, M., Touvelle, M., Hudson, C., Klein, D., Baird, W. (2002). Selective ring opening of naphthalin molecules. *J. Catal.*, 210(1), 137–148.

202. Demirel, B., Wiser, W. H. (1998). Thermodynamic probability of the conversion of multiring aromatics to isoparaffins and cycloparaffins. *Fuel Proc. Technol.*, 55, 83–91.

203. Sapre, A. V., Gates, B. C. (1981). Hydrogenation of aromatic hydrocarbons catalyzed by sulfided $CoO–MoO_3/\gamma–Al_2O_3$: reactivities and reaction networks. *Ind. Eng. Chem. Process De. Dev.*, 20, 68–73.

204. Kubicka, D., Salmi, T., Tiitta, M., Murzin, D. Y. (2009). Ring opening of decalin—kinetic modeling. *Fuel*, 88, 366–373.

205. Lapinas, A. T., Klein, M. T., Gates, B. C., Macris, A. (1991). Catalytic hydrogenation and hydrocracking of fluorene: reaction pathways, kinetics, and mechanisms. *Ind. Eng. Chem. Res.*, 30(1), 42–50.

206. Korre, S. (1995). Polynuclear aromatic hydrocarbons hydrogenation. 1. Experimental reaction pathways and kinetics. *Ind. Eng. Chem. Res.*, 34, 101–117.

207. Beltramone, A. R., Resasco, D. E., Alvarez, W. E., Choudhary, T. V. (2008). Simultaneous hydrogenation of multiring aromatic compounds over NiMo catalyst. *Ind. Eng. Chem. Res.*, 47, 7161–7166.

208. Kokayeff P. (1993). Aromatics saturation over hydrotreating catalysts: reactivity and susceptibility to poison. In *Catalytic Hydro Processing of Petroleum and Distillates* (eds. M. C. Oballa and S. S. Shih). Dekker, New York, 253–278.

209. Santana, R. C., Do, P. T., Santikunaporn, M., Alvarez, W. E., Taylor, J. D., Sughrue, E. L., Resasco, D. E. (2006). Evaluation of different reaction strategies for the improvement of cetane number in diesel fuels. *Fuel*, 85, 643–656.

210. Vrinat, M., Bacaud, R., Laurenti, D., Cattenot, M., Escalona, N., Gamez, S. (2005). New trend sin the concept of catalytic sites over sulfide catalysts. *Catal. Today*, 107(10), 570–577.

211. Houalla, M., Nag, N. K., Sapre, A. V., Broderick, D. H. (1978). Hydrodesulfurization of dibenzothiophene catalyzed by sulfided CoO–MO$_3$/γ-Al$_2$O$_3$: the reaction network. *AIChE J.*, 24(6), 1015–1021.

212. Laredo, G. C., Saint-Martin, R., Martinez, M. C., Castillo, J., Cano, J. L. (2004). High quality diesel by hydrotreating of atmospheric gas oil/light cycle oil blends. *Fuel*, 83, 1381–1389.

213. Marchal, N., Mignard, S., Kasztelan, S. (1996). Aromatics saturation by sulfided nickel-molybdenum hydrotreating catalysts. *Catal. Today*, 29, 203–207.

214. Ali, S. A., Siddiqui, A. B. (1997). Dearomatization, cetane improvement and deep desulfurization of diesel feedstock in a single-stage reactor. *React. Kinet. Catal. Lett.*, 61(2), 363–368.

215. Macaud, M., Milenkovic, A., Schulz, E., Lemaire, M., Vrinat, M. (2000). Hydrodesulfurization of alkylbenzothiophenes: evidence of highly unreactive aromatic sulfur compounds. *J. Catal.*, 193, 255–263.

216. Marin, C., Escobar, J., Galván, E., Murrieta, F., Zárate, R., Vaca, H. (2004). Light straight run gas oil hydrotreatment over sulfided CoMoP/Al$_2$O$_3$–USY zeolite catalysts. *Fuel Proc. Technol.*, 86, 391–405.

217. Leliveld, B. (2009). Increasing ULSD production with current assests. *Petrol. Technol. Quart. Catal.*, 51–54.

218. Lee, S. L. (2006). Latest developments in Albemarle hydroprocessing catalysts. *Albemarle Catal. Courier*, 63, 8–9.

219. Sigurdson, S., Sundaramurthy, V., Dalai, A. K., Adjaye, J. (2008). Phosphorus promoted trimetallic NiMoW/γ-Al$_2$O$_3$ sulfide catalysts in gas oil hydrotreating. *J. Molecul. Catal. A.*, 291, 30–37.

220. Huirache-Acuna, R., Albiter, M. A., Espino, J., Ornelas, C., Alonso-Nunez, G., Paraguay-Delgado, F., Rico, J. L., Martinez-Sánchez, R. (2006). Synthesis of Ni–Mo–W sulphide catalysts by ex situ decomposition of trimetallic precursors. *Appl. Catal. A*, 304, 124–130.

221. Chiou, J. F., Huang, Y. L., Lin, T. B., Chang, J. R. (1995). Aromatics reduction over supported platinum catalysts. 1. Effect of sulfur on the catalyst deactivation of tetralin hydrogenation. *Ind. Eng. Chem. Res.*, 34, 4277–4283.

222. Lin, T. B., Jan, C. A., Chang, J. R. (1995). Aromatics reduction over supported platinum catalysts. 2. Improvement on sulfur resistance by addition of palladium to supported platinum catalysts. *Ind. Eng. Chem. Res.*, 34, 4284–4289.

223. Yoshimura, Y., Toba, M., Matsui, T., Harada, M., Ichihashi, Y., Bando, K. K., Yasuda, H., Ishihara, H., Morita, Y., Kameoka, T. (2007). Active phases and sulphur tolerance of bimetallic Pd–Pt catalysts used for hydrotreatment. *Appl. Catal. A*, 322, 152–171.

224. Simone, A., Castellon, E. R., Livi, M., López, A. J., Vaccari, A. (2004). Hydrogenation and hydrogenolysis/ring-opening of naphthalene on Pd/Pt supported on zirconium-doped mesoporous silica catalysts. *J. Catal.*, 228, 218–224.

225. Varga, Z., Hancsók, J., Tolvaj, G., Wáhl Horvath, I., Kalló, D. (2002). Hydrodearomatization, hydrodesulfurization and hydrodenitrogenation of gas oils in one step on Pt, Pd/H–USY. *Proceeding of 2nd FEZA Conference*, September 1–5, Taormina, Giardini Naxos, Italy.

226. Varga, Z., Hancsók, J., Nagy, G., Kalló, D. (2005). Hydrotreating of gasoils on bimetallic catalysts: effect of the composition of the feeds. *Studies Surf. Sci. Catal.—Molecul. Sieves: From Basic Research to Industrial Applications*, 158, 1891–1898.

227. Fujikawa, T., Idei, K., Ohki, K., Mizuguchi, H., Usui, K. (2001). Kinetic behavior of hydrogenation of aromatics in diesel fuel over silica-alumina-supported bimetallic Pt–Pd catalyst. *Appl. Catal. A.*, 205, 71–77.

228. Nagy, G., Hancsók, J., Varga, Z., Polczmann, Gy., Kalló, D. (2007). Investigation of hydrodearomatization of prehydrogenated gas oil fractions on Pt–Pd/H–USY catalysts. *Topics Catal.*, 45(1–4), 195–201.

229. Nagy, G., Hancsók, J. (2010). Hydrodearomatization of different prehydrogenated gas oils on noble metal/zeolite catalysts. In *Silica and Silicates in Modern Catalysis* (ed. I. Halász). Transworld Research Network, Kerala, 457–476.

230. Nagy, G., Pölczmann, Gy., Kalló, D., Hancsók, J. (2009). Investigation of hydrodearomatization of gas oils on noble metal/support catalysts. *Chem. Eng. J.*, 154, 307–314.

231. Yoshimura, Y., Toba, M., Farag, H., Sakanishi, K. (2004). Ultra instence hydrodesulfurization of gas oils over sulfide and/or noble metal catalysts. *Catal. Surveys Asia*, 8(1), 47–60.

232. Kaufmann, T. G., Kaldor, A., Stuntz, G. F., Kerby, M. C., Ansell, L. L. (2000). Catalysis science and technology for cleaner transportation fuels. *Catal. Today*, 62, 77–90.

233. Lee, S. L., de Wind, M. (1992). Cetane improvement of diesel fuels with single and dual stage hydrotreating. *Preprints, A.C.S., Div. Petr. Chem.*, 37(3), 718–728.

234. Suchanek, A. J., Davé, D., Gupta, A., Van Stralen, H., Karlsson, K. (1993). SnySat[SM] HDS/HDA—a revamp at Scanraff. *Proceedings of 1993 NPRA Annual Meeting*, March 21–23, San Antonio.

235. Dath, J. P. (2001). Deep HDA and very low sulfur in diesel fuels with the new KF 200. *Proceedings of Ecological and Economic Challenges for the Transportation Sector and Related Industries in the Next Decade*, June 11–13, Nordwijk Aan Zee, Holland.

236. Peries, J. P. (1991). IFP deep hydrodesulfurization and aromatics hydrogenation on straight run and pyrolysis middle disstillates. *Proceedings of NPRA Annual Meeting*, San Antonio.

237. Lonka, S., Toppinen, S., Markkanen, V., Aittamaa, J. (1997). New approach to reduce aromatics in refinery products. *Hydrocarb. Eng.*, 10, 83–85.

238. Patel, R. H. (2003). How are refiners meeting the ultra low sulfur diesel challenge? 2003 NPRA Annual Meeting, March 23–25, San Antonio.

239. de la Fuente, E., Christensen, P., Johansen, M. K. (1999). Options for meeting EU year 2005 fuels specifications. *Proceedings of the European Refining Technology Conference*, November 22–24, Paris.

240. Rutherford, F. (1998). Refinery issues and options. European Diesel Seminar, June 24, Prague.

241. Hargrove, J. D., Elkes, G. J., Richardson, A. H. (1979). New dewaxing process proven in operations. *Oil Gas J.*, 77(3), 103–105.

242. Köhler, E., Weyda, H. (2000). Catalysts of the future. *Hydrocarb. Eng.*, 5(7), 49–51.

243. Ireland, H. R., Redini, C., Raff, A. S., Fava, L. (1979). Distillate dewaxing in operation. *Hydrocarb. Proc.*, (May), 58(5), 119–122.

244. Rolfe, J. R. K. (1984). Development in BP refining processes inducing catalytic dewaxing and isomerisation. JPI Petroleum Refining Conference, October, Tokyo.

245. Chen, N.Y., Garwood, W. E., Dwyer, F. G., eds. (1989). *Shape Selective Catalysis in Industrial Applications*, Dekker, New York.

246. Gergely, J., Perger, J., Szalmás-Pécsvári, G. (1997). Hydrodewaxing process at Danube refinery. *Petrol. Technol. Quart.*, (1), 35–41.

247. Pappal, D. A., Hilbert, T. L. (1996). Isomerization dewaxing a new selective process, *Petrol. Technol. Quart.*, (2), 35–41.

248. Tracy, W. J., Chitnis, G. K., Novak, W. J., Helton, N. E. (2002). ExxonMobil MIDW process: innovative solutions for production of low sulfur distillates using selective dewaxing and advanced hydrotreating catalysts. *Proceedings of 3rd European Catalyst Technology Conference*, February 26–27, Amsterdam.

249. Sastre, G. Chica, A., Corma, A. (2000). On the mechanism of alkane isomerization (isodewaxing) with unidirectional 10-member ring zeolites: a molecular dynamics and catalytic study. *J. Catal.*, 195(2), 227–236.

250. Pappal, D. A., Tracy W. J., Weinstabl, H. (1998). MAKFining technology—pathway to enhanced diesel fuel. *Proceedings of ERTC*, November 16–18, Berlin.

251. Ertl, G., Knözinger, H., Weitkamp, J., eds. (1997). *Handbook of Heterogeneus Catalysis*. Wiley, Weinheim.

252. McKetta, J. J., Cunningham, W. A., eds. (1982). *Encyclopedia of chemical processing and Design*. Dekker, New York.

253. Geng, C. H., Zhang, F., Gao, Z. X., Zhao, L. F., Zhou, J. L. (2004). Hydroizomerization of n-tetradecan over Pt/SAPO-11 catalyst. *Catal. Today*, 485–491.

254. Ioannou, M. G., Hagan, A. P., Hofmann, E. J., Huizinga, T. (1999). Optimisation of the hydrocracker in a complex refinery. *European Refinery Technology Conference*, November 22–24, Paris.

255. Law, D. V. (2000). Hydrocracking: past, present and future. *Petrol. Technol. Quart.*, 55–63.

256. Hennico, A., Billon, A., Bigeard, P.-H. (1992). Hydrocracking—new market demands. *Hydrocarb. Techn. Int.*, 19–29.

257. Pölczmann, Gy., Hancsók, J. (2010). Investigation of catalytic production of modern base oils and fuels. *Proceedings of 17th International Colloquium Tribology*, January 19–21, Stuttgart/Ostfildern.

258. Clark, R. H., Unsworth, J. F. (1999). The Performance of diesel fuel manufactured by the shell middle distillate synthesis process. *Fuels 1999, 2th International Colloquium*, January 20–21, Esslingen, Germany.

259. Zhao, T.-S. et. al. (2005). Selective synthesis of middle isoparaffins via a two-stage Fischer–Tropsch reaction: activity investigation for a hybrid catalyst. *Ind. Eng. Chem. Res.*, 44, 769–775.

260. Wu, T. et. al. Physical and chemical properties of GTL-diesel fuel blends and their effects on performance and emissions of a multicylinder DI compression ignition engine. *Energy Fuels*, 21, 1908–1914.

261. Bouchy, C., Hastoy, G., Guillon, E., Masters, J. A., (2009). Fischer–Tropsch upgrading via hydrocracking and selective hydroisomerisation. *Oil Gas Sci. Technol.*, 64, 91–112.

262. Pappal, D. A., Miyauchi, Y. A., Plantenga, F. L. (2001). Next generation hydrocracking catalysts. *Ecological and Economic Challenges for the Transportation Sector and Related Industries in the Next Decade*, June 11–13, Nordwijk aan Zee, Holland.

263. Shakeel, A., Mazen, A. S. (2000). Preparation and characterization of CoMo/SiO$_2$–Al$_2$O$_3$ catalysts, *Preprints of Symposia*, 45(3), 612–616.

264. Belato, D. A. S., Lima, J. R. D., Oddone, M. R. R. (2002). Hydrocracking—a way to produce high quality low sulphur middle distillates. *Proceedings of 17th World Petroleum Congress*, September 1–5, Rio de Janeiro.

265. Marcilly, C. R. (2000). Where and how shape selectivity of molecular sieves operates in refining and petrochemistry catalytic process. *Topics Catal.*, 31(4), 357–366.

266. Wade, R., Vislocky, J., Maesen, T., Torchia, D. (2009). Hydrocracking catalyst and processing developments. *Petrol. Technol. Quart.*, (3), 86.

267. Wasse, G., Perry, S., Wu, K., Adarme, R., Martinez, L. (2010). Expand hydrocracker operating window through process equipment, catalyst system and operational design. *2010 NPRA Annual Meeting*, March 21–23, Phoenix.

268. Peng, Q. (2005). New approaches to hydrocracking technologies. *18th World Petroleum Congress*, September 25–29, Johannesburg, South Africa.

269. Abdo, S. F. (2006). Hydrocracking for clean fuels production. *AIChE Spring National Meeting*, April 23–27, Orlando, FL.

270. Rispoli, G., Sanfilippo, D., Amoroso, A. (2009). Advanced hydrocracking technology upgrades extra heavy oil. *Hydrocarb. Proc.*, 2009, 88 (12), 42.

271. Mukherjee, U., Mayer, J., Srinivasan, B. (2005). Hydroprocessing revamp configurations. *Petrol. Technol. Quart. Catal.*, 49–56.

272. Pappal, D. A., et al. (1997). Converting VGO HDS units to moderate pressure hydrocracking. *Petrol. Technol. Quart.*, (2), 45.

273. Hilbert, T., Patel, V. (2009). Economic conversion to maximize distillate production. *Proceedings of 2009 NPRA Annual Meeting*, March 22–24, San Antonio.

274. Hilbert, T. L., Chitnis, G. K., Umansky, B. S., Kamienski, P. W., Patel, V., Subramanian, A. (2008). Consider new technology to produce "clean" diesel. *Hydrocarb. Proc.*, 87 (2), 47–56.

275. Marion, P., Koseoglu, P. (1998). Build flexible hydrocracking configurations: consider the cost-effective routes to hydrocracking pressure management. *Hart's Fuel Technol. Manag.*, (1–2), 51–54.

276. Gosselink, W., Stork, W. H. J. (1997). Coping with catalyst deactivation in hydrocracking: catalyst and process development. *Ind. Eng. Chem. Res.*, 36(8), 3354–3359.

277. Reno, M. E., Gala, H. B., Antos, G. J. (1997). Hydrocracking for maximum profitability. *Proceedings of 1997 NPRA Annual Meeting*, March 16–18, San Antonio.

278. Thakkar, V. (2004). Increase the value of your light cycle oil. *Technology & More*. UOP Company Brochure, 2004(Autumn).

279. Martindale, D. et al. (1996). Continuing innovation in hydrocracking technology. *Hydrocarb. Asia*, (Sept.), 88.

280. Sarrazin, P., Bonnardot, J., Wambergue, S., Morel, F. (2005). New mild hydrocracking route produces 10-ppm-sulfur diesel. *Hydrocarb. Proc.*, 84 (2), 57–60.

281. Dahlberg, A., Mukherjee, U., Olsen, C.W. (2007). Consider using integrated hydroprocessing methods for processing clean fuels. *Hydrocarb. Proc.*, 86(9), 111–120.

282. Morel, F., Bonnardot, J., Benazzi, E., (2009). Hydrocracking solutions squeeze more ULSD from heavy ends. *Hydrocarb. Proc.*, (4), 79–87.

283. Ushio, M., Kamiya, K., Yoshida, T., Honjou, I. (1992). Production of high VI base oil by VGO deep hydrocracking. *Am. Chem. Soc., Div. Pet. Chem. Prepr. Pap.*, 1493–1302.

284. Farrell, T. R., Zakarian, J. A. (1986). Lube facility makes high-quality lube oil from low-qualiy feed. *Oil Gas J.*, 84(20), 47–51.

285. Wenzel, F. W. (1989). A commercial route for bottom of the barrel upgrading. NPRA Annual Meeting, March 19–21, San Francisco.

286. Waugh, W. B., Muir, G., Skripek, M. (1995). Advanced hydroprocessing. of heavy oil and refinery residues. 13th World Petroleum Congress, October 20–25, Buenos Aires.

287. Colyar, J. J. (1998). Ebullated bed reactor technology. *Hydrocarb. Eng.*, 3(5), 86–94.

288. Fujita, K., Abe, S., Inoue, Y., Plantegna, F.L., Leliveld, B. (2002). New development in resid hydroprocessing. *Petrol. Technol. Quart.*, 51–58.

289. Ross, J., Kressmann, S., Harlé, V., Tromer, P. (2000). Maintaining on-spec products with residue hydroprocessing. IFP Refining Seminar, November 13, Rome.

290. James, R. D., Nee, P. A., Diddams, S. S, Paloumbis, G. D. (2001). FCC catalist technology for maximum residue upgrading. *Petrol. Technol. Quart.*, 6(2), 37–41.

291. Letzsch, W. S., Lauritzen, J. (2004). A range of options for resid conversion. *Petrol. Technol. Quart.*, (2), 63–66.

CHAPTER 4

Alternative Fuels

In the previous chapter, we considered materials suitable for energy transmission into the alternative fuels along with the crude oil based gasoline, kerosene, and diesel gasoil. In this chapter, we discuss fuels derived from alternative sources that can be used alone or with the conventional based fuels.

The reasons for using nonconventional or alternative fuels are the following [1–15]:

- Contribute to meeting increasing energy demands
- Decrease crude oil dependence
  - Reduce unequal distribution of the crude oil resources (decrease dependence on imports)
  - Prevent forecast depletion of the crude oil reserves (ca. 100–120 years)
  - Limit periodic jumps in crude oil prices
- Limit deleterious effects on the environment and human health (lower CO_2 emission during the whole life cycle, decrease green house effect, prevent acid rains, eliminate ozone depletion in atmosphere.)
- Replace environment-contaminating fossil energy carriers
- Apply renewable energy resources
- Improved quality control
- Lower investment cost
- Secure political independence of foreign fuel sources (production of own industrial feedstock, increase employment, ensure safety, keep population in land, etc.)
- Utilize uncultivated fields
- Contribute to the soil and water protection, and to lifestyle improvements.

However, the alternative fuels for internal combustion engines must meet requirements that overlap with those for gasolines and diesel gas oils [1,4,8,12,15,16]:

- Be applicable for energy transmission under internal combustion conditions in engines

Fuels and Fuel-Additives, First Edition. S. P. Srivastava and Jenő Hancsók.
© 2014 John Wiley & Sons, Inc. Published 2014 by John Wiley & Sons, Inc.

- Easy exploitation and production
- No or low harmful emissions during the retrieval and production
- Available in a high and constant quantity
- Not contain compounds proved to be environmental hazards, or contain very low amounts (e.g., sulfur content <10 mg/kg)
- High energy content
- Required evaporation rate
- Not be corrosive
- Thermal Stability
- Suitable chemical resistance
- Not be toxic
- Fuel waste and combustion products must not be environment contaminants, corrosion initiators, or erode cylinders and other engine parts
- Suitable distillation curve (if the boiling point of the constutents is in a wide range)
- Low vapor pressure
- Suitable lubricity (feeding pump, engine parts)
- Suitable level of additives (if it is necessary)
- Compatibility with motor oils
- Easy and safe handling
- Acceptable price.

The alternative fuels must meet further requirements beside previously detailed issues [1,10]:

- *Be applicable in existing engines*
- *Call for little modification of existing engines*
- *Comply with environmental protection and human health concerns same as with conventional fuels*
- *Operate at lowest additional expense*
- *Be abundant and of high quality*

Alternative fuels can arbitrarily be classified as follows [1]:

- By origin:
  - Fossil (e.g., alcohols, gasoils produced from coal through synthesis gas)
  - Biomass (e.g., alcohols)
  - Others (solar, hydro, water, geothermic, wind, nuclear energy, hydrogen produced by water electrolysis)
- By availability:
  - Exhaustible
  - Renewable (fuels obtained with transforming of biomass)

◦ Renewable (solar energy, water, geothermic, wind, nuclear energy, hydrogen)
- By type of engine:
  ◦ Alternative fuels for Otto engines
  ◦ Alternative fuels for diesel engines
  ◦ Alternative fuels for other engines

Different types of alternative fuels can be produced from different raw materials, and from a single raw material (see Table 4.1, Figure 4.1) [1].

In the this chapter, we review the quality characteristics, production, application possibilities (advantages and disadvantages), and international requirements (if they exist) of each alternative fuel.

4.1 LIGHT (GASEOUS) HYDROCARBONS

Light hydrocarbons like methane, propane and the propane-butane belong to the category of alternative fuels. However, ethane, whose boiling point is between that of methane and propane, occurs a low amount.

Methane is used mostly as an alternative fuel in compressed form, and it is called compressed natural gas (CNG). The maximum pressure of compressed natural gas is ordinarily 200 bar; it is used in the gas phase as a fuel in vehicles (ISO 14532). (The allowed maximum pressure of the natural gas that is stored in the fuel tank is 250 bar; the volume of gas tank is about five times as large as a conventional fuel tank.) Extended research has also been carried out on the use of the liquefied methane. This fuel is called liquefied natural gas (LNG), and operates at pressure 4–5 bar, storage temperature of −162 °C, with the insulated tank of a volume being 1.5 times larger than that of a conventional fuel tank. These two nonconventional fuels are simply called natural gas, because methane is the main component of the different natural gases [1,17–19].

Methane occurs as a natural gas in nature, in the hydrocarbon gases of coal [20–24] and oil shale beds [25–27], and in natural gas hydrates [28–31]. It can further be produced from biogas (methane content > 60–65 v/v%), too [32,33]. The principle scheme of methane production from natural gas is shown in Figure 4.2 [1].

The properties of vehicle fuels drived with compressed natural gas (CNG), for example, vehicles with modified diesel engines, are summarized in Table 4.2.

4.2 PROPANE-BUTANE GAS

The propane-butane (PB is another name for liquefied petroleum gas LPG). PB is a gas mixture that accompanies crude oil and natural gas but is also an accompanying product of different processes in petroleum refining [34,35]. PB is similarly a mixture of hydrocarbons. Its main components are propane and n-butane; some other components are isobutane, pentane, ethane, propene, and butenes. Additionally

TABLE 4.1 Important primary energy resources of motor fuels

	Energy Source									
	Primary									Derived
Motor Fuels	Natural gas[a]	Coal	Oil Shale, Sand	Nuclear Energy	Wind/Water Energy	Renewable (Biomass)				Wastes
						Sugar Plants	Grains	Oil Seeds	Wood	
Synthetic gasoline	x	x	x				x		x	x
Biogas oil	x	x						x		x
Synthetic (bio)gas oil	x	x	x			x	x	x	x	x¹
CNG	x	x	x							x
Ethanol/buthanol	x	x	x			x	x		x	x
FAME	–	–	–	–	–	–	–	x	–	–
DME	x	x	x						x	x
Methanol	x	x	x				x		x	x
Hydrogen	x	x	x	x	x				x	x
Electrical energy	x	x	x	x	x				x	x

Note: ¹CNG: compressed natural gas; FAME: fatty acid methyl esther; DME: dimethyl-ether
All are productable from crude oil.
ᵃincludes methane gas of coalbeds, oil shale, and biogas.

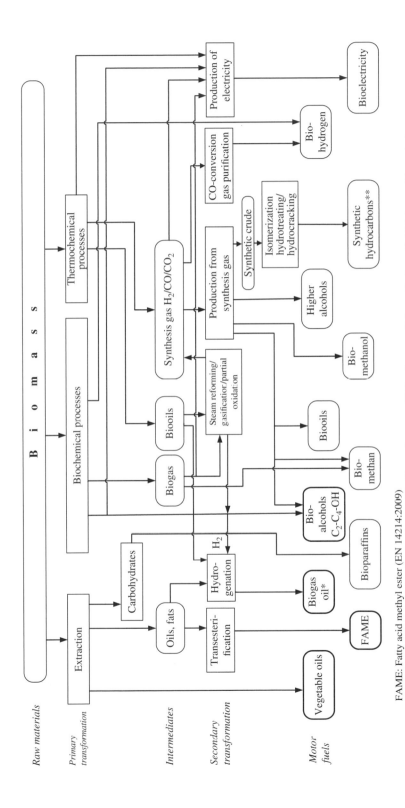

FAME: Fatty acid methyl ester (EN 14214:2009)
*iso-and n-paraffin containing fraction in gas oil boiling point range; **Gasoline, gasoil, base oil, etc., with high isoparaffin content

FIGURE 4.1 Renewable raw materials and production processes of motor fuels

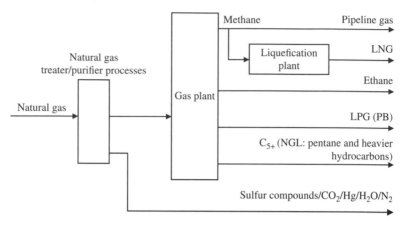

FIGURE 4.2   Simplified scheme of natural gas processing

TABLE 4.2   Main features of natural gas for motor fuel (EN ISO 15403–1:2006)

Properties	Value
Carbon content, %	73.3[a]
Hydrogen content, %	23.9[a]
Oxygen content, %	0.4[a]
Carbon/hydrogen atomic ratio	0.25–0.33
Relative density (1 bar, 15 °C)	0.72–0.81
Boiling point (1 bar), °C	−162
Autoignition temperature, °C	540–650
Octane number (RON/MON)	115–135
Methane number	69–99
Stoichiometric air/fuel mass ratio	17.2
Heating value, MJ/kg	30.2–47.2
Methane content, v/v%	80–99
Sulfur content, mg/kg	<5
Wobbe number[b], MJ/m^3	37.8–56.5

[a] Typical North Sea gas composition.
[b] Wobbe index (MJ/m^3): heating value, on a volumetric basis, at specified reference conditions, divided by the square root of the relative density at the same specified reference conditions of measurements.

sulfur-containing odorants are mixed into the gas to produce a smell for safety reasons [1,36,37]. The PB is in the gas phase under ambient conditions, but can be liquefied at low pressure (4–5 bar).

The composition of the commercial PB differs across countries, depending on the availability, local market prices, the production facilities and the climate conditions. LPG contains mainly propane in the US, Germany, and Finland, mostly in a concentration of 98%. This is likely due to the natural occurrence of propane in these regions. In asain countries, it is mainly butane (C4-hydrocarbons).

Vehicles with Otto engines can be operated, after a modification, with a standardized quality LPG (see Table 4.3).

TABLE 4.3 **Quality requirements of propane-butane gas (EN 589:2004)**

Properties	Limit Value Minimum	Maximum	Test Method
Motor octane number, MON	89.0		[a]
Total dienes content[b], mole%		0.5	EN 27941
Hydrogen sulfide	Negative		EN ISO 8819
Total sulfur content		50	EN 24260
(after stenching), mg/kg			ASTM D 3246
			ASTM D 6667
Copper strip corrosion	Class 1		EN ISO 6251
(1 h at 40 °C), grade			
Evaporation residue, mg/kg		60	EN 15470
			EN 15471
Vapor pressure, gauge, at		1550	EN ISO 4256
40 °C, kPa			EN ISO 8973
Vapor pressure, gauge, min. 150			
kPa at a temperature of, °C			
For grade A		−10	
For grade B		−5	
For grade C		0	EN ISO 8973
For grade D		+10	
For grade E		+20	
		(from 2005)	
Odor	Unpleasant and distinctive at 20% of lower explosive limit		

[a] According to the composition of GC (gas chromatograph).
[b] If the concentration of 1,3-butadiene is higher than 0.1%, it must be signed according to the Directive for subtitles of the European Union

4.3 MIXTURES OF SYNTHETIC LIQUID HYDROCARBONS

Synthetic liquid hydrocarbons are produced directly or indirectly from "synthetic crude oil" or from other primary energy carriers. These mixtures consist mainly of carbon- and hydrogen-containing compounds, that can be transformed, with the suitable processing, or refined to products with quality very similar to that of gasolines and diesel gasoils obtained from crude oil. The production possibilities are the following [1]:

- From synthesis gas obtained from natural gas, coal, hydrocarbon condensates, oil shale or oil sand, biomass, polyolefin wastes, or rubber [15,38–42]
- Transformation of methanol to gasoline (the methanol can be produced from a synthesis gas of natural gas, coal, biomass, etc., or directly from the biomass) [42–47]
- Coupling of methane [48–49]

- Conversion of C_2–C_4 hydrocarbons to gasoline and gas oil (see Chapter 3)
- Distillation of coal [15,50]
- Liquefaction of coal [15,50–55]
- Super critical extraction of coal [15,50]
- Distillation and pyrolysis of biomass [15,46,56–59]
- Production of hydrocarbons from carbondioxide and hydrogen [60,61]
- Thermal and catalytic cracking of polyolefin wastes [62–64]
- Pyrolysis of rubber [65]

From these liquid hydrocarbon mixtures, with suitable technologies (usually the different hydrotreating processes), fuels can be produced with similar compositions and of better quality than conventional gasolines and diesel gasoils. These fuels are applicable in Otto and diesel engines alone or as blending components, with or without additives, so that the produced fuels may satisfy the quality requirements discussed in Chapter 2.

Some of the synthetic gasolines and gasoil are being produced from synthesis gas and methanol are shown in Figure 4.3 [1]. Other potential fuels produced from synthesis gas are also indicated in this figure.

Not all the production possibilities of synthetic liquid hydrocarbons can be reviewed here. Also the technologies used to produce synthesis gas such as the biogas oil production from triglycerides discussed here, may be short-lived. What is important is that these alternative liquid fuels have been proved to be applicable in internal combustion engines without requiring significant engine redesigns. The logistics of storage, transportation, and distribution, as well as quality control have not presented any problems. These are all close to the logistics in place for the conventional gasoline and diesel gas oil. Additional investments, if any, are apparently not yet needed.

4.3.1 Liquid Synthetic Hydrocarbon Mixtures from Synthesis Gas

The production possibilities of liquid synthetic hydrocarbons from synthesis gas are the following:

- GtL (gas to liquid; liquid hydrocarbon mixture from natural gas) [66–71]
- CtL (coal to liquid; liquid hydrocarbon mixture from coal) [70,72–74]
- BtL (bio to liquid; liquid hydrocarbon mixture from biomass) [75–80]
- RtL (residue to liquid; liquid hydrocarbon mixture from residue) [81,82]
- WtL (waste to liquid; liquid hydrocarbon mixture from wastes) [81,82]

It should be noted that gasoline from synthesis gas as obtained from coal is not a new discovery. Liquid synthesis gas was produced between the two World Wars. It is fundamentally a Fischer–Tropsch synthesis process. As the molecules are built up by – CH_2– units, the obtained product mixture is the synthetic crude oil.

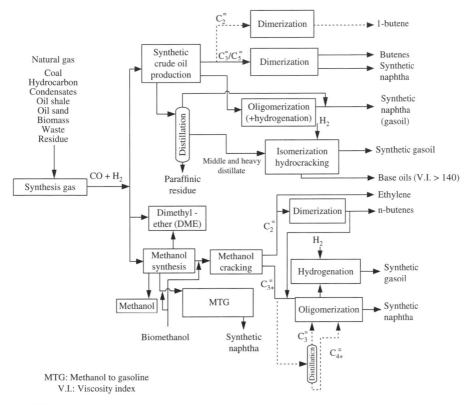

FIGURE 4.3   Scheme of alternative fuels production (The intermediate product is synthesis gas)

The feedstocks of the process producing liquid hydrocarbon mixtures are synthesis gases consisting of mixtures of hydrogen and carbon monoxide in various ratios. These can be sourced from different raw materials, including natural gases, coals, naphthas, heating oils, bitumenes, heavy crude oils, residues of petroleum refining, petrol coke, biomass, wastes containing hydrogen, and coal, among others. [1,15,38,39,43,83–86].

The technologies for synthesis gas production from all these raw materials are not being discussed here. Still, in treating the production of hydrogen as an alternative fuel, some of the more typical processes are being covered in this chapter.

The integrated Fischer–Tropsch processes consist of three main units (Figure 4.4) [1,64,66,69–82]:

- Synthesis gas production
- Synthesis of mainly high molecular weight paraffins (including product separation)

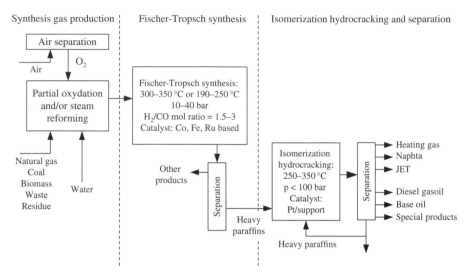

FIGURE 4.4   Integrated synthetic crude oil producing and processing plant

- Conversion of high molecular weight paraffins (hydrocracking, isomerization, isomerizer hydrocracking)

The production of synthetic crude oil from synthesis gas in the Fischer–Tropsch synthesis involves the following reactions [1,15,87–89]:

Main Reactions
Formation of paraffins:

$$(2n+1)H_2 + nCO \rightarrow C_nH_{2n+2} + nH_2O \quad \Delta H = -165 kJ/mol$$

Formation of olefins:

$$2nH_2 + nCO \rightarrow C_nH_{2n} + nH_2O$$

Formation of alcohols:

$$2nH_2 + nCO \rightarrow C_nH_{2n+1}OH + (n-1)H_2O$$

Side Reactions
Water–gas reaction:

$$H_2O + CO \rightleftharpoons CO_2 + H_2 \quad \Delta H = -41 kJ/mol$$

Coke formation on the catalyst surface:

$$\left(\frac{x+y}{2}\right)H_2 + xCO \rightarrow C_xH_y + xH_2O$$

Boudouard disproportionation:

$$CO + CO \rightarrow CO_2 + C$$

Oxidation-reduction of the catalyst:

$$yH_2O + xM \leftrightarrow M_xO_y + yH_2$$
$$yCO_2 + xM \leftrightarrow M_xO_y + yCO$$

Carbide formation:

$$yC + xM \rightarrow M_xC_y$$

The reaction rates depend on the physical-chemical properties of the catalyst, the type of the reactor(s) and their construction, the technological setup, process parameters, and so on.

These reactions are enhanced by iron-, cobalt-, and ruthenium-based catalysts, which contain—beside the support alumina and silica—different promoters (e.g., K_2O, Cu in the case of iron, a small amount of ruthenium in the case of cobalt, etc.) to increase the selectivity and the stability [90–100]. The ruthenium-based catalysts are the most active, so they operate at the lowest temperatures (150–200 °C). On these catalysts the highest average molecular weight paraffins can be produced. Recent results show the catalyst developments of the Fe- and Co/zeolite catalyst families [101–105] to form relatively short chain paraffins (maximum carbon numbers are 20–26). Catalysts of other compositions are being studied too [106,107].

Among the major companies that use integrated technology are Exxon/Mobil [108–110], Shell [67,69,111–113], Sasol(+Chevron) [109,114–119], Syntroleum [109,120,121], Rentech [122,123], Intevep [109], and ENI-IFP [109,124,125]. Each integrated technology has critical points. For example, the choice of the synthesis gas-producing technology must be suitable for the production of high quantities of the synthesis gas with the desired ratio of H_2/CO. The separation of the heavy paraffins from the catalyst is important as well.

Of course, the composition of the products obtained by different Fischer–Tropsch technologies is different, and is determined by the factors mentioned above. For example, a typical average hydrocarbon composition is C_1–C_2: 4–18 %; C_3–C_5 hydrocarbons: 2–17%; C_5–C_{11} (gasoline) hydrocarbons: 20–50%; C_{12}–C_{20} hydrocarbons: 15–25%; C_{21} and higher hydrocarbons: 5–58 %; the whole amount of olefins: 5–35%; oxygen-containing compounds: 1–4 % [1,108]. The newer catalysts promote the formation of medium (C_{10}–C_{20}) and long (>C_{20}) hydrocarbon chains, with minimum amount of light products.

Figure 4.5 shows a comparison on between the product yields of crude oil processing and of an arbitrarily chosen integrated GtL process (included the isomerizing hydrocracking of heavy paraffins too) [126–130] to middle distillate. As is apparent, in the latter case significantly more middle distillate can be produced (by ca. 22 absolute%), beside the formation of naphtha and other products [1,131].

Notice that to increase the octane number of the gasoline boiling point range light fraction, obtained by the integrated Fischer–Tropsch synthesis (including the isomerizing hydrocracking and product separation), the best solution is catalytic isomerization. In the case of the heavier fraction, the appropriate reforming corresponds to the product demand, or it may be used as feedstock for steam pyrolysis (together with the light naphtha) to produce light olefins (ethylene, propylene).

The diesel gasoil fraction has excellent quality. In particular, it is free of sulfur and aromatics (≤ 2 mg/kg and <0.05%), with cetane numbers: 70–80; density: 0.775–0.785 kg/m³. This is one of the big advantages of GtL technologies [109,126–138].

The advantages and disadvantages of Fischer-Tropsch gasoils are the following [1,64,69–82,108–138]:

Advantages

- Several raw materials (conventional and renewable) can be used
- Products practically free of sulfur, nitrogen, and aromatics
- Applicable in conventional vehicles and engines
- Applicable in existing logistical system
- Lower undesirable effect on the activity of the aftertreatment catalyst of vehicles

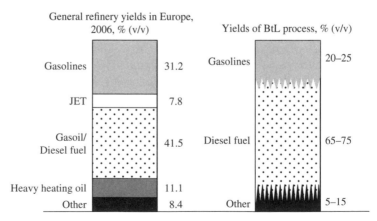

FIGURE 4.5 Comparison of typical product yields from crude oil with integrated GtL technology

- Lower toxic effect during treatment, transportation, distribution
- Better biodegradation compared to the crude based diesel gasoils
- Lower general emission
- Blending with crude oil based diesel gasoils improves the emission properties

Disadvantages
- Available in low quantity (minimum 1.5×10^6 t/year in 2010), but capacity data of ongoing and future investments very promising—by 2020 between about 27 and 95×10^6 t/year ($0.54–1.9 \times 10^6$ b/d), and by 2030 between about 48 and 202×10^6 t/y ($0.96–4.04 \times 10^6$ b/d)] [139]
- Lower volumetric energy content (by 7–8 %)
- Do not satisfy the density specification in actual diesel gasoil standards (marketing only together with crude oil based diesel gasoil)
- The acceptable lubricity only with additives
- Production cost higher than the cost of conventional gasoils

4.3.2 Biogas Oils from Triglycerides

In a wider sense biogas oils can be classified as the synthetic liquid hydrocarbons. These oils are likewise mixtures of iso- and -paraffins [140,141] and are produced from different feedstocks having a high triglyceride content (and fatty acid content) by one or moresteps of catalytic hydrogenation (a special form of hydrocracking), which can, if necessary, be accomplished with isomerization. The biogas oils (actually HVOs: hydrogenated vegetable oils) belong to the second-generation agricultural alternative fuels (see Table 4.4) [140–142].

Biogas oil production starts with the elimination of disadvantageous properties consisting of the fatty acid-alkyl-esters bond in triglycerides, such as in vegetable oils, used frying oils, and greases. Triglycerides are undesirable because of their thermal and oxidation instabilities due to olefinic double bonds. Their hydrolysis sensitivity is due to the ester bond, which forms corrosive acids. Then the phosphorus, alkali, and alkaline metal content must be removed, lowering the energy content by about 10% compared to the gasoil; and poor cold flow, among other negative properties [4,8,143–149].

Biogas oils are nevertheless being developed based on the fact that the long unsaturated hydrocarbons chains ($C_{12}–C_{18}$) derived from triglycerides (vegetable oils, used frying oils and greases, slaughter house greases, "brown fat" of sewage works, algae oils, etc.) have a very high cetane number, around 95–110 (see Table 3.5 and Figure 3.18) in saturated form. Furthermore the high freezing points (ca. between 18 and 32 °C) of the n-paraffins can be decreased at least by 15–30 °C with the transformation to isoparaffins (see Table 3.6 and Figure 3.19.).

$CoMo/Al_2O_3$ [150–153], $NiMo/Al_2O_3$ [153–162], $CoNiMo/Al_2O_3$ [163,164] and other catalysts used for desulphurization of the middle distillates have been suggested

TABLE 4.4 Biomotor fuels classification according to the chronology of their recognition and application

First	Second	Third	Fourth
		Generation	
• Bioethanol	• **Biogas oils** (hydrogenation/ isomerization of vegetable oils)	• Synthetic biofuels from biosynthesis gas	• Biohydrogen • Synthetic biomethane
• Vegetable oils	• Bioethanol from lignocelluloses	• Biogasoline and biogasoil (hydro- cracking of biooils produced by biomass pyrolysis)	• Biomethanol
• Biodiesels	• Biobutanol • Biocomponents as molecular constituents	• Bioparaffins from lignocelluloses carbohydrates	• Bio electricity (indirectly for fuel cells)
• Blends of the previous and conventional petroleum-based fuels	• Biomethan (biogas)	• Biodimethyl ether (DME)	• γ-Valero-lactone • Unknowns

for the purpose of hydrocracking triglyerides. The main reactions proceeding on these catalysts are besides the formation of fatty acids, saturation of the olefinic double bonds and deoxygenation reactions [155,165,166]:

For example, in the case of a triglyceride containing only an oleic acid, hydrogenation takes three steps:

1) Deoxygenation with reduction (classic hydrodeoxygenation; HDO)

$$\xrightarrow[\text{cat., p, t}]{+ 15H_2} 3C_{18}H_{38} + C_3H_8 + 6H_2O$$

2) Decarboxylation

$$CH_2-O-\overset{O}{\overset{||}{C}}-CH_2(CH_2)_6-CH=CH-(CH_2)_7-CH_3$$

$$CH-O-\overset{O}{\overset{||}{C}}-CH_2(CH_2)_6-CH=CH-(CH_2)_7-CH_3 \quad\xrightarrow[\text{cat., p, t}]{+6H_2}\quad 3C_{17}H_{36}+C_3H_8+3CO_2$$

$$CH_2-O-\overset{O}{\overset{||}{C}}-CH_2(CH_2)_6-CH=CH-(CH_2)_7-CH_3$$

3) Decarbonylation

$$CH_2-O-\overset{O}{\overset{||}{C}}-CH_2(CH_2)_6-CH=CH-(CH_2)_7-CH_3$$

$$CH-O-\overset{O}{\overset{||}{C}}-CH_2(CH_2)_6-CH=CH-(CH_2)_7-CH_3 \quad\xrightarrow[\text{cat., p, t}]{+9H_2}\quad 3C_{17}H_{36}+C_3H_8+3CO+3H_2O$$

$$CH_2-O-\overset{O}{\overset{||}{C}}-CH_2(CH_2)_6-CH=CH-(CH_2)_7-CH_3$$

The first step of deoxygenation is the elimination of propane by hydrogenolysis, which results in the formation of carboxyl acid. Then a consecutive hydrodeoxygenation of the carboxylic acid must take place. The main isomerization and cracking of n-paraffins reactions may be followed by a water–gas shift, methanization, and even cyclization and aromatization, depending on the applied catalyst(s) and the reaction conditions. The suggested process parameter combinations are $T = 280–380\,°C$; $P = 20–80\,bar$; $LHSV = 0.8–1.5\,h^{-1}$; $H_2/HC = 450–600\,Nm^3/m^3$ [151–166]. For example, on the conventional transition metal-support catalysts, a product mixture can be obtained from sunflower oil that has a cetane number of 96–102, but high CFPP value (between +22 and +28 °C) [152]. Decarboxylation catalysts for deoxygenation of natural tryglicerides were also suggested (Pt, Pd, Ni, Ir, Ru, etc., supported on Al_2O_3, SiO_2, zeolite, or activated carbon) [167–169];. The isomerization of paraffinic products includes, for example, Pt/SAPO-11 and Pt/HZSM-22 catalysts ($T = 320–350\,°C; P = 35–60\,bar; LHSV = 1.0–2.5\,h^{-1}; H_2/HC = 250–300\,Nm^3/m^3$) [140,141,170]. The CFPP values of obtained products range between −5 °C and −25 °C, but the cetane number is a bit lower only (80–90), though significantly higher than the generally specified 50–55 unit (e.g., EN 590:2009 A1 + 2010, ASTM D 975–11). Currently, only three plants are in operation for the production of biogas oil (two in Finland and one in the Netherlands) [171]. These plants include hydrodeoxygenation and n-paraffin isomerization blocks (Figure 4.6) [162].

Over the past decade several studies focused on the simultaneous catalytic transformation of tryglicerides and different gasoils [145,172–176]. In summary, they found that catalytic coprocessing could be successfully accomplished in an existing HDS plants. The modifications were small, suggesting advantages of lower investment cost and the production of the biocomponent containing gasoil in one

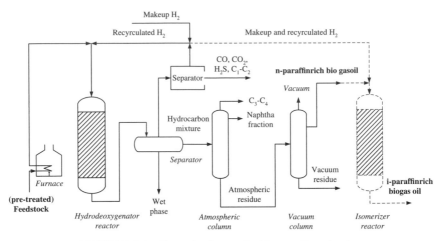

FIGURE 4.6 General scheme of a biogas oil production

step. The cetane number, and sulfur, nitrogen, and aromatic contents of the final product were very favorable.

4.3.3 Production of Bioparaffins from Lignocellulose and Carbohydrates

The bioparaffins produced from lignocellulose and carbohydrates may be utilized in the future as jet fuel and/or diesel fuel blending components [177,178]. If necessary, their quality may even be improved by catalytic isomerization. Glucose and xylose are produced from the biomass by acidic hydrolysis, and the water-soluble organic compounds of higher molecular weight are dehydrated in the acidic media followed by aldol condensation (when carbon–carbon bonds are formed on solid basic catalyst). In the last step in a multi-phase dehydration/hydrogenation unit on a bifunctional catalyst (containing metallic and acidic sites, for example Pt/zeolite), long-chain alkanes are formed. Hydrogen gas is produced from some of the glucose and xylose by aqueous-phase reforming. The reactions are carried out at 60°C to 250°C and 30–35 bar [177,178].

4.4 OXYGEN-CONTAINING ENGINE FUELS AND BLENDING COMPONENTS

Oxygen-containing fuels and blending components contain at least one oxygen atom by molecule beside the carbon and hydrogen atoms. The presence of oxygen promotes burning without emitting high amounts of inert material (e.g., into the air N_2), and thus add to the harmful material (NO_x) emissions at the operation conditions of internal combustion engines. The other important property of oxygen-containing compounds is that they have high octane and cetane numbers required for Otto and diesel engines. For example the research octane number of ethanol is around. 111, while the cetane number of dimethyl-ether is around 55–65. Both have excellent burning properties in their corresponding engines, and they can improve the properties of some less effective components [179,180].

Oxygen-containing fuels and blending components can be classified as follows [1]:

- Type of the applied engines:
  - Otto-engines (methanol, ethanol, methyl-tertiary-butyl-ether, etc.)
  - Diesel-engines (dimethyl-ether, fatty-acid-methyl-ester, etc.)
  - Fuel cells (e.g., methanol)
- Type of the compound:
  - Alcohols (e.g., tertiary-butyl-alcohol)
  - Ethers (e.g., bio ethyl-tertiary-butyl-ether)
  - Esters (vegetable oil, synthetic fatty-acid-methyl-esters, etc.)
- Type of substituted fuel:
  - Instead of gasoline, or use of gasoline with oxygen containing compounds
  - Diesel gas oil substituents and blending components
  - Suitable for both gasoline and diesel gasoil substitute (e.g., ethanol)
- Extent of fuel substitution:
  - Solely applicable
  - Applicable as a blending component (additive)
  - Solely applicable as a blending component

In the following sections, the main types of alcohols, ethers and esters, their properties, production, advantages and disadvantages of the application in engines, including their effect on the environment will be discussed.

4.4.1 Alcohols

Monovalent alcohols having from one to four carbon atoms can be used as fuels in internal combustion engines. The main properties of the top candidates are summarized in Table 4.5 [1]. The roles of isopropyl and tertiary butyl alcohols are negligible compared to the methyl and ethyl alcohols, and the butanols will be discussed later.

Methanol   Among the alcohols, methanol is the cheapest to use as a fuel (in internal combustion engine, in fuel cell) or mixed, such as with 15 v/v% in gasoline or diesel gas oil, and it may also be mixed with dimethyl-ether (DME) [44,179–184]. Its solubility in gasolines is limited, so method is used together with higher carbon number alcohols [44,185,186]. Its use in MTBE manufacturing has decreased due to the toxicity and undesirable odor (especially in drinking water) of MTBE and the spread of bio-ETBE [187–191].

TABLE 4.5 Characteristics of alcohols applicable as motor fuels

Properties	Methyl Alcohol	Ethyl Alcohol	Isopropyl Alcohol (IPA)	Tertiary butyl Alcohol (TBA)	Butanol	Gasoline
Density (at 20 °C), g/cm³	0.791	0.789	0.786	0.791	0.812	0.720–0.775
Boiling point, °C	65.0	77.8	82.0	82.9	118	20–210
Freezing point, °C	−117.3	−97.8	—	−25.6	−89	—
Flash point, °C	13	11	11.7	1–5	35	−43–(−)39
Ignition temperature, °C	446	423	—	470	325	460–495
Vapor pressure, at 37.8 °C (Reid), kPa	35.5	16.2	13.2	14.2	18.0	50–70
Blending vapor pressure, kPa	413.7	124.1	89.0	82.7	—	—
Azeotrope formation with hydrocarbons	Yes	Yes	—	Yes	Yes	—
Oxygen content, wt%	49.9	34.8	26.6	21.6	21.6	—
Stoichiometric air/fuel mass ratio	6.45	8.97	—	11.3	11.2	14.7
Research octane number (RON)	108	111	112	109	94	95
Motor octane number (MON)	89	92	93	93	81	85
$(R+M)/2$	98.5	101.5	101.5	101	87.5	90
Energy content, MJ/kg	19.9	27.7	30.3	32.7	35.9	43
Heat of vaporization, kJ/kg	1.150	0.845	0.720	0.598	0.349	—
Solubility in water			Miscible			
Component in water, v/v %						
Water in component, v/v %						

Methanol may contain sulfur, nitrogen, aromatics, and olefins but only as contaminants. It has a high octane number; consequently it has a low cetane number. It is moreover extremely toxic and corrosive. If methanol gets into the groundwater near the surface it can be a disastrous for the human population. The toxicity of methanol is however, lower for fish and animals than for people. Water-containing methanol causes corrosion of zinc, aluminum, and magnesium, but water-free methanol is not so corrosive. A major hazard of methanol as a fuel is that toxic formaldehyde is formed during combustion, causing allergies to flare up and skin irritation. In the use of M85 (85 v/v% methanol and 15 v/v% gasoline), the emission of formaldehyde can be 10 times higher than that of gasoline [182,192,193].

The methanol is produced from synthesis gas made of different feedstocks (e.g., coal or natural gas). The main reaction is:

$$CO + 2H_2 \rightleftharpoons CH_3OH \quad \Delta H = -92\,kJ/mol$$

which is very exothermic.

Actually the most economical way to produce methanol is to synthetize it from the synthesis gas obtained from fossil sources. In the synthesis gas produced from methane-rich natural gas, the CO/H_2 molar ratio is 1:3, which differs from the stoichiometric ratio required for the methanol production. Hence, in the processes that apply the synthesis gas containing excess hydrogen, carbon dioxide is added to the gas mixture, and the following reaction for methanol formation takes place:

$$CO_2 + H_2 \rightleftharpoons CH_3OH + H_2O \quad \Delta H = -50\,kJ/mol$$

The methanol-producing processes can be classified according to the catalysts being applied, the process parameters (especially pressure and temperature), and the configuration of the reactors [1,15,181,194,195].

The general scheme of methanol producing technology on synthesis gas basis is the following:

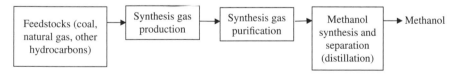

The pressures applied are based on arbitrarily chosen ranges: high (100–350 bar), medium (50–100 bar), and low (<50 bar) pressure processes have been tried [15,181,194,195]. The development of more active and selective catalysts has made it possible to decrease the pressure range. So Methanol synthesis can be carried out at 50–100 bar and at 200–280 °C on $CuO–ZnO–Al_2O_3$ catalysts [195–198].

Methanol production based on biomass is carried out nowadays only by the thermal method, though theoretically it is possible to use biochemicals. An economical fermentation process for the production of methanol on an industrial scale has yet to be developed.

The quality of methanol fuel (>85% alcohol content) is specified in the standards of countries where their use promises to be significant, such as in the United States. Of course, methanol in gasoline presents substantially different properties that must be considered as well (see Table 4.5). Generally, in the so- called flexible fuel vehicles (FFV), these differences do not pose a problem. These vehicles' computer-controlled sensors control the methanol content (alcohol content) of the gasoline, and adjust the operating parameters accordingly [199,200].

The use of methanol as motor fuel has the following advantages and disadvantages [1,15,44,182,201,202]:

Advantages
- Obtained from well-known fossil or from renewable energy sources
- Improves combustion (lower total emission compared even to ethanol)
- Lower volatility than of gasoline
- High research and motor octane number
- Relatively low ozone-forming activity
- Potential to be transported via pipeline
- Cheaper than ethanol
- NO_x and particle emission lower than of conventional engine fuels
- Flames can be extinguished with water
- Lower combustion temperature than gasoline (sparing structural materials)
- Can be applied in fuel cells
- Carbon dioxide emission reduced when produced from biomass
- Several elastomers suitable for seals (e.g., fluorine silicone rubber, styrene butadiene rubber, ethylene propylene terpolymer rubber).

Disadvantages
- Energy balance of production negative compared to that of gasoline
- Low caloric value (18 MJ/dm^3)
- Cold start difficulties
- Relatively high methanol emission, very high aldehyde emission
- Toxic
- Solubility in water
- High vapor pressure
- Because of its limited solubility, according to European Standard EN 228:2010, only 3 v/v% methanol can be blended into motor fuels in the absence of a co-solvent

Methanol is an important raw material for several engine fuels (e.g., gasoline from methanol, dimethyl ether from methanol) and engine fuel additives (methyl-tertiary-butyl-ether: MTBE; tertiary-amyl-methyl-ether: TAME). It is hard to predict the extent to which methanol can become an engine fuel or blending component because of its high toxicity and water solubility. Then again, interest in bioethanol is becoming widespread.

Ethanol   Ethanol (ethyl-alcohol: CH_3-CH_2-OH) is one of the oldest applied fuels (since 1877). Ethanol is called bioethanol when it is distinguished from the ethanol obtained through chemical synthesis. The nearly 16 million tons of bioethanol used as fuel in 2000 grew to nearly 51 million tons in 2012. Nowadays it is around 5.7% of the gasoline use of the world [203].

For the production of ethanol several processes were developed [1,15,175,181]:

Chemical Synthesis
- Direct hydration of ethylene
- Indirect hydration of ethylene
- Homologization of methanol
- Partial oxidation of ethane
- Processes based on synthesis gas

Biochemical Processes
- From sugar via fermentation—grape-sugar (glucose), malt-sugar (maltose), beet sugar (saccharose) etc.
- From starch via fermentation after hydrolysis (the best-known technology world over)
- From cellulose via fermentation after special pre-treating and hydrolysis (on the basis of agricultural by-products, wood, etc.; still in phase of research or pilot plant).

Chemical syntheses will not be discussed, because these are fully covered in petrochemical books [204,205]. The biochemical processes are of increasing importance, so these will be reviewed briefly.

Several processes have been developed for the production of bioethanol, because of the chemical compositions of the different feedstocks. The major feedstocks for bioethanol production are the following:

- Sugarplants (beetroot, sugarcane, fodder beet, sugar sorghum)
- Crops with high starch content (corn, wheat, barley, potato)
- Crops with high inulin content (sweetpotato or sunroot)
- Lignocelluloses (cornstalk, straw, woody plant parts)
- Industrial by-products (molasses, whey, refuse paper, sawdust, etc.)

In North America the raw material of motor fuel grade ethanol (MFGE) is the starch obtained from different grains, primarily from corn, which is around 90%. In Central and South America it is produced from sugar vegetables, mostly from sugarcane. (Ethanol can be obtained directly from the sugar syrup of chopped sugar-cane—from independent manufacturers—or from treacle, which is the by-product of sugar production—from ethanol manufacturers connected with the sugar industry.) In Europe grains and sugar carrot have the dominating role, but the sweet sorghum can become important, as well. Generally, the choice of feedstock unequivocally follows the structure of agricultural production.

Bioethanol is a product of a microbiological transformation—fermentation—of a biomass. Fermentation is an enzyme-catalyzed, energy-producing chemical reaction by which complex organic molecules are transformed into lower molecular weight organic compounds. Consider the following example [1,181,206]:

$$(C_6H_{10}O_5)_n + nH_2O \rightarrow nC_6H_{12}O_6$$
$$nC_6H_{12}O_6 \rightarrow 2nCH_3CH_2OH + 2nCO_2 \quad \text{heat release}: 92.2\,kJ/mol\,(anaerobic)$$
$$C_6H_{12}O_6 = 6CO_2 + 6H_2O \quad 2.82\,MJ/mol\,(aerobic)$$

From the formula above, theoretically 51.1 kg alcohol and 48.9 kg CO_2 are formed from 100 kg glucose. However, the effective yield is lower than this; ordinarily it is 90% of the theoretical yield.

After fermentation, the mash contains 8–12% ethanol depending on the technology and the sugar content of the feedstock. The recovery of ethanol is performed by distillation. The ethanol concentration of the final product is roughly 96%. Water-free, "absolute" alcohol can be obtained by azeotropic distillation, or by separation with a molecular sieve or membrane. The product is of 99–99.9% ethanol content.

The most frequently used starch-containing plant crops are corn, wheat, barley, and potato. The starch content of the first three is around 60%, while that of the last one is 12–20%. Their common property is that the sugar is present in polymeric form.

The **hydrolysis of starch** (liquefaction, conversion to sugar) takes place in three steps [1,15,179,206,207]:

$$Starch + Water \xrightarrow{\alpha\text{-Amylase}} Maltose$$
$$(C_6H_{10}O_5)_{2n} + nH_2O \rightarrow nC_{12}H_{22}O_{11} + By\text{-products}$$
$$C_{12}H_{22}O_{11} + H_2O \xleftarrow{Glucomylase} 2C_6H_{12}O_6$$
$$\text{Maltose} \qquad\qquad \text{Glucose}$$
$$C_{12}H_{22}O_{11} + H_2O \xrightarrow{Invertase} C_6H_{12}O_6 + C_6H_{12}O_6$$
$$\text{Sacharose} \qquad\qquad \text{Glucose} \quad \text{Fructose}$$

During fermentation, glucose and fructose transform to ethanol and CO_2 in an anaerobic reaction in the presence of enzyme complexes:

$$C_6H_{12}O_6 \rightarrow 2C_2H_5OH + 2CO_2$$
$$\text{Glucose, fructose} \qquad \text{Ethanol}$$

The fermentation is followed by the purification of the product mixture. First around 95% concentrated ethanol is produced by conventional distillation.

The ethanol concentration can be increased using azeotropic distillation and adding a demolisher to the mixture (e.g., benzene or cyclohexane). An alternative is to use pressure swing adsorption (PSA, e.g., over zeolites). Yet another solution could be membrane separation. Production of economical, industrial-scale ethanol from lignocelluloses (e.g., from soft and hard woods, sawdust, cornstalks, and corn cobs) has not been realized and currently is only at the experimental stage [1,10,179,208–211].

Ethanol is used as a blending component of gasolines (ca. 5–15 v/v %) or in E85 (ca. 85 v/v% ethanol and ca. 15 v/v% gasoline) [212–214]. Ethanol has performed well in so-called FFV (flexible fuel vehicles). The operating conditions are based on the ethanol concentration of the fuel, which is controlled by a computer [199,200,215].

Based on the experiences of bioethanol programs worldwide, and numerous other research and development activities connected to ethanol, the advantages and disadvantages of this fuel in Otto engines are unambiguous [1,10,179,182,206,216–218].

Advantages:
- Reduction in use of fossil energy sources
- High octane number (octane number increasing with additive)
- Contributes to the reduction of aromatic content of gasolines
- Improved combustion (oxygen-containing compound)
- Higher compression ratio, and thus higher efficiency (38% reached in some tests in Brazil)
- Enhanced performance and increased torque
- Mixtures have higher volumetric energy content
- Longer engine life due to lower operating temperatures
- Environmental advantages in reducing emissions of some carbon compounds (e.g., ca. 60–65% carbon dioxide, ca. 50% carbon monoxide), and also of solid particles
- Biodegradability
- Lower damage to ozone layers than gasoline (but higher than that of methanol)
- Less toxic than methanol
- Performance with lean mixture, resulting in lower carbon monoxide and hydrocarbon emissions (hydrogen/carbon ratio higher than that of gasoline),
- No raining of topsoil or groundwater

Disadvantages:
- Energy content lower by mass
- Energy consumption highest with pure ethanol (i.e., performance declines as purity increases)
- New gasoline blends not yet efficient at high blending vapor pressures
- High hydrocarbon evaporization emissions (e.g., in E15G blends the vapor pressure of gasoline must be decreased)

- Cold start problems of pure ethanol (low vapor pressure due to high evaporation heat)
- Higher ethane, ethylene, and acetaldehyde, but lower formaldehyde emission
- Poor lubricity
- Corrosion damage to engines to iron, steel, zinc, etc., so more corrosion inhibitor needed
- Absorption of water resulting in phase separation of gasoline mixture
- Gaskets (elastomers), plastics may be damaged
- Need for tank of special material, etc.

4.4.2 Ethers

Ethers produced from alternative sources can be used as fuels alone or as blending components in diesel or Otto engines. The most important ethers are the dimethylether (DME), methyl-tertiary-butyl-ether (MTBE), ethyl-tertiary-butyl-ether (ETBE, bio-ETBE), and tertiary-amyl-methyl-ether (TAME) [1,219–224].

Ethers Used in Gasolines   In the Otto engines, ethers can be used as gasoline-blending components. Such ethers are methyl-tertiary-butyl-ether (MTBE), ethyl-tertiary-butyl-ether (ETBE), tertiary-amyl methyl-ether (TAME), and heavier (higher carbon number) ones. These ethers are considered as bioengine fuels when they are produced from bioalcohols such as from biomethanol and bioethanol. The production and properties of these ethers were covered in the previous chapter.

Dimethyl-ether (DME)   Dimethyl-ether is suitable alone to operate diesel engines (cetane number: 60–65). An essential fact is that the miscibility with the diesel gas oil is good [1,225–227]. Dimethyl-ether is produced from methanol (manufactured from synthesis gas) or directly from synthesis gas [228,229]:

$$CO + 2H_2 \rightarrow CH_3 - OH$$
$$2CH_3 - OH \rightarrow CH_3 - O - CH_3 + H_2O$$

This is a slightly exothermic reaction. The catalyst can be amorphous alumina with ca. 10% silica. The main reactions during the direct production from synthesis gas are [228,230,231]:

$$3CO + 3H_2 \rightarrow CH_3 - O - CH_3 + CO_2$$
$$2CO + 4H_2 \rightarrow CH_3 - O - CH_3 + H_2O$$
$$2CO + 4H_2 \rightarrow 2CH_3 - OH$$
$$2CH_3 - OH \rightarrow CH_3 - O - CH_3 + H_2O$$
$$CO + H_2O \rightarrow CO_2 + H_2$$

The considerable heat released during the highly exothermic reaction must be removed. Catalysts were developed (by methanol synthesis of a copper-zinc-zirconium based catalyst on solid acids like γ-Al_2O_3 or ZSM-5 zeolite) that generate hydrogen formation from the water and carbon monoxide dissolved in the solvent [229,231,232].

The favorable process parameters are 260–270 °C and roughly 40–60 bar [231,232]. DME may contain less than 0.3% other oxygen-containing compounds, and a lubrication improver additive is needed in a concentration of 500–2000 mg/kg [233,234].

The advantages of the DME application are [1,235–237]:

- Low ignition temperature (high cetane number), therefore suitable for compression ignition engines
- High vapor pressure at ambient temperature, therefore evaporates almost immediately after injecting into the cylinder
- Very high oxygen content, low carbon/hydrogen ratio and free of carbon–carbon bond, therefore the soot formation during combustion is lower and the particle emission is very low
- Can be used as a hydrogen source for the fuel cells
- Ignition delay is short because of its high cetane number and good volatility
- Noise of the engine is low
- Has good cold start properties

The Disadvantages are:

- Requires special fuel-injection system, but so far only a few engine and fleet experiments have been carried out with DME-fueled engines
- Due to the poor lubrication properties, abrasion may occur on friction surfaces in fuel-injection system
- Chemically and physically attacks conventional gaskets and plastic components
- Compressibility of DME is much higher than that of the diesel gas oil, therefore the compression energy demand of the DME-fuel injection is much higher
- At the same energy content, approx. 1.8 times higher volume of DME must be injected than diesel gas oil to combustion chamber, thus timing of the injection must be improved

4.4.3 Vegetable Oils and Their Oxygen-Containing Derivatives

As we noted earlier, the fuels produced from plants or crops will increase in importance worldwide because the carbon dioxide emitted during burning—as in case of other biofuels—does not burden the environment and will likely be assimilated by cultivation of the next crop of vegetation.

The suitability of vegetable oils for operating diesel engines has been known for a long time. In 1912 Rudolph Diesel presaged that the use of vegetable oils could someday become as important to engine operation as the crude oil and coal tar products.

These materials have come to be called biodiesels. Biodiesels range widely due to the different vegetable oils and natural fats. Their oxygen-containing derivatives are also applicable in diesel engines. In a sense, transesterificated derivatives of plant triglycerides, are the fatty acid alkyl-esters.

The vegetable oils are obtained mostly from the seeds of plants, and sometimes from other plant parts. The oils used are: in the United States, soybean oil and some sunflower oil; in Canada, rapeseed oil and mainly pine pulp oil ("tall oils"), in Europe, rapeseed, sunflower oils; and in southern countries, palm oil.

The compositions of vegetable oils are very similar, because 95–97% of them are triglycerides that are built up of similar fatty acids; the compositions differ only in the ratios of the individual fatty acids. In addition, the raw vegetable oil contains free fatty acids and their oxidized products, phosphatides (e.g., lecithin), vitamins (e.g., tocopherols), water, mono- and diglycerides, colors, free sugar, glucolypides, hydrocarbons, resins, sterols, waxes, taste and smell materials, traces of metals having oxidation catalyzing effect (e.g., copper), among other things. (The emulgeated materials in vegetable oils are called slime [1,4,8,238–243].)

The oils are retrieved from suitably prepared acids and alcoholscrops of vegetables by extruding (e.g., screw extruders) or extruding milling (oil mills) and/or with extraction (e.g., with hexane) [1,4,8]. Extruding milled vegetable oils and their derivatives are used as engine fuels. (In the food industry more refining steps are necessary.) A valuable by-product, called "extruded cake" (animal food), is obtained in this operation, which has 4–8 % oil content [244–246].

The oil content of the various seeds is different (eg., rapeseed has ca. 40%; sunflower has ca. 47%; soybean has ca. 18%; palm has ca. 47%; peanut has ca. 45%) The compositions (Table 4.6) and qualities (Table 4.7) of obtained oil differ as well [1].

The structures of vegetable oil molecules fundamentally determine their physical-chemical properties. For example, double bonds impair the thermal and oxidation stability of molecules [1,4,8,247–249]. The molecules are miscible with the water because the ester groups hydrolyze easily to acids and alcohols. In both processes the acids that form, cause corrosion in the fuel supply lines and engine parts, and they also damage gaskets. In some parts these acidic components interact directly with the engine oil. As a consequence the basic reserve of engine oil decreases, and antagonistic interactions with the additives take place. The engine oil change times are shortened significantly (by 30–50%) [1,4,8,161,162]. Most of the esters have relatively unstable bonds that are effectively destroyed by microorganisms inherent to them, but during storage as well [1,4,8,249–252]!

Vegetable oils can be used in diesel engines alone or with diesel gasoils in various rated mixtures, and after chemical transformations. The chemical transformations are structural changes to the oxygen-containing compound(s), which are different fatty acid alkyl esters in biodiesels. Biogas oils contain mixtures of iso- and n-paraffins produced from vegetable oils of high triglyceride content and other raw materials by catalytic hydrogenation. Very little, if any, oxygen-containing compounds are present, as we explained in our discussion of synthetic liquid hydrocarbons (see Chapter 4).

TABLE 4.6 Fatty acid composition of vegetable oils and fats

Oil	C8:0; C10:0	C12:0	C14:0	C16:0	C16:1	C18:0	C18:1	C18:2	C18:3	C20:0 C22:0	C20:1 C22:1	Other	Iodine Value gI₂/100 g
Rapeseed oil													
Low erucic acid content	—	—	—	3–5	—	1–2	55–65	20–26	8–10	0.5	1–2	<0.5	96–117
High erucic acid content	—	—	—	2–4	—	1–2	14–18	13	8–10	1	45–52	<0.5	98–108
High oleic acid content	—	—	—	2–4	—	0–2	70–80	12	1–5	0–1	0–1	<0.5	85–100
Sunflower oil													
Conventional	—	—	—	6	—	3–5	17–22	67–74	—	0.6	—	<0.5	127–142
High oleic content	—	—	—	4	—	4	70–95	1–13	—	—	—	<0.5	85–90
Soybean oil	—	—	0–1	7–12	0–1	3–6	22–34	49–60	2–11	0–10	—	<0.5	121–143
Palm oil	—	0–1	1–6	32–48	—	1–6	37–52	2–11	0–1	0–1	—	<0.5	53–57
Babassu oil	4–6 6–8	44–48	15–20	5–11	—	2.5–5.5	10–16	1–3	—	—	—	C6:0: 0–1	10–18
Coconut oil	5–9 4–10	44–51	13–20	7–10	—	1–4	5–8	1–3	—	—	—	C6:0: 0–2	6–12
Corn oil	—	—	0–2	9–19	1–2	1–5	26–50	34–62	0–1	—	0–2	<0.5	109–140
Cottonseed oil	—	—	0–3	17–28	—	1–3	13–41	34–59	—	—	2–3	<0.5	90–117
Olive oil	—	—	—	7–16	0–1	1–3	64–85	4–15	0.5–4	0–1	0–1	<0.5	78–94
Palm seed oil	2–5 3–7	45–55	14–19	3–9	0–1	1–3	10–21	1–2	—	1–2	—	C6:0: 0–2	12–16

[a]The first number represents the number of carbon atoms, and the second the number of double bonds in the molecule.

TABLE 4.7 **Physical-chemical properties of different vegetable oils**

Properties	Rapeseed Oil	Sunflower Oil	Soybean Oil	Palm Oil	Len Oil
Density, at 15 °C, g/cm^3	0.915	0.925	0.930	0.920	0.935
Flash point, °C	320	315	330	265	—
Cloud point, °C	0	−15	−10	30	−20
Pour point, °C	−12	−18	−18	25	−25
Kinematic viscosity, at 20 °C, mm^2/s	98	66	65	—	51.5
Iodine number, g I$_2$/100 g	115	130	135	—	185
Saponification value, mg KOH/g	175	190	190	—	190
Heating value, MJ/kg	40.5	40.0	40.0	35.0	39.5
Cetane number	50	35	38	40	—

Tables 4.8 and 4.9 list the properties of the most widely used rapeseed and soybean oils as fuels and the fatty-acid-alkyl-esters produced from them [1]; they also summarize the most important properties of EU and US diesel gasoils.

Recall that in the production of the vegetable oils and diesel gas oils listed Tables 4.8 and 4.9 fatty-acid-alkyl(methyl)-esters from the vegetable oil are produced by transesterification (alcoholysis) of suitably prepared vegetable oils (triglyceride content >95 %) [1,4,8,244,245], and they are also produced from used frying oils [1,4,8,252] with low molecular weight alcohols (e.g., methanol, bioethanol). Bases (KOH, NaOH, K- or Na-methylate, etc.) [1,4,8,245,252–257], acids (e.g., sulfuric acid) [1,4,8,245,256–259], supported bases (IFP) [260–263], solid acids, or enzymes [264–268] are used as catalysts. Supercritical alcohols (e.g., methanol, ethanol) have been suggested for transesterification [1,4,8245,269–272]; their chemical transformation (Figure 4.7) occurs at 35–250 °C, with their 3.1–3.8:1.0 alcohol:triglyceride mol ratio depending on the catalyst and on how the technology for the transesterification is constructed [1,4,8,245,256,257]. The fatty-acid-methyl-ester phase is separated from the glycerin phase with decantation, and its quality can be improved by different finishing operations, before it is additivated.

Purification of the transesterificated product can be done in any of the following ways [1,4,8,245,256,273–278]:

- Washing with water, followed by water removal with distillation.
- Separating the ester by distillation (e.g., film evaporation requires high energy, but the product is very pure, and methanol gets separated in the process).
- Applying a surface active additive (that sufficiently dissolves methanol, leaving water and catalyst residue).

Fatty-acid-methyl-esters can be produced not only from vegetable oils but also from used frying oils [1,4,8,252,279–282], animal fats [283–285], algae oils

TABLE 4.8   **Properties of rapeseed oil, RME, and diesel gasoil (European Union)**

Properties	Rapeseed Oil	Methyl Ester (RME)	Diesel Gasoil [EN 590: 2010]
Density, at 15 °C, g/cm^3	0.905–0.930	0.860–0.900	0.820–0.845 (0.830)
Kinematic viscosity, at 40 °C, mm^2/s	33–43 (39)	3.5–5.0	2.0–4.5
Heating value,	37.0–37.2	36.7–37.7	40.0–44.0
MJ/kg	33.7–34.0	32.4–33.1	35.7–35.8
MJ/dm^3			
Cetane number	40–44 Refined 50–56	≥51	≥51
Polyaromatic content, (di+), v/v %	<0.01	—	≤8.0
Sulfur content, mg/kg	10–120	≤10	≤10
Flash point, °C	315–325	≥120	≥55
CFPP, °C	5–18	≤−20	≤−20
Conradson-number, %	0.15–0.5	≤0.3	≤0.3
Initial boiling point, °C	~210	~320	~180
End boiling point, °C	Breaking from ~320 °C	~360	360–370
Sulfate ash content, %	—	≤0.02	<0.01
Total contaminant content, mg/kg	—	≤24	≤24
Copper corrosion, grade	—	1	1
Water content, mg/kg	100–1000	≤500	≤200
Iodine number, g/100 g	110–120	≤120	*
Total glycerin content, %	—	≤0.25	*
Free glycerin content, %	—	≤0.02	*
Methanol content, %	—	≤0.2	*
Phosphor content, mg/kg	—	≤7.0	*
I. Metal group (Na + K), mg/kg		≤5.0	*
II. Metal group (Ca + Mg), mg/kg		≤5.0	*
Acid number, mg KOH/g		<0.50	*
Polyunsaturated (≥4 double bond)		<1.0	*
Oxidation stability	—	EN 14112:2003 min. 6 h	EN 15751:2009 min. 20 h

*Note: The*indicates no limitation.*

[a] Determined from the 10% residue

TABLE 4.9 **Properties of soybean oil, SME, and diesel gasoil (United States)**

Properties	Soybean		Diesel Gasoil
	Oil	Methyl Ester	
Density, at 15 °C, g/cm³	0.920	0.886	0.8495
Kinematic viscosity, at 40 °C, mm²/s	33	3.891	2.98
Heating value	39.3	39.77	45.42
MJ/kg	36.2	35.24	38.58
MJ/dm³			
Cetane number	38	55	49
Polyaromatic content, (di+), v/v %			
Sulfur content, mg/kg	≤100	≤120	≤360
Flash point, °C	—	188	74
CFPP, °C			
Conradson number, %	—	0.068	0.16

[286–289] and from other non edible oils [290–293]. Before the transesterification the appropriate pre-treatment of these raw materials is necessary.

Data in Table 4.8 clearly show that the properties of vegetable oils, their fatty acid methyl esters (B100), and diesel gasoils are fundamentally different in their molecular structures. The undesirable properties of vegetable oils do not disappear by blending them into gasoils, but they deteriorate the gasoil quality (depending on the blending ratio—B5, B7, B10, B20, etc.). The application of pure vegetable oils alone or as diesel gasoil substituents is not suggested.

The use of vegetable oils as an engine fuel has been approached from the point of view of redesigning engines to make biogas oils more suitable as fuels. However, the development and investment costs are still very high, and as yet, attempts have not been successful.

Theoretically, suitable components for diesel engines could be produced from the different natural triglycerides that result after they are transformed into molecules similar to those in gasoils, but with lower average molecular weights and a lower content of unsaturated compounds. So far, the only ways known to accomplish the hydrogenation of triglycerides and the transesterification of triglycerides are with monovalent alcohols. This former process was discussed previously in this chapter.

During esterification, as is evident in the scheme of Figure 4.7, the molecular structure of fatty-acid-methyl-ester mixtures, which are formed in consecutive (multiple-step) reactions, becomes very similar to that of the paraffin hydrocarbons in gasoils (except for the ester group and the double bond(s) in hydrocarbon chain).

On comparing the main properties of vegetable oils and fatty-acid-methyl-esters (FAME) to similar data for gasoils (Tables 4.8 and 4.9), it can be seen that the chemical transformation can raise the quality of FAME and even exceed that of gas oils. In effect, on average molecular structure of FAME is closer to the structure of diesel gas oils than to that of rapeseed oil, for example. However, the density, viscosity, cold filter plugging point, and Conradson number values of FAME are definitely not as good as those of

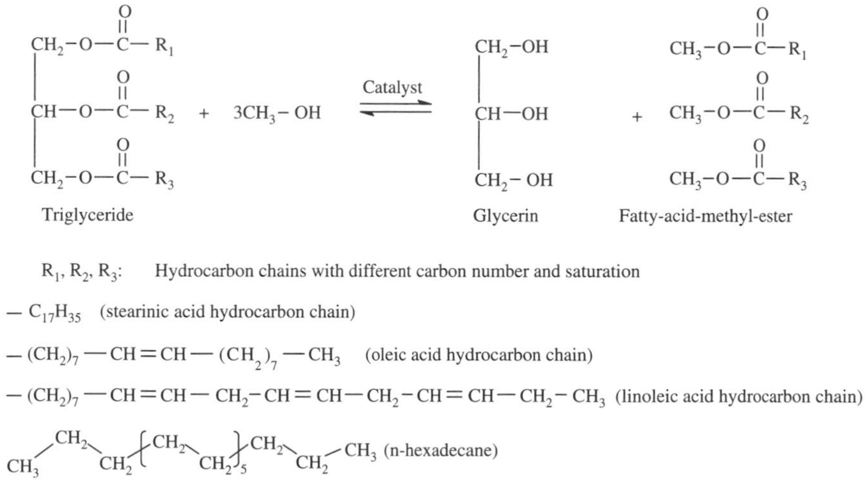

$$
\begin{array}{ccc}
\text{Triglyceride} & \text{Glycerin} & \text{Fatty-acid-methyl-ester}
\end{array}
$$

CH$_2$—O—C(=O)—R$_1$
CH—O—C(=O)—R$_2$ + 3CH$_3$–OH ⇌ (Catalyst)
CH$_2$—O—C(=O)—R$_3$

CH$_2$—OH
CH—OH
CH$_2$—OH

+ CH$_3$—O—C(=O)—R$_1$
CH$_3$—O—C(=O)—R$_2$
CH$_3$—O—C(=O)—R$_3$

R$_1$, R$_2$, R$_3$:   Hydrocarbon chains with different carbon number and saturation

— C$_{17}$H$_{35}$ (stearinic acid hydrocarbon chain)

— (CH$_2$)$_7$ — CH = CH — (CH$_2$)$_7$ — CH$_3$ (oleic acid hydrocarbon chain)

— (CH$_2$)$_7$ — CH = CH — CH$_2$–CH = CH — CH$_2$–CH = CH — CH$_2$– CH$_3$ (linoleic acid hydrocarbon chain)

CH$_3$—CH$_2$—(CH$_2$)$_5$—(CH$_2$)$_x$—CH$_2$—CH$_3$ (n-hexadecane)

FIGURE 4.7   Gross reaction of transesterification of vegetable oils (triglyceride) and the molecule structure of n-hexadecane

diesel gasoil; the cetane numbers are approximately the same in range, while the higher flashpoint and the better lubricity of FAME can be considered two advantages. The water, glycerin, methanol, and phosphorus contents of FAME are clear disadvantageous, and they can be sources of serious trouble. The presence of the toxic methanol is extremely undesirable and that of phosphorus is environmentally disastrous, since it deactivates the exhaust catalysts. The heating value of FAME is definitely lower than that of diesel gasoils (by 8–14%), which is the main reason for the higher fuel consumption (by 10–15 v/v%) of engines operating with FAME. Because of the double bonds, FAME has weak thermal and oxidation stability (storage), and because of the ester bonds, it is sensitive to hydrolysis. In addition, because of high solubility of FAME, special sealants (e.g., fluoro elastomers) must be used.

To reduce or eliminate the above-mentioned disadvantages of FAME, different additives are necessary (flow improvers, corrosion inhibitors, detergent-dispersant, and cetane improvers) [294–301].

The chemical structure of FAME compounds, while differing from that of gasoil and posing disadvantage, can also by the lower heating value be a significant advantage. Fatty-acid-methyl-ester molecules have excellent lubrication properties. The abrasion to the fuel pump is consequently less than that of diesel gasoils, due to the low sulfur and reduced aromatic content. A further blending with diesel gasoils in small concentrations (e.g., 0.5–2%) can improve significantly the lubricating, properties of gasoils [302–304].

The consumption of biodiesels from different triglyceride-containing feedstocks has grown significantly over the last decade. The mentioned preferable properties and the significant greenhouse gas emission reductions over a vehicle's life cycle (40–55%) outweight the mentioned disadvantages. But there is also an indication of decrease of use, such as in Germany. Nevertheless, the data show biodiesel production to rise from around 800 thousand tons in 2000 to around 15.7 million tons in 2009 worldwide, with around 680 thousand tons in 2000, around 8.8 million tons in 2011 and 7.8 million tons in 2011 being in the European Union [305].

4.5 HYDROGEN

In 1820 there was already a trial for developing hydrogen-fueled engines, but the prototype engine was not built until in 1860. Otto used approximately 50% hydrogen-containing gas in the 1860s and 1870s, when he tested his internal combustion engines [306–308].

Since then, hydrogen has been used in large quantites in the US space program, since hydrogen has the highest energy content per mass. Hydrogen is best fuel for rockets, and these have included the *Apollo* missions to the Moon, the *Skylab* and *Viking* missions to Mars, and the *Voyager* mission to Saturn. The main driving gear of a spacecraft operates with hydrogen. In the 1950s hydrogen was even tried in a *Martin B-57* bomber, to estimate the advantages of hydrogen in jet engine (increase of flying distance and flying height) [307].

When hydrogen is used as an engine fuel, after the combustion only a very small amount of impurities is emitted. If it is burned in an internal combustion engine, the main combustion product is water, CO_2 does not form at all, and NO_x-compounds form in a small amount. If we consider only the burning process, hydrogen is the most pure fuel for internal combustion engines. Hydrogen will likely be used in high quantities in the near future, directly or indirectly, in the practically emission-free fuel cells [1,306,307,311].

4.5.1 Production of Hydrogen

The most practical ways of achieving hydrogen production are shown in Figure 4.8 [1]. The principles of considered technologies are discussed below.

Steam Methane Reforming (SMR) [306]   During steam reforming, hydrogen is produced by the reaction of hydrocarbons (e.g., mainly methane) with steam. For example, for methane [1,309–313]:

$$CH_4 + H_2O = 3H_2 + CO \qquad \Delta H = +206\,kJ/mol\,H_2$$

Or in general,

$$C_nH_m + nH_2O = (n + m/2)H_2 + nCO$$

The reaction is carried out ordinarily at 800–850 °C, 25–35 bar, on a nickel catalyst inserted in a heated pipe reactor. Due to the severe operating conditions, the catalyst is supported on mechanically strong alumina spheres of high surface area and of appropriate porosity, ensuring the required heat and material transfer. The main catalyst poisons are the sulfur and chlorine compounds (≤0.1 mg/kg) [314,315]

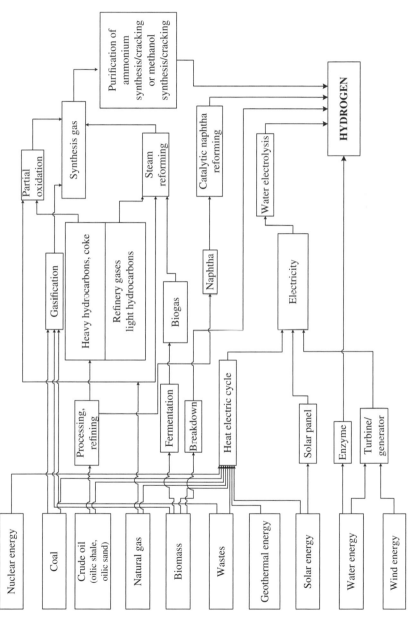

FIGURE 4.8 Top raw materials considered in the production of hydrogen

Cracking and coke formation limit the application of hydrocarbons with a higher boiling point than 180 °C. For their application a pre-reforming step is included, where a low temperature catalyst bed is used [306,316–318].

The CO content of the yielded gas mixture is reacted with steam. By this "water-gas shift" reaction the hydrogen output is increased:

$$CO + H_2O = CO_2 + H_2 \quad \Delta H = -41 kJ/mol H_2$$

After the shift reaction, the product mixture contains mostly hydrogen and carbon dioxide. The CO_2 is then removed. In some technologies the residual CO and CO_2 are eliminated by methanization (which is the reverse of reforming):

$$CO + 3H_2 = CH_4 + H_2O \qquad \Delta H = -206 kJ/mol H_2$$
$$CO_2 + 4H_2 = CH_4 + 2H_2O$$

H_2 can be enriched with pressure swing adsorption, membrane separation, and CO_2 absorption [309,320]. In recent years, several studies have explored water-free, so-called dry reforming, processes [1,306,307,314,321,322]:

$$CH_4 + CO_2 \rightarrow 2H_2 + 2CO$$

Partial Oxidation (POX)   Hydrogen can be produced by the partial oxidation of different hydrocarbons [1,306,307,314,323–325]. In this process, hydrocarbons are reacted with oxygen at a high temperature:

$$CH_4 + \tfrac{1}{2}O_2 = CO + 2H_2 \quad \Delta H = -38 kJ/mol$$

Water–gas shift reaction takes place to some extent due to the unavoidably formed water, so the product contains CO and CO_2 besides hydrogen. For partial oxidation, the catalyst is not needed, so light and pure hydrocarbons without purification can be used as feedstocks. The expenses of partial oxidation are high. In the case of light olefins, steam reforming is preferred. For heavy feedstocks, this is indeed the only suitable and economic process. Partial oxidation is also applicable for the conversion of residues.

The typical reaction conditions are the following: pressure 50–60 bar, temperature 1300–1400 °C. The methane content in the output is very low due to the high temperatures. The gas mixture is cooled with water quenching or in heat exchangers connected to the steam generation.

Autothermal Refoming (ATR)   Autothermal (self-supporting) reforming is similar to steam reforming. The only difference is that the catalyst is not placed in hard alloyed tubes, but in a reactor having well-isolated walls. The process is called autothermal reforming because the reaction heat of the reforming is covered partly by the partial combustion of the hydrocarbon feedstock. In this technology the desulphurized feedstock is preheated, mixed with steam, and then introduced in the ATR reactor, where oxygen is added into a special burner [1,306,307,326].

Production of Hydrogen from Alcohols with Steam Reforming   Alcohols (methanol, ethanol, etc.) and steam are reacted in the presence of a catalyst (e.g., copper-zinc). Hydrogen, carbon dioxide, carbon monoxide, and water are formed [327–332]. The typical reaction conditions are the following: temperature 250–300 °C, pressure 10–25 bar. This process is profitable only when the price of the alcohols is low.

Production of Hydrogen with Decomposition Hydrogen-Sulfide   The decomposition of hydrogen-sulfide is an equilibrium reaction. It can be carried out, for example, on Pt/cobalt-oxide, on molybdenum- and tungsten-disulfides, and on transition metal-sulfides supported on aluminum-oxide carriers [333,334]. The equilibrium of this reaction is shifted to decomposition only at higher temperatures. Recycling and recovery of hydrogen-sulfide from the product gas mixture is rather expensive. Several other methods are being developed to produce hydrogen from hydrogen-sulfide. Among these are plasma technologies and thermal cracking processes.

Hydrogen Production with Electrolysis of Water   Hydrogen production with electrolysis of water is a very old technology. Except for some special applications, it cannot compete economically with the steam reforming or with partial oxidation. Hydrogen production with electrolysis can be competitive in regions where the production of electric energy is cheap and the production environmentally compatible, for example, by using energy of water or wind [307,335,336]. The electrolysis of water is well known; a survey here would be superfluous [337].

Production of Hydrogen as a By-product of Catalytic Naphtha Reforming   Catalytic naphtha reforming serves to produce aromatic-rich hydrocarbon fractions. A high amount of hydrogen is then formed as a by-product of the dehydrogenation and dehydrocyclization reactions. This technology is the most economic hydrogen source in refineries. The petroleum industry consumes this large amount of hydrogen, however, and it is often not enough to satisfy the demand (see Chapter 3).

Hydrogen Production in Biological Way   There are different biological methods to produce hydrogen:

- Direct biophotolysis
- Indirect biophotolysis
- Photo fermentation
- Dark fermentation

Unfortunately, due to limitations of space, their discussion must be omitted.

TABLE 4.10   Typical properties of hydrogen and other fuels

Properties	Hydrogen	Gasoline	Methanol	Ethanol
Burning heat, MJ/kg	120.1	43.4	19.9	27.7
Stochiometric hydrogen/air mass ratio	34.3	14.5	6.45	8.97
Maximal laminar flame velocity, ms^{-1}	2.91	0.37	0.52	—
Adiabatic flame temperature, °C	2756	2637	2576	2594
Research octane number	>130	91–100	108	111
Motor octane number		82–94	89	92
Density, g/cm³	0.0709(−252 °C)	0.720–0.775	0.791	0.789

Obviously hydrogen can be obtained from the gas mixtures produced in coal and biomass gasification [306,307].

From the hydrogen-rich gas mixtures produced by the above-mentioned technologies, the hydrogen thus produced has impurities, such as CO, CO_2, light hydrocarbons, and water, that must be eliminated to get fuel quality hydrogen. Following possibilities are being explored [307,320]:

- Pressure swing adsorption
- Membrane separation
- Cryogen technology
- Absorption

4.5.2   Main Characteristics of Hydrogen

The density of hydrogen is very low (gas at 0 °C: 0.09 kg/m³; liquid at −252 °C: 70.9 kg/m³) [307]. This entails to two problems:

- A very large storage volume is necessary for the suitable running distance of the vehicle
- The energy density of hydrogen/air mixture is low, so the achievement is low too.

The flame velocity of burning hydrogen is nearly one order of magnitude higher than that of gasoline. This means that in stochiometric mixtures, the hydrogen engines approach better the thermodynamically ideal motorcycle. In lean mixtures, the flame velocity is significantly lower [307,338,339].

Some important properties of hydrogen fuel are summarized in Table 4.10.

The performances of gasoline and different hydrogen engines can be increased with modern fuel-injection methods or with the use of liquid hydrogen. In the latter case the quantity of the injected hydrogen can be increased by about 33%, and the performance by about. 37%. The performance can be further increased if the hydrogen is directly injected into the cylinder (combustion space) [340].

The use of hydrogen in *compression ignition engines* is limited due to its high self-ignition temperature. A hydrogen–air mixture does not ignite easily without an external ignition source. So the majority of experiments for the development of pure compression ignition hydrogen engines has failed [338–340].

Harmful Material Emissions   An important advantage of hydrogen engines is that they have lower harmful material emissions than gasoline engines. The components of exhaust gas are only water, nitrogen, and nitrogen oxides (NO_x). Some hydrogen is exhausted, but hydrogen is not toxic and does not contribute to smog-generating reactions [307,340–342].

NO_x emission is unavoidable in hydrogen engines. Nitrogen oxides can thin the ozone layers of the atmosphere and produce acid rain. The nitrogen oxides form because air is used for combustion. Air contains 79% nitrogen. so if the temperature of the combustion is high enough, then it forms NO_x with oxygen. During combustion NO_x formation basically depends on the following factors [307,340]:

- Temperature of combustion
- Time of reaction
- Quantity of available oxygen

If one of these factors increases, then NO_x emission increases too. When hydrogen engines operate in the lean mixture mode, the combustion temperature decreases [341], and so do NO_x emissions. This state can be attained as exhaust gas is recirculated, or by injecting water and liquid fuel.

Over the life cycle of a vehicle, the greenhouse gas emission of hydrogen fuel is around one-third that of new composited gasoline. The use of hydrogen from natural gas is a bit more expensive, by about 20% [343].

The efficiency of hydrogen-fueled internal combustion engines seems to be satisfactory, but it is more unfavorable with respect to the WTW balance than the use of natural gas in CNG form. Hydrogen in liquid form requires more energy than in compressed gas form. Liquid hydrogen is, however, a good alternative for fuel cell use. If hydrogen could be produced from natural gas, the total GHG emission can only be decreased by use of fuel cell vehicles [308,343,344].

4.5.3   Hydrogen Storage on Vehicle and Reloading

A huge challenge in engineering hydrogen-driven engines is the construction of a hydrogen tank that meets modern technical and safety requirements. Because of the very low density of hydrogen, hydrogen storage systems have to be larger and heavier than the tanks storing gasoline and gasoil. Hydrogen is stored as a liquid in cryogenic tanks onboard the vehicle, in hydride form (combined with some metals), or as a compressed gas at high pressure. Table 4.11 shows the properties of this system [345].

TABLE 4.11   Comparison of on-board hydrogen storage systems

Properties	Gasoline	Liquefied Hydrogen	Hydride (FeTi) (1.2% H$_2$)	Compressed Hydrogen (20.7–69.0 MPa)
Energy content, MJ[a]	665	665	665	665
Fuel mass, kg	13.9	4.7	4.7	4.7
Tank mass, kg	6.3	18.6	547.5	63.3–86.0
Full fuel system mass, kg	20.2	23.3	552.2	68–90.7
Tank volume, dm^3	18.9	177.9	189.3	408.8–227.2

[a] The heating value of 18.9 liters of gasoline is 665 MJ.

Liquid hydrogen is stored at 2 bar pressure and 20 K (–253 °C) temperature. To maintain these conditions, an extra insulated tank with a duplex wall is needed. Hydrogen gas can be removed from these tanks. The daily vaporization loss is ordinarily less than 2%. It is should be noted that the energy demand of the liquefaction process is roughly 40% of the heating value of the obtained liquid hydrogen [340–345].

Compared to the other onboard storage systems, an advantage of using liquid hydrogen is that the mass of the fuel system is just a little larger than that of the conventional gasoline fuel system. Its volume demand is 6 to 10 times higher, which reduces the storage and/or passenger space. Another advantage of the liquefied hydrogen is that it can be injected into the engine in cold form (–80 °C). As the thermal efficiency of the engine increases, the NO$_x$ emission decreases, because cold hydrogen behaves as a good heat absorber, and decreases the temperature in the burning chamber [343,345].

The basis of the metal hydride storage system is that the gas-phase hydrogen adsorbs on a metal surface by weak bonding. The metal hydrides are granulated or in powder form, have a high specific surface area and hydrogen storage capacity. Hydrogen gas is produced as the hydride is warmed up in some extent. It can be emitted directly or an indirectly with the exhaust gas. The advantage of the hydride storage systems is that high pressure or an extremely low temperature is not needed. The greatest disadvantage of the system is the low energy density *per* mass, so high mass is needed. This limits the run distance of vehicle to 150 to 300 km. A further disadvantage is the volume of the hydride storage system, which is high, around 100 to 300 dm^3 [345,346].

The high-pressure storage tanks of compressed hydrogen are usually aluminum cylinders covered with glass fiber. Other tanks are also being investigated, such as aluminum cylinders covered with Kevlar (plastics reinforced with graphite fiber) and high-strength aluminum rolls. The storage of compressed hydrogen has disadvantages from the point of view of weight and volume too. A car with a 160-km efficient mileage needs to have onboard two 45-kg hydrogen tanks [343,345]. Other hydrogen storage possibilities are activated carbon, and carbon nanotubes, and fullerenes, for example. [307,340,344–346]

Hydrogen may be the most promising fuel of the future, but today it is still much more expensive than crude-derived gasoline and gasoil. Environmentally it is very advantageous based on a full life cycle, and if the electricity for the electrolysis of the water becomes cheaper from renewable sources. The use of hydrogen could be enhanced by resolving the issue of economic and safe storage in its liquid state.

REFERENCES

1. Hancsók, J. (2004). *Fuels for Engines and JT Engines. Part III: Alternative Fuels*, Veszprém University Press, Veszprém.

2. European Parliament and the Council of the European Union. (2003). Directive 2003/30/EC of the European Parliament and of the Council. *Official J. EU*, 31(13), 188–192.

3. European Parliament and the Council of the European Union. (2009). Directive 2009/30/EC of the European Parliament and of the Council. *Official J. EU*, 140, 88–110.

4. Mittelbach, M., Remschmidt, C., eds. (2004). *Biodiesel. The Comprehensive Handbook*. Mittelbach, Graz

5. Crabbe, E., Ishizaki, A. (2004). Biodiesel production and application. *Concise Encyclopedia of Bioresource Technology*, 475–481.

6. Demirbas, A. (2008). Comparison of transesterification methods for production of biodiesel from vegetable oils and fats. *Energy Convers. Manag.*, 49, 125–130.

7. Sharma, Y. C., Singh, B., Upadhyay, S. N. (2008). Advancements in development and characterization of biodiesel a review. *Fuel*, 87, 2355–2373.

8. Knothe, G., van Gerpen, J., Krahl, J., eds. (2005). *The Biodiesel Handbook*. AOCS Press, Springer, New York.

9. Oosterveer, P., Mol, A. P. J. (2010). Biofuels, trade and sustainability a review of perspectives for developing countries. *Biofuels, Bioproducts & Biorefining*, 4, 66–76.

10. Bartz, W. J., ed. (2011). How sustainable are biofuels for transportation? *Proceedings of the 8th International Colloquium*, January 19–20, Technische Akademie Esslingen, Ostfildern, Germany.

11. Hancsók, J., Varga, Z., eds. (2003). *The Biogasoline, Vegetable Based Fuels and Their Blending Components of Otto-Engines*. BME OMIKK, Budapest.

12. Hancsók, J., Kovács, F., eds. (2002). *The Biodiesel*. BME OMIKK, Budapest.

13. Krár, M., Thernesz, A., Tóth, Cs., Kasza, T., Hancsók, J. (2010). Investigation of catalytic conversion of vegetable oil/gas oil mixtures. In *Silica and Silicates in Modern Catalysis* (ed. I. Halász). Transworld Research Network, 435–455, Kerala.

14. Eller, Z., Hancsók, J. (2011). Reduced aromatic Jet fuel, *Proceedings of 8th International Colloquium Fuels 2011*, January 19–20, Stuttgart/Ostfildern.

15. Lee, S., Speight, J. G., Loyalka, S. K. (2007). *Handbook of Alternative Fuels Technologies*. Taylor & Francis, New York.

16. Hancsók, J., Kovács, F., Krár, M. (2005). Investigation of the production of vegetable oil derivates with high cetane number. *Proceedings of 5th International Symposium on Materials made from Renewable Resources*, September 1–2, Erfurt, Germany.

17. Korakianitis, T., Namasiyavam, A. M., Crookes, R. J. (2011). Natural-gas fueled spark-ignition (SI) and compression-ignition (CI) engine performance and emissions. *Progr. Energy Combust. Sci.*, 37, 89–112.

18. Di Pascoli, S., Femia, A., Luzzati, T. (2001). Natural gas, cars and the environment. A (relatively) "clean" and cheap fuel looking for users. *Ecol. Econ.*, 38, 179–189.

19. Hekkert, M. P., Hendriks, F. H. J. F., Faaij, A. P. C., Neelis, M. L. (2005). Natural gas an alternative to crude oil in automotive fuel chains well-to-wheel analysis and transition strategy development. *Energy Pol.*, 33, 579–594.

20. Trotter, G., Rhodes, Z. (2009). Catalytic oxygen removal from coal mine methane, *Petrol. Technol. Quart.: GS*, 41–44.

21. Reeces, S. R. (2003). Enhanced CBM recovery, coalbed CO_2 sequestration assessed. *Oil Gas J.*, 101 (27), 49–53.

22. Petzet, A. (2007). Water issues overshadow Powder River coal gas play. *Oil Gas J.*, 105 (4), 30–36.

23. Tambach, T. J., Mathews, J. P., van Bergen, F. (2009). Molecular exchange of CH_4 and CO_2 in coal: enhanced coalbed methane on a nanoscale. *Energy Fuel*, 23, 4845–4847.

24. Squarek, J., Dawson, M. (2006). Coalbed methane expands in canada. *Oil Gas J.*, 104 (28), 37.

25. Mohr, S. H., Evans, G. M. (2007). Model proposed for world conventional, unconventional gas. *Oil Gas J.*, 105 (47), 46–51.

26. Wylie, G., Hyden, R., Parkey, V., Grieser, B., Middaugh, R. (2007). Unconventional gas technology—2. Custom technology makes shale resources profitable. *Oil Gas J.*, 105 (48), 41–49.

27. Lobato, F. S. (2011). Shale gas, oil, minerals processing offer synergies in Brazil's Amazon basins. *Oil Gas J.*, 109(10), 54.

28. Demirbas, A., ed. (2010). *Methane Gas Hydrate*. Springer, Dordrecht.

29. Kida, M., Suzuki, K., Kawamura, T., Oyama, H., Nagao, J., Ebinuma, T., Narita, H., Suzuki, H., Sakagami, H., Takahashi, N. (2009). Characteristics of natural gas hydrates occuring in pore-spaces of marine sediments collected from the eastern nanaki trough, off Japan. *Energy Fuels*, 23, 5580–5586.

30. Borowski, W. S. (2004). A review of methane and gas hydrates in the dynamic, stratified system of the Blake Ridge region, offshore southeastern North America. *Chem. Geol.*, 205, 311–346.

31. Sassen, R., Roberts, H. H., Carney, R., Milkov, A. V., DeFreitas, D. A., Lanoil, B., Zhang, C. (2004). Free hydrocarbon gas, gas hydrate, and authigenic minerals in chemosynthetic communities of the northern Gulf of Mexico continental slope: relation to microbial processes. *Chem. Geol.*, 205, 195–217.

32. *A Biogas Road Map for Europe*. (2009). European Biomass Association, Brussels.

33. Kettl, K.-H., Niemetz, N., Sandor, N., Eder, M., Narodoslawsky, M. (2010). Ecological evaulation of biogas feedstock from intercrops. *Chem. Eng. Trans.*, 21, 433–438.

34. Häring, H.-W. (2008). *Industrial Gases Processing*. Wiley, Chichester.

35. Brok, T. K. (2007). Integrated treating options for sour natural gases. *Petrol. Technol. Quart.*, 12(3), 67–71.

36. Brown, W. G. (2009). Selecting gas treating technologies. *GAS*, 13–19.

37. Tan, J. A., Pelletier, D. (2009). Canadian gas plant handles NORM in replacing C_3 treater's mol sieve. *Oil Gas J.*, 107 (25), 59–65.

38. Knifton, J. F. (1993). New synthesis gas chemistry. *Catal. Today*, 18, 355–384.

39. Yamashita, K., Barreto, L. (2005). Energyplexes for the 21th century: coal gasification for co-producing hydrogen, electricity and liquid fuels. *Energy*, 30, 2453–2473.

40. Arena, U., Zaccariello, L., Mastellone, M. L. (2007). Gasification of a plastic waste in a pilot fluidized bed reactor, *Chem. Eng. Trans.*, (12), 641–646.

41. Shen, J. (2000). Ultra clean transportation fuels for the 21th century: the Fischer–Tropsch option—an overview. *Proceedings of Symposium on Advances in Fischer–Tropsch Chemistry.* American Chemical Society, March 26–31, San Francisco.

42. Ohira, T. (2005). Status and prospects for the development of synthetic liquid fuels. *Sci. Technol. Trends, Quart. Rev.*, 17, 48–62.

43. Hartmann, M. (2007). Production of synthetic fuels from natural gas or biomass—status of commercial realization. *Erdöl, Erdgas Kohle*, 123(10), 362–369.

44. Oláh György, Goeppert, A., Prakash, G. K. S., eds. (2006). *After Crude Oil and Natural Gas: The Methanol Economics.* Better, Budapest.

45. Keil, F. J. (1999). Methanol-to-hydrocarbons: process technology. *Microporous Mesoporous Materials*, 29, 49–66.

46. Kaltschmitt, M., Brdigwater, A. V., eds. (1997). *Biomass Gasification and Pyrolysis.* CPL Press, Chippenham.

47. Liu, Z., Liang, J. (1999). Methanol to olefin conversion catalysts. *Solid State Materials Sci.*, 4, 80–84.

48. Salerno, D., Arellano-Garcia, H., Wozny, G. (2011). Techno-economic analysis for ethylene and oxygenates products from the oxidative coupling of methane process. *Chem. Eng. Trans.*, 24, 1507–1512.

49. Holmen, A. (2009). Direct conversion of methane to fuels and chemicals. *Catal. Today*, 142, 2–8.

50. Talbiersky, J. (2007). Der Chemietohstoff Kohle: gestern, heute und morgen. *Erdöl Erdgas Kohle*, 123(1), 32–37.

51. Dong, J., Hu, Z., Zhang, Y., Xin, J., Nie, H., Gao, X. (2009). Development and application of coal liquified oil hydroupgrading technology, *Prepr. Pap.—Amer. Chem. Soc. Div. Petr. Chem.*, 54(2), 116–117.

52. Sun, Q., Fletcher, J. J., Zhang, Y., Ren, X. (2005). Comparative analysis of costs of alternative coal liquefaction processes. *Energy Fuels*, 19, 1160–1164.

53. Deem, K., Sayles, S. (2009). Coal liquids as refinery feedstock. *Petrol. Quart. Technol.*, (3), 69–79.

54. Williams, R. H., Larson, E. D. (2003). A comparison of direct and indirect liquefaction technologies for making fluid fuels from coal. *Energy Sustain. Dev.*, 6(4), 103–129.

55. Höök, M., Aleklett, K. (2010). A review on coal to liquid fuels and its coal consumption. *Int. J. of Energy Res.*, 34(109), 848–864.

56. Holmgren, J., Marinangeli, R., Marker, T., McCall, M., Petri, J. (2007). Opportunities for biorenewables. *Hydrocarb. Eng.*, 12 (6), 75–82.

57. Boateng, A. A., Daugaard, D. E., Goldberg, N. M., Hicks, K. B. (2007). Bench-scale fluidized-bed pyrolysis of switchgrass for bio-oil production. *Ind. Eng. Chem. Res.*, 46, 1891–1897.

58. Wee, S. K., Chok, V. S., Srinivasakannan, C., Chua, H. B., Yan, H. M. (2008). Fluidization quality study in a compartmented fluidized bed gasifier (CFBG). *Energy Fuels*, 22, 61–66.

59. Zhang, H., Xiao, R., Wang, D., Zhong, Z., Song, M., Pan, Q., He, G. (2009). Catalytic fast pyrolysis of biomass in a fluidized bed with fresh and spent fluidized catalytic cracking (FCC) catalysts. *Energy Fuels*, 23, 6199–6206.

60. Zhang, Y., Jacobs, G., Sparks, D. E., Dry, M. E., Davis, B. H. (2001). CO and CO_2 hydrogenation study on supported cobalt Fischer–Tropsch synthesis catalysts. *Catal. Today*, 71(3–4), 411–418.

61. Dorner, R. W., Hardy, D. R., Williams, F. W., Davis, B. H., Willauer, H. D. (2009). Influence of gas feed composition and pressure on the catalytic conversion of CO_2 to hydrocarbons using a traditional cobalt-based Fischer–Tropsch catalyst. *Energy Fuels*, 23, 4190–4195.

62. del Remedio Hernandez, M., Garcia, A.N., Gómez, A., Agulló, J., Marcilla, A. (2006). Effect of residence time on volatile products obtained in the HDPE pyrolysis in the presence and absence of HZSM-5. *Ind. Eng., Chem. Res.*, 45, 8770–8778.

63. Arandes, J. M., Torre, I., Castano, P., Olazar, M., Bilbao, J. (2007). Catalytic cracking of waxes produced by the fast pyrolysis of polyolefins. *Energy Fuels*, 21, 561–569.

64. Sultanov, A. S., Berdiev, S. A. (2007). Recycling of polyethylene production wastes. *Petrol. Technol. Quart.*, (3), 143–145.

65. Zabaniotou, A., Oudenne, P. D., Jung, C.G., Fontana, A. (2005). Activated carbon production from char issued from used tyres pyrolysis: industrial improvement. *Erdöl Erdgas Kohle*, 121(4), 160–162.

66. Schubert, P. F. (2001). Expanding markets for GTL fuels and specialty products. *Studies Surf. Sci. and Catal.*, 136, 459–464.

67. Van Wechem, V. M. H., Senden, M. M. G. (1994). Conversion of natural gas to transportation fuels via the shell middle distillate synthesis process (SMDS). *Natural Gas Conver. II*, 1994, 43–71.

68. de Graaf, W., Schrauwen, F. (2002). World scale GTL. *Hydrocarb. Eng.*, 7(5), 55–58.

69. Lüke, H.-W. (2003). Shell GTL—introducing a synthetic gas-to-liquid fuel into the market. *VDI-Berichte, Nr.* 1808, 241–252.

70. Rahmim, I. I. (2008). GTL, CTL finding roles in global energy supply. *Oil Gas J.*, 106(12), 22–31.

71. Steynberg, A. P., Dry, M. E., eds. (2004). *Fischer–Tropsch Technology*. Elsevier, Amsterdam.

72. Robinson, K. K., Tatterson, D. F. (2007). Fischer–Tropsch oil-from-coal promising as transport fuel. *Oil Gas J.*, 105(8), 20–31.

73. Mantripragada, H. C., Rubin, E. S. (2011). Techno-economic evaluation of coal-toliquids (CTL) plants with carbon capture and sequestration. *Energy Pol.*, 39, 2808–2816.

74. Robinson, K. K., Tatterson, D. F. (2008). Economics on Fischer–Tropsch coal-to-liquids method updated. *Oil Gas J.*, 106(40), 22–26.

75. Zwart, R. W. R., Boerrigter, H. (2005). High efficiency co-production of synthetic natural gas (SNG) and Fischer–Tropsch (FT) transportation fuels from biomass. *Energy Fuels*, 19, 591–597.

76. Hamelinck, C. N. (2004). Production of FT transportation fuels from biomass; technical options, process analysis and optimisation, and development potential. *Energy*, 29, 1743–1771.

77. Tijmensen, M. J. A. (2002). Exploration of the possibilities for production of Fischer–Tropsch liquids and power via biomass gasification. *Biomass Bioenergy*, 23, 129–152.

78. Blades, T. (2005). BtL SunFuel from CHOREN's carbo-V process. 1st International Biorefinery Workshop, July 20–21, Washington, DC.

79. Sunde, K., Brekke, A., Solberg, B. (2011). *Environmental impacts and costs of woody biomass-to-liquid (BTL) production and use—a review*. Forest Pol. Econ.

80. McDaniel, J., Litt, R., Weidert, D., Kilanowski, D. (2008). Microchannel Fischer–Tropsch for waste-to-liquids. The 2008 Spring Natural Meeting

81. Ohira, T. Status and prospects for the development of synthetic liquid fuels. *Sci. Technol. Trends*, 48–62.

82. Blades, T. (2006). BTL—High yield, low CO_2. Second Generation Biofuel, 1st European Fuels Conference, March 14–15, Paris.

83. Young, G. C., ed. (2010). *Municipal Solid Waste to Energy Conversion Processes*. Wiley, Hoboken, NJ.

84. Rostrup-Nielsen, J. R. (1993). Production of synthesis gas. *Catal. Today*, 18, 305–324.

85. Bonneau, R. (2010). Upgrade syngas production. *Hydrocarb. Proc.*, 89(4), 33–42.

86. Demirbas, A., ed. (2009). *Biofuels*. Springer, London.

87. Srivastava, S. P. (2000). Study of the temperature and enthalpy of thermally induced phase-transitions in n-alkanes, their mixtures and Fischer–Tropsch waxes. *Petrol. Sci. and Technol.*, 18 (5–6), 493–518.

88. Van Santen, R. A., Shetty, S. G., van Steen, E. (2008). Elementary reaction steps of the Fischer–Tropsch reaction. *Prepr. Pap.—Amer. Chem.Soc., Div. Petr. Chem.*, 53(2), 70–74.

89. Schulz, H., Claeys, M. (1999). Reactions of olefins of different chain length added during Fischer–Tropsch synthesis on a cobalt catalyst in a slurry reactor. *Appl. Catal.*, 186(1–2), 71–90.

90. Perego, C., Bortolo, R., Zennaro, R. (2009). Gas to liquids teschnologies for natural gas reserves valorization: the Eni experience. *Catal. Today*, 142, 9–16.

91. Ernst, B. (1999). Preparation and characterization of Fischer–Tropsch active Co/SiO_2 catalysts. *Appl. Catal. A: General*, 186 (1–2), 145–168.

92. Zennaro, R. (2001). Quantitative comparison of supported cobalt and iron Fischer–Tropsch synthesis catalysts. *Studies Surf. Sci. Catal.*, 136, 513–518.

93. Geerlings, J. J. C. (1999). Fischer–Tropsch tcchnology—from active site to commercial process. *Appl. Catal. A: General*, 186 (1–2), 27–40.

94. Jacobs, G., Das, K. T., Zhang, Y., Li, J., Racoillet, G., Davis, H. B. (2002). Fischer–Tropsch synthesis: support, loading, and promoter effects on the reducibility of cobalt catalysts. *Appl. Catal.*, 233, (1–2), 263–281.

95. Jacobs, G. (2004). Fischer–Tropsch synthesis: influence of reduction promoters on cluster size and stability of Co/Al_2O_3 catalysts for GTL. *Prepr. Pap—Amer. Chem. Soc., Div. Pet. Chem.*, 49(2), 186–191.

96. Jacobs, G., Zhang, Y., Das, T. K., Li, J., Patterson, P. M., Davis, B. H. (2001). Deactivation of a Ru promoted Co/Al_2O_3 catalyst for FT synthesis. *Studies Surf. Sci. and Catal.*, 139, 415–422.

97. Zhang, Y., Koike, M., Yang, R., Hinchiranan, S., Vitidsant, T., Tsubaki, N. (2005). Multi-functional alumina-silica bimodal pore catalyst and its application for Fischer–Tropsch synthesis. *Appl. Catal. A: General*, 292, 252–258.

98. Zhao, T.-S., Chang, J., Yoneyama, Y., Tsubaki, N. (2005). Selective synthesis of middle isoparaffins via a two-stage Fischer–Tropsch reaction: activity investigation for a hybrid catalyst. *Ind. Eng. Chem. Res.*, 44 (4), 769–775.

99. Rytter, E. (2008). Catalyst particle size of cobalt/rhenium on porous alumina and the effect on Fischer–Tropsch catalytic performance. *Ind. Eng. Chem. Res.*, 46, 9032–9036.

100. Li, S., Krishnamoorthy S., Anwu Li, A., Meitzner, D. G., Iglesia, E. (2002). Promoted iron-based catalysts for the Fischer–Tropsch synthesis: design, synthesis, site densities, and catalytic properties. *J. Catal.*, 206, 202–217.

101. Botes, F. G. (2005). The effect of a higher operating temperature on the Fischer–Tropsch/ HZSM-5 bifunctional process. *Appl. Catal. A: General*, 284(1–2), 21–29.

102. Pour, A. N., Zamani, Y., Tavasoli, A., Shahri, S. M. K., Taheri, S. A. (2008). Study on products distribution of iron and iron-zeolite catalysts in Fischer–Tropsch synthesis. *Fuel*, 87, 2004–2012.

103. Li, Y. (2008). Gasoline-range hydrocarbon synthesis over cobalt-Based Fischer–Tropsch catalysts supported on SiO_2/HZSM-5. *Energy Fuels*, 22(3), 1897–1901.

104. Cano, L. A., Cagnoli, M. V., Bengoa, J. F., Alvarez, A. M., Marchetti, S. G. (2011). Effect of the activation atmosphere on the activity of Fe catalysts supported SBA-15 in the Fischer–Tropsch synthesis. *J. of Catal.*, 278, 310–320.

105. Mu, S., Li, D., Hou, B., Jia, L., Chen, J., Sun, Y. (2010). Influence of ZrO_2 loading on SBA-15-supported cobalt catalysts for Fischer–Tropsch synthesis. *Energy Fuels*, 24, 3715–3718.

106. Madikizela, N. N., Coville, N. J. (2002). A study of Co/Zn/TiO_2 catalysts in the Fischer–Tropsch reaction. *J. Molecul. Catal.*, 181(1–2), 129–136.

107. Tavasoli, A., Trépanier, M., Abbaslou, R. M. M., Dalai, A. K., Abatzoglou, N. (2009). Fischer–Tropsch synthesis on mono- and bimetallic Co and Fe catalysts supported on carbon nanotubes. *Fuel Proc. Technol.*, 90, 1486–1494.

108. Fiato, R. A., Iglesia, E., Soled, S. L. (2002). Gas-to-liquids technology—FT catalysis in ExxonMobil's AGC-21. *Fuel Chem. Div. Prepr.*, 47(1), 235–236.

109. Wender, I. (1996). Reactions of synthesis gas. *Fuel Proc. Technol.*, (48), 189–297.

110. Lopez, A. M., Fiato, R. A., Ansell, L. L., Quinlan, C. W., Ramage, M. P. (2003). ExxonMobil's advanced gas-to-liquids technology. *Hydrocarb. Asia*, (July–Aug)., 56–63.

111. Senden, M., Mcewan M. (2000). The Shell middle distillate synthesis (SMDS) experience. *Proceedings of 16th World Petroleum Congress*, June 11–25, Calgary, Canada.

112. Clark, R. H., Battersby, N. S., Palmer, A. P., Stradling, R. J., Whale, G. F., Louis, J. J. J. (2003). The environmental benefits of shell GTL diesel. *Proceedings of 4th International Colloquium Fuels*, January 15–16, Ostfildern, Germany.

113. Eilers, J., Posthuma, S. A., Sie, S. T. (1990). The Shell middle distillate synthesis process. *Catal. Lett.*, 7, 253–270.

114. Leckel, D. (2009). Diesel production from Fischer–Tropsch: the past, the present, and new concepts. *Energy Fuels*, 23, 2342–2358.

115. Dancuart, L. P., Steynberg, A. P. (2007). Fischer–Tropsch based GTL technology a new process? In *Fischer–Tropsch Synthesis, Catalysts and Catalysis* (eds. B. H. Davis and M. L. Occelli). Elsevier, Amsterdam 379–399.

116. Tait, B. (2006). Sasol's activities on synthetic fuels. 2nd International BtL Congress, October 13.

117. Halstead, K. (2008). Oryx GTL from conception to reality. *Nitrogen + Syngas*, 292, (Mar–Apr), 43–50.

118. Fleisch, T. H., Sills, R. A., Briscoe, M. D. (2002). 2002—Emergence of the gas-to-liquids industry: a review of global GTL developments. *J. Nat. Gas Chem.*, 11, 1–14.

119. van Dyk, J. C., Keyser, M. J., Coertzen, M. (2006). Syngas production from South African coal sources using Sasol–Lurgi gasifiers. *Int. J. Coal Geol.*, 65, 243–253.

120. Knott, D. (1997). Gas-to-liquids projects gaining momentum as process list grows. *Oil Gas J.*, 95(25), 16–21.

121. Syntroleum pursues global GTL opportunities (2002). *Eur. Chem. News*, 77(2016), 26.

122. Yakobson L. D. (2000). Gas-to-liquids markets and opportunities. Howard Weil 28th Annual Energy Conference, New Orleans, USA.

123. Caprani, E. (2008). Gasel™: the XTL technology suite for the conversion of syngas to diesel. OAPEC—IFP Joint Seminar, June 18, Rueil-Malmaison, France.

124. Fedou, S., Douziech, D., Caprani, E. (2008). Synthesis gas to high cetane diesel: the Gasel process. *J. Petrotech Soc.*, (Jan), 26–31.

125. Marion, M.-C., Hugues, F. (2008). Diesel yield according to Fischer–Tropsch process conditions. *DGMK-Tagungsbericht*, (3), 117–124.

126. Möller, K., le Grange, P., Accolla, C. (2009). A two-phase model for the hydrocracking of Fischer–Tropsch-derived wax. *Ind. Eng. Chem. Res.*, 48, 3791–3801.

127. Bouchy, C., Hastoy, G., Guillon, E., Martens, J. A. (2009). Fischer–Tropsch waxes upgrading via hydrocracking and selective hydroisomerization. *Oil Gas Sci. Technol.*, 64(1), 91–112.

128. Calemma, V., Gambaro, C., Parker Jr., W. O., Carbone, R., Giardano, R., Scorletti, P. (2010). Middle distillates from hydrocracking of FT waxes: composition, characteristics and emission properties. *Catal. Today*, 149, 40–46.

129. Hancsók, J., Kovács, S., Pölczmann, Gy., Kasza, T. (2011). Investigation the effect of oxygenic compounds on the isomerization of bioparaffins over Pt/SAPO-11. *Topics Catal.*, 54, 1094–1101.

130. Pölczmann, Gy., Hancsók, J. (2012). Production of high stability base oil from Fischer–Tropsch wax. *Proceedings of 18th International Colloquium Tribology*, January 10–12, Stuttgart/Ostfildern.

131. Nagy, G., Hancsók, J. (2009). Key factors of the production of modern diesel fuels. *Proceedings of 7th International Colloquium Fuels, Mineral Oil Based and Alternative Fuels*, January 14–15, Stuttgart/Ostfildern.

132. Schaberg, P. W., Morgan, P. M., Myburgh, I. S., Roets, P. N. J., Botha J. J. (1999). An overview of the producion, properties, and exhaust emissions performance of sasol slurry phase distillate diesel fuel. *Proceedings of 2nd International Colloquium Fuels*, January 20–21, Esslingen, Germany.

133. Vogel, A., Mueller-Langer, F., Kaltschmitt, M. (2008). Technical and economic assessment of existing and future BtL-plants—state of knowledge 2008. *DGMK-Tagungsbericht*, (2), 23–34.

134. Clark, R. H., Louis, J. J. J., Stradling, R. J. (2005). Shell gas to liquids in the context of future engines and future fuels. *Proceedings of 5th Internaional Colloquium Fuels*, January 12–13, Esslingen, Germany.

135. Leckel, D. (2009). Hydroprocessing Euro 4-type diesel from high-temperature Fischer–Tropsch vacuum gas oils. *Energy Fuels*, 23, 38–45.

136. Kamara, B. I., Coetzee, J. (2009). Overview of high-temperature Fischer–Tropsch gasoline and diesel quality. *Energy Fuels*, 23, 2242–2247.

137. Leckel, D. (2009). A closer look at the intrinsic low aromaticity in diesel hydrocracked from low-temperature Fischer–Tropsch wax. *Energy Fuels*, 23, 32–37.

138. Prada Silvy, R., Belizario dos Santos, A. C., Sousa-Aguiar, E. F. (2010). Surging GTL promises lube-market opportunities. *Oil Gas J.*, (August), 98–105.

139. Rahmim, I. I. (2008). GTL, CTL finding roles in global energy supply. *Oil and Gas J.*, 106(12), 22–31.

140. Hancsók, J., Krár, M., Magyar, Sz., Boda, L., Holló, A., Kalló, D. (2007). Investigation of the production of high cetane number biogasoil from pre-hydrogenated vegetable oils over Pt/HZSM-22/Al$_2$O$_3$. *Microporous Mesoporous Materials*, 101(1–2), 148–152.

141. Hancsók, J., Krár, M., Magyar, Sz. Boda, L., Holló, A., Kalló, D. (2007). Investigation of the production of high quality biogasoil from pre-hydrogenated vegetable oils over Pt/SAPO-11/Al$_2$O$_3$. *Studies Surf. Sci. Catal. 170 B—From Zeolites to Porous MOF Materials*, 170, 1605–1610.

142. Hancsók, J., Kasza, T. (2011). The importance of isoparaffins at the modern engine fuel production. *Proceedings of 8th International Colloquium Fuels 2011*, January 19–20, Stuttgart/Ostfildern.

143. Mäki-Arvela, P., Kubičková, I., Eränen, K., Murzin, D.Y. (2007). Catalytic deoxygenation of fatty acids and their derivatives. *Energy Fuels*, 21 (1), 30–41.

144. Kubičková, I., Snåre, M., Eränen, K., Murzin, D.Y. (2005). Hydrocarbons for diesel fuel via decarboxylation of vegetable oils. *Catal. Today*, 106, 197–200.

145. Mikulec, J., Cvengros, J., Joríková, L., Banic, M., Kleinová, A. (2009). Diesel production technology from renewable sources—second generation biofuels. *Chem Eng. Trans.*, 18, 475–480.

146. Krár, M., Kovács, S., Kalló, D., Hancsók J. (2010). Fuel purpose hydrotreating of sunflower oil on CoMo/Al$_2$O$_3$ catalyst. *Biores. Technol.*, 101(23), 9287–9293.

147. Rantanen, L., Linnaile, R., Aakko, P., Harju, T. (2005). NExBTL— biodiesel fuel of the second generations. *SAE Paper*, 1, 1–17.

148. Donnis, B., Egeberg, R. G., Blom, P., Knudsen, K. G. (2009). Hydroprocessing of bio-oils and oxygenates to hydrocarbons, understanding the reaction routes. *Topics Catal.*, 52, 229–240.

149. Hancsók, J., Baladincz, P., Kasza, T., Kovács, S., Tóth, Cs., Varga, Z. (2011). Bio gas oil production from waste lard. *Biomed. Biotechnol. J.*, article ID 384184, p 9.

150. Veldsink, J. W., Bouma, M. J., Schöön, N.-H., Beenackers, A. A. C. M. (1997). Heterogeneous hydrogenation of vegetable oils a literature review. *Catal. Rev.*, 39(3), 254–318.

151. Kubička, D., Šimáček, P., Žilková, N. (2009). Transformation of vegetable oils into hydrocarbons over mesoporous-alumina-supported CoMo catalysts. *Topics Catal.*, 52, 161–168.

152. Krár, M., Kasza, T., Kovács, S., Kalló, D., Hancsók, J. (2011). Bio gas oils with improved low temperature properties. *Fuel Proc. Technol.*, 92(5), 886–892.

153. Şenol, O. I., Viljava, T.-R., Krause, A. O. I. (2005). Hydrodeoxygenation of aliphatic esters on sulphided NiMo/γ-Al$_2$O$_3$ and CoMo/γ-Al$_2$O$_3$ catalyst: the effect of water. *Catal. Today*, 106(1–4), 186–189.

154. Toba, M., Abe, Y., Kuramochi, H., Osako, M., Mochizuki, T., Yoshimura, Y. (2011). Hydrodeoxygenation of waste vegetable oil over sulfide catalysts, *Catal. Today*, 164, 533–537.

155. Liu, Y., Sotelo-Boyás, R., Murata, K., Minowa, T., Sakanishi, K. (2009). Hydrotreatment of jatropha oil to produce green diesel over trifunctional Ni–Mo/SiO$_2$–Al$_2$O$_3$ catalyst. *Chem. Lett.*, 38, 552–553.

156. Laurent, E., Delmon, B. (1994). Study of the hydrodeoxygenation of carbonyl, carboxylic and guaiacyl over sulfided CoMo/ γ-Al$_2$O$_3$ and NiMo/γ-Al$_2$O$_3$ catalysts. I. Catalytic reaction schemes. *Appl.Catal. A: General*, 109, 77–96.

157. Wang, X., Ozkan, U. S. (2005). Effect of pre-treatment conditions on the performance of sulphided Ni–Mo/γ-Al$_2$O$_3$ catalysts for hydrogenation of linear aldehydes. *J. Molecul. Catal. A: Chemical*, 232, 101–112.

158. Şenol, O. İ., Ryymin, E.-M., Viljava, T.-R., Krause, A. O. I. (2007). Effect of hydrogen sulphide on the hydrodeoxygenation of aromatic and aliphatic oxygenates on sulphided catalysts. *J. Molecul. Catal. A: Chemical*, 277, 107–112.

159. Furimsky, E. (2000). Catalytic hydrodeoxygenation. *Appl. Catal. A: General*, 199, 147–190.

160. Huber, G. W., Iborra, S., Corma, A. (2006). Synthesis of transportation fuels from biomass: chemistry, catalysts, and engineering. *Chem. Rev.*, 106, 4044–4098.

161. Kovács, S., Boda, L., Leveles, L., Thernesz A., Hancsók, J. (2010). Catalytic hydrotreating of triglycerides for the production of bioparaffin mixture. *Chem. Eng. Trans.*, 21, 1321–1326.

162. Hancsók, J., Kasza, T., Kovács, S., Solymosi, P., Holló, A. (2012). Production of bioparaffins by the catalytic hydrogenation of natural triglycerides. *J. Cleaner Product*, 34, 76–81.

163. Benyamna, A., Bennouna, C., Leglise J., van Gestel, J., Duyme, M., Duchet, J. C., Dutartre, R., Moreau, C. (1998). Effect of the association of Co and Ni on the catalytic properties of sulfided CoNiMo/Al$_2$O$_3$. *Prepr. Pap.—Am. Chem. Soc. Div. Petr. Chem.*, 43(1), 39–42.

164. Sanghavi, K., Torrisi, S. (2007). Converting a DHT to ULSD service. *Petrol. Technol. Quart.*, (1), 45–54.

165. da Rocha Filho, G. N., Brodzki, D., Djéga-Mariadassou, G. (1993). Formation of alkanes, alkylcycloalkanes and alkylbenzenes during the catalytic hydrocracking of vegetable oils. *Fuel*, 72(4), 543–549.

166. Snåre, M., Kubičková, I., Mäki-Arvela, P., Eränen, K., Murzin, D.Yu. (2006). Continuous deoxygenation of ethyl stearate: a model reaction fot production diesel fuel hydrocarbons. In *Catalysis of Organic Reactions* (ed. S. R. Schmidt). CRC Press, Taylor & Francis Group, Boca Raton, FL, 415–425.

167. Smith, B., Greenwell, H. C., Whiting, A. (2009). Catalytic upgrading of tri-glycerides and fatty acids to transport biofuels. *Energy Environ. Sci.*, 2(3), 262–271.

168. Lestari, S., Mäki-Arvela, P., Simakova, I., Beltramini, J., Lu, G. Q. M., Murzin, D. Y. (2009). Catalytic deoxygenation of stearic acid and palmitic acid in semibatch mode, *Catal. Lett.*, 130 (1–2), 48–51.

169. Snare, M., Kubickova, I., Maki-Arvela, P., Chichova, D., Eranen, K., Murzin, D. Y. (2008). Catalytic deoxygenation of unsaturated renewable feedstocks for production of diesel fuel hydrocarbons, *Fuel*, 87, 933–945.

170. Geng, C.-H., Zhang, F., Gao, Z.-X., Zhao, L.-F., Zhou, J.-L. (2004). Hydroisomerization of n-tetradecane over Pt/SAPO-11 catalyst. *Catal. Today*, 93–95, 485–491.

171. Egeberg, R., Michaelsen, N., Skyum, L., Zeuthen, P. (2010). Hydrotreating in the production of fuels from biomass. *Angew. Chem. Int. Ed.*, 46, 7184–7201.

172. Huber, G. W., O'Connor, P., Corma, A. (2007). Processing biomass in conventional refineries: production of high quality diesel by hydrotreating vegetable oils in heavy vacuum oil mixtures. *Appl. Catal. A: General*, 329, 120–129.

173. Templis, Ch., Vonortas, A, Sebos, I., Papayannakos, N. (2011). Vegetable oil effect on gas oil HDS in their catalytic co-hydroprocessing. *Appl. Catal. B: Environmental*, 104, 324–329.

174. Bezergianni, S., Kalogianni, A., Vasalos, I. A. (2009). Hydrocracking of vacuum gas oil-vegetable oil mixtures for biofuels production. *Biores. Technol.*, 100, 3036–3042.

175. Tóth, Cs., Baladincz, P., Kovács, S., Hancsók, J. (2011). Producing clean diesel fuel by co-hydrogenation of vegetable oil with gas oil. *Clean Techn. Environ. Pol.*, 13, 581–585.

176. Hancsók, J., Krár, M., Kasza, T., Kovács, S., Tóth, Cs., Varga, Z. (2011). Investigation of hydrotreating of vegetable oil-gas oil mixtures. *J. Environ. Sci. Eng.*, 5, 500–507.

177. Huber, G. W., Chedda, J. N., Barre, Ch. J., Dumesic, J. A. (2005). Production of liquid alkanes by aqueous-phase processing of biomass-derived carbohydrates. *Science*, 308, 1446–1450.

178. Xing, R., Subrahmanyam, A. V., Olcay, H., Qi, W., van Walsum, P., Pendse, H., Huber, G. W. (2010). Production of jet and diesel fuel range alkanes from waste hemicellulose-derived aqueous solutions. *Green Chem.*, 12, 1933–1946.

179. Jacques, K. A., Lyons, T. P., Kelsall, D. R., eds. (1999). *The Alcohol Textbook*. Nottingham University Press, Nottingham, UK.

180. API. *(2001)*: *Alcohols and Ethers, Publication No. 4261 3rd ed.* American Petroleum Institue, Washington, DC.

181. Jordan, J. (2008). World methanol overview: 2007 and beyond, *Hydrocarb. Proc.*, 87(4), 61–68.

182. J. C. Guibet, ed. (1999). *Fuels and Engines Technology Energy Environment*. Editions Technip, Paris.

183. Sayin, C., Ozsezen, A. N., Canakci, M. (2010). The influence of operating parameters, on the performance and emissions of a DI diesel engine using methanol-blended-diesel fuel. *Fuel*, 89, 1407–1414.

184. Shenghua, L., Clemente, E. R. C., Tiegang, H., Yanjv, W. (2007). Study of spark ignition engine fueled with methanol/gasoline fuel blends, *Appl. Therm. Eng.*, 27, 1904–1910.

185. Bahadori, A., Vuthaluru, H. B., Mokhatab, S. (2008). Estimation of methanol vaporization loss and solubility in hydrocarbon liquid phase. *Oil Gas Eur. Mag.*, 3, 149–151.

186. Bahadori, A. (2007). Model accurately predicts HC solubility in methanol. *Oil Gas J.*, 105 (33), 40–42.

187. Schmidt, T. C. (2003). Analysis of methyl tert-butyl ether (MTBE) and tert-butyl alcohol (TBA) in ground and surface water. *Trends Anal. Chem.*, 22(10), 776–784.

188. Achten, C., Kolb A., Püttmann W. (2001). Methyl tert-butyl ether [MTBE] in urban and rural precipitation in Germany. *Atmospheric Environment J.*, 35(36), 6337–6345.

189. Lee, A. K. K., Al-Jarallah, A. (2001). Treatibility of MTBE-contaminated groundwater by ozone and peroxone. *J. Amer. Water Works Assoc.*, 93(6), 110–120.

190. Concawe (1997). The Health Hazards and Exposures associated with gasoline containing MTBE. Concawe Report 97/54, Brussels.

191. Moolenaar, R. L., Hefflin, B. J., Ashley, D. L., Middaugh, J. P., Etzel, R. A. (1994). Methyl tertiary butyl ether in human blood after exposure to oxygenated fuel in Fairbanks, Alaska. *Arch. Environ. Health*, 49, 402–409.

192. Agarwal, A. K. (2007). Biofuels (alcohols and biodiesel) applications as fuels for internal combustion engines. *Prog. Energy Combust. Sci.*, 33, 233–271.

193. Kelly, K. J., Bailey, B. K., Coburn, T. C., Clark, W., Eudy, L., Lissiuk, P. (1996). FTP emissions test results from flexible-fuel methanol Dodge spirits and ford econoline vans. SAE International Spring Fuels and Lubricants Meeting, May 6–8, 1996, Dearborn, MI.

194. Netzer, D., Wallsgrove, C. (2011). Flow scheme envisions producing methanol for gasoline blending. *Oil Gas J.*, 109 (13), 114–120.

195. Cybulski, A. (1994). Liquid phase methanol synthesis: catalysts, mechanism, kinetics, chemical equilibria, vapor–liquid equilibria, and modeling—a catalytic review. *Sci. Eng.*, 36 (4), 557–615.

196. Bae, J.-W., Potdar, H. S., Kang, S-H. (2008). Coproduction of methanol and dimethyl ether from biomass-derived syngas on a Cu–ZnO–Al$_2$O$_3$/γ-Al$_2$O$_3$ hybrid catalyst. *Energy Fuels*, 22, 223–230.

197. Kumabe, K., Fujimoto, S., Yanagida, T. (2008). Environmental and economic analysis of methanol production process via biomass gasification. *Fuel*, 87, 1422–1427.

198. Luyben, W. L. (2010). Design and control of a methanol reactor/column process. *Ind. Eng. Chem. Res.*, 49, 6150–6163.

199. *Ohio's First Ethanol-Fueled Light-Duty Fleet: Final Study Results* (1998). Battle for the US Department of Energy's National Renewable Energy Laboratory and the State of Ohio Department of Administrative Services, Columbus.

200. de Serves, C. (2005). *Emissions from Flexible Fuel Vehicles with Different Ethanol Blends*. Swedish Road Administration, Stockholm.

201. Song, R., Liu, J., Wang, L., Liu, S. (2008). Performance and emissions of a diesel engine fuelled with methanol. *Energy Fuels*, 22, 3883–3888.

202. Bilgin, A., Sezer, İ. (2008). Effects of methanol addition to gasoline on the performance and fuel cost of a spark ignition engine. *Energy Fuels*, 22, 2782–2788.

203. *Biofuels Business, Buyer's Guide 2011*, 5(2), pp. 17.

204. Meyers, R. A. (2005). *Handbook of Petrochemicals Production Processes*. McGraw-Hill, New York.

205. Wells, G. M. (1991). *Handbook of Petrochemicals and Processes*. Gower Publishing, Aldershot, UK.

206. Minteer, S. (2006). *Alcoholic Fuels*. Taylor & Francis, New York.

207. White P. J., Johnson L. A., eds. (2003). *Corn: Chemistry and Technology*, American Association of Cereal Chemists, St. Paul, MN.

208. Demirbas, A. (2008). Products from lignocellulosic materials via degradation processes. *Energy Sources*, 30, 27–37.

209. Lennartsson, P. R., Karimi, K., Edebo, L., Taherzadeh, M. J. (2008). Ethanol production from lignocelluloise by the dimoprhic fungus mucor indicus. *Proceedings of World Bioenergy 2008*, May 27–29, Jönköping. Sweden.

210. Lingitz, A., Jungmeier, G., Reith, H., van der Linden, R., von Weymarn, N., Penttila, M., Vehlow, J., Dimitriou, I., O'Donohue, M., Peck, P., Kolodziejczyk, K. (2008). Perspectives of wood to bioethanol. *Proceedings of World Bioenergy 2008*, May 27–29, Jönköping, Sweden.

211. Yu, Y., Lou, X., Wu, H. (2008). Some recent in hydrolysis of biomass in hot-compressed water and its comparisons with other hydrolysis methods. *Energy Fuels*, 22, 46–60.

212. Drevna, C. T. (2011). Consumer protection is a key issue for E15. *Hydrocarb. Proc.*, 90 (2), 33–36.

213. Macduff, M., Marshall, G., Vilardo, J., Nilsson, M., Danielsson, D., Johansson, K. (2009). Field study on the performance of deposit control additives in E85 fuel. Fuels 2009 7th International Colloquium, January 14–15, 2009, Esslingen, Germany.

214. Persson, H. (2008). Fuel vapour composition and flammability properties of E85. World Bioenergy 2008, May 27–29, 2008, Jönköping, Sweden.

215. Blondy, J. (2007). *FFVs E85 The French Case*, EFC, March 15., 2007.

216. Dobrev, D., Stratiev, D., Kirilov, K., Petkov, P. (2007). Effect of gasoline hydrocarbon composition on the properties of the blend gasoline/bioethanol. IPC2007—43rd International Petroleum Conference, September 24–26, Bratislava, Slovak Republic.

217. Smokers, R., Smit R. (2004). Compatibility of pure and blended biofuels with respect to engine performance, durability and emissions—a literature review. *Report 2GAVE04.01.* Senter Novem, Amsterdam.

218. Merkl, Von J., Jungmeier, G. (2008). Die Energie- und Treibhausgas-Bilanz von Bioethanolanlagen am Beispiel der AGRANA-Bioethanolanlage in Pischelsdorf. *Erdöl Erdgas Köhle*, 124 (7–8), 319–322.

219. Unzelmann, G. H. (1989). Ethers have good gasoline-blending attributes. *Oil Gas J.*, 87(15), 33–37.

220. Piel, W. J. (1989). Ether will play key role in "clean" gasoline blends. *Oil Gas J.*, 87(49), 40–44.

221. Palkhe, G, Leonhard, H., Tappe, M. (2000). Mögliche Umweltelastungen durch die Nutzung von MTBE als Kraftstoffzusatz in Deutschland und Westeuropa, *Erdöl Erdgas Kohle*, 116(10), 498–504.

222. Koehl, W. J., Gorse, R. A. Jr., Knepper, J. C., Rapp, L. A., Benson, J. D., Hochhauser, A. M., Leppard, W. R., Reauter, R. M., Burns, V. R., Painter, L. J., Rutherford, J. A. (1993). Comparison of effects of MTBE and TAME on exhaust and evaporative emissions—Auto/Oil Air Quality Improvement Research Program. Paper No. 932730. Society of Automotive Engineers, Troy, MI.

223. Gold, R. B., Lichtblau, J. H., Goldstein, L. (2002). MTBE vs. ethanol: sorting through the oxigenate issues. *Oil Gas J.*, 100(2) 18–32.

224. Chase, J. D., Woods, H. J., (1979). MTBE and TAME—a good octane boosting combo. *Oil Gas J.*, 77(15), 149–152.

225. Semelsberger, T. A., Borup, R. L., Greene, H. L. (2006). Dimethyl ether (DME) as an alternative fuel. *J. Power Sources*, 156, 497–511.

226. Ying, W., Genbao, L., Wei, Z., Longbao, Z. (2008). Study on the application of DME/diesel blends in a diesel engine. *Fuel Proc. Technol.*, 89, 1272–1280.

227. Yoon, H., Bae, C. (2011). The comperative study on dme and diesel PCCI engine combustion with two-stage fuel injection strategy. *Proceedings of 8th International Colloquium Fuels*, January 19–20, Stuttgart/Ostfildern.

228. Ohno, Y., Ogawa, T., Shikada, T., Hayashi, H. (2001). Development of dimethyl ether synthesis technology and associated diesel engine test. *Oil Gas Eur. Mag.*, 27 (2), 35–39.

229. Yaripour, F., Baghaei, F., Schmidt, I., Perregaard, J. (2005). Catalytic dehydration of methanol to dimethyl ether (DME) over solid-acid catalysts. *Catal. Commun.*, 6 (2) 147–152.

230. Moser, F. X. (2000). Dimethyl ether, the diesel fuel from natural gas—a key to clean technologies and resources. 2nd International Symposium on Fuels and Lubricants, March 10–12, New Delhi.

231. Omata, K., Watanabe, Y., Umegaki, T., Ishiguro, G., Yamada, N. (2001). Low-pressure DME synthesis with Cu-based hybrid catalysts using temperature-gradient reaktor. *Fuel*, 81, 1605–1609.

232. Ge, Q., Huang, Y., Qui, F. (1998). A new catalyst for direct synthesis of dimethyl ether from syngas. *React. Kinet. Catal. Lett.*, 63 (1), 137–142.

233. Sivebaek, I. M., Sorenson, S. C. (2001). Assessment of lubricity of dimethyl ether using the medium frequency reciprocating rig. 3rd International Colloquium on Fuels, January 17–18, Esslingen, Germany.

234. Wain, K. S., Perez, J. M., Chapman, E., Boehman, A. L. (2005). Alternative and low sulfur fuel options: boundary lubrication performance and potential problems. *Tribol. Int.*, 38 (3), 313–319.

235. Sorenson, S. C. (1999). Dimethyl ether as a fuel for diesel engines. 2nd International Colloquium on Fuels, January 20–21, Esslingen, Germany.

236. Teng, H., McCandless, J. C., Schneyer, J. B. (2001). Thermochemical characteristics of dimethyl-ether—an alternative fuel for compression-ignition engines. *SAE Trans.*, 177–187.

237. Bhide, S. V., Boehman, A. L., Perez, J. M. (2001). Viscosity of DME-diesel fuel blends. *Fuel Chem. Div. Preprints*, 46 (2), 400–401.

238. Karmakar, A., Karmakar, S., Mukherjee, S. (2010). Properties of various plants and animals feedstocks for biodiesel production, *Biores. Technol.*, 101, 7201–7210.

239. Alcantara, R., Amores, J., Canoira, L., Fidalgo, E., Franco, M. J., Navarro A. (2000). Catalytic production of biodiesel from soy-bean oil, used frying oil and tallow. *Biomass Bioenergy*, 18(6), 515–527.

240. Canakci, M., Sanli, H. (2008). Biodiesel production from various feedstocks and their effects on the fuel properties. *J. Ind. Microbiol. Biotechnol.*, 35, 431–441.

241. Durrett, T. P., Benning, C., Ohlrogge, J. (2008). Plant triacylglycerols as feedstocks for the production of biofuels. *The Plant J.*, 54, 593–607.

242. Lang, X., Dalai, A. K., Bakhshi, N. N., Reaney, M. J., Hertz, P. B. (2001). Preparation and characterization of bio-diesels from various bio-oils. *Biores. Technol.*, 80, 53–62.

243. Shah, S., Gupta, N. M. (2007). Lipase catalyzed preparation of biodiesel from Jatropha oil in a solvent free system. *Proc. Biochem.*, 42, 409–414.

244. Bockisch, M., ed. (1998). *Fats and Oils Handbook*. AOCS Press, Hamburg.

245. Demirbas, A. (2008). *Biodiesel—A Realistic Fuel Alternative for Diesel Engines*. Springer, Hamburg.

246. Achten, W. M. J., Verchot, L., Franken, Y. J., Mathijs, E., Singh, V. P., Aerts, R., Muys, B. (2008). Jatropha bio-diesel production and use (a literature review). *Biomass Bioenergy*, 32(12), 1063–1084.

247. Sharma, M., Dohen, K. C., Puri, S. K., Sarin, R., Gupta, A. A., Malhotra, R. K. (2008). Influence of antioxidans on biodiesel oxidation sability. *Proceedings of International symposium on fuels and lubricants* (ISFL-2008), New Delhi.

248. Schober, S., Mittelbach, M. (2004). The impact of antioxidants on biodiesel oxidation stability. *Eur. J. Lipid Sci. Technol.*, 106(6), 382–389.

249. McCormick, R. L., Westbrook, S. R. (2010). Storage stability of biodiesel and biodiesel blends. *Energy Fuels*, 24, 690–698.

250. Ciric, L., Philp, J. C., Whiteley, A. S. (2010). Hydrocarbon utilization within a diesel-degrading bacterial consortium. *FEMS Microbiol. Lett.*, 303, 116–122.

251. Bento, F. M., Gaylarde, C. C. (2001). Biodeterioration of stored diesel oil: studies in Brazil. *Int. Biodeter. Biodegrad.*, 47, 107–112.

252. Hancsók, J., Kovács, F., Krár, M. (2004). Production of vegetable oil fatty acid methyl esters from used fying oil by combined acidis/alkali transesterification. *Petrol. Coal*, 46(3), 36–44.

253. Dias, J. M., Alvim-Ferraz, M. C. M., Almeida, M. F. (2008). Comparison of the performance of different homogeneous alkali catalysts during transesterification of waste and virgin oils and evaluation of biodiesel quality. *Fuel*, 87, 3572–3578.

254. Mittelbach, M., Trathnigg, B. (1990). Kinetics of alkaline catalyzed metanolysis of sunflower oil. *Fat Sci. Technol.*, 92(4), 145–148.

255. Vicente, G., Martínez, M., Aracil, J. (2004). Integrated biodiesel production: a comparison of different homogeneous catalysts systems. *Biores. Technol.*, 92, 297–305.

256. Ma, F., Hanna, M. A. (1999). Biodiesel production a review. *Biores. Technol.*, 70, 1–15.

257. Meher, L. C., Vidya Sagar, D., Naik, S. N. (2006). Technical aspects of biodiesel production by transesterification—a review, *Renew. Sustain. Energy Rev.*, 10, 248–268.

258. Canakci, M., Gerpen, J. V. (1999). Biodiesel production via acid catalysis. *Trans. ASAE*, 42(5), 1203–1210.

259. Zheng, S., Kates, M., Dubé, M. A., McLean, D. D. (2006). Acid-catalyzed production of biodiesel from waste frying oil. *Biomass Bioenergy*, 30, 267–272.

260. Bournay, L., Casanave, D., Delfort, B., Hillion, G., Chodorge, J.A. (2005). New heterogeneous process for biodiesel production: a way to improve the quality and the value of the crude glycerin produced by biodiesel plants. *Catal. Today*, 106, 190–192.

261. Sharma, Y. C., Singh, B., Korstad, J. (2010). Application of an efficient nonconventional heterogeneous catalyst for biodiesel synthesis from Pongamia pinnata oil. *Energy Fuels*, 24, 3223–3231.

262. Melero, J. A., Bautista, L. F., Morales, G., Iglesias, J., Briones, D. (2009). Biodiesel production with heterogeneous sulfonic acid-functionalized mesostructured catalysts. *Energy Fuels*, 23, 539–547.

263. Fu, B., Gao, L., Niu, L., Wei, R., Xiao, G. (2009). Biodiesel from waste cooking oil via heterogeneous superacid catalyst SO_4^{2-}/ZrO_2. *Energy Fuels*, 23, 569–572.

264. Bélafi-Bakó, K., Kovács, F., Gubicza, L., Hancsók, J. (2002). Enzimatic biodiesel production from sunflower oil by *Candida antarctica* lipase in solvent-free system. *Biocatal. Biotransform.*, 20(6), 437–439.

265. Shimada, Y., Watanabe, Y., Sugihara, A., Tominaga, Y. (2002). Enzymatic alcoholysis for biodiesel fuel production and application of the reaction to oil processing. *J. Molecul. Catal. B: Enzymatic* 17(3–5), 133–142.

266. Shieh, C. J., Liao, H. F., Lee, C. C. (2003). Optimization of lipase-catalyzed biodiesel by response surface methodology. *Biores. Technol.*, 88(2), 103–106.

267. Kovács, F., Hancsók, J., Krár, M., Nagy, G., Neményi, M. (2005). Enzymatic transesterification of high quality sunflower oils. *Proceedings of 14th European Biomass Conference and Exhibition. Biomass for Energy, Industry and Climate Protection*, October 17–21., Paris.

268. Ranganathan, S. V., Narasimhan, S. L., Muthukumar, K. (2008). An overview of enzymatic production of biodiesel, *Biores. Technol.*, 99, 3975–3981.

269. Demirbas, A. (2003). Biodiesel fuels from vegetable oils via catalytic and noncatalytic supercritical alcohol transesterifications and other methods a survey, *Energy Convers. and Manag.*, 44(13), 2093–2109.

270. Vieitez, I., da Silva, C., Alckmin, I., Borges, G. R., Corazza, F. C., Oliviera, J. V., Grompone, M. A., Jachmanián, I. (2009). Effect of temperature on the continous synthesis of soybean esters under supercritical ethanol. *Energy Fuels*, 23, 558–563.

271. Patil, P. D., Gude, V. G., Deng, S. (2010). Transesterification of *Camelina Sativa* oil using supercritical and subcritical methanol with cosolvents. *Energy Fuels*, 24, 746–751.

272. Yin, J.-Z., Ma, Z., Hu, D.-P., Xiu, Z.-L., Wang T.-H. (2010). Biodiesel production from subcritical methanol transesterification of soybean oil with sodium silicate. *Energy Fuels*, 24, 3179–3182.

273. Cooke, B. S. (2006). Adsorbent purification of biodiesel. Eastern Biofuels Conference and Expo II, May 30–June 1, Budapest.

274. Demirbas, A. (2009). Progress and recent trend sin biodiesel fuels. *Energy Convers. Manag.*, 50, 14–34.

275. He, H. Y., Guo, X., Zhu, S. L. (2006). Comparison of membrane extraction with tradition extraction methods for biodiesel production. *JAOCS*, 83, 457–460.

276. Atadashi, I., M., Aroua, M., K., Abdul Aziz A. (2011). Biodiesel separation and purification: a review. *Renew. Energy*, 36, 437–443.

277. Vicente, G., Martínez, M., Aracil, J. (2007). Optimisation of integrated biodiesel production. Part I. A study of biodiesel purity and yield. *Biores. Technol.*, 98, 1724–1733.

278. Suppalakpana, K., Ratanawilai, S.B., Tongurai, C. (2010). Production of ethyl ester from mesterified crude palm oil by microwave with dry washing by bleaching earth. *Appl. Energy*, 87, 2356–2359.

279. Chin, L. H., Hameed, B. H., Ahmad, A. L. (2009). Process optimization for biodiesel production from waste cooking palm oil (*Elaeis guineensis*) using response surface methodology. *Energy Fuels*, 23, 1040–1044.

280. Araujo, V. K. W. S., Hamacher, S., Scavarda, I. F. (2010). Economic assessment of bio diesel production from waste frying oils. *Biores. Technol.*, 101, 4415–4422.

281. Kovács, S., Krár, M., Hancsók, J. (2007). Comparison of conventional (combined acidic/alkali-catalyzed) and enzyme-catalyzed transesterification of used frying oil. *Proceedings of 43rd International Petroleum Conference*, September 24–26, Bratislava, Slovakia.

282. Kovács, S., Krár, M., Beck, Á., Hancsók, J. (2007). Enzymatic transesterification of used frying oils. *Proceedings of 15th European Biomass Conference and Exhibition. Biomass for Energy, Industry and Climate Protection*, May 7–11, Berlin.

283. Hsu, A.-F., Jones, K., Marmer, W. N. Foglia, T.A. (2001). Production of alkyl esters from tallow and grease using lipase immobilized in a phyllosilicate sol-gel. *J. Amer. Oil Chem. Soc.*, 78(6), 585–588.

284. Lapuerta, M., Rodríguez-Fernández, J., Oliva, F., Canoira, L. (2009). Biodiesel from low-grade animal fats: biodiesel engine performance and emissions. *Energy Fuels*, 23, 121–129.

285. Montefrio, M. J., Xinwen, T., Obbard, J. P. (2010). Recovery and pre-treatment of fats, oil and grease from grease interceptors for biodiesel production, *Appl. Energy*, 87, 3155–3166.

286. Levine, R. B., Pinnarat, T., Savage, P. E. (2010). Biodiesel production from wet algal biomass through in situ lipid hydrolysis and supercritical transesterification. *Energy Fuels*, 24, 5235–5243.

287. Vijayaraghavan, K., Hemanathan, K. (2009). Biodiesel production from freshwater algae. *Energy Fuels*, 23, 5448–5453.

288. Mata, T. M., Martins, A. A., Caetano, N. S. (2010). Microalgae for biodiesel production and other applications: a review. *Renew. Sustain. Energy Rev.*, 14, 217–232.

289. Gouveia, L., Oliveira, A. C. (2008). Microalgae as a raw material for bio fuels production. *J. Ind. Microbiol. Biotechnol.* 1367–5435.

290. Koh, M. Y., Ghazi, I. M. (2011). A review of biodiesel production from *Jatropha curcas L.* oil. *Renew. Sustain. Energy Rev.*, 15, 2240–2251.

291. Ong, H. C., Mahlia, T. M. I., Masjuki, H. H., Norhasyima, R. S. (2011). Comparison of palm oil, *Jatropha curcas* and *Calophyllum inophyllum* for biodiesel: a review. *Renew. Sustain. Energy Rev.*, 15, 3501–3515.

292. Gui, M. M., Lee, K. T., Bhatia, S. (2008). Feasibility of edible oil vs. non-edible oil vs. waste edible oil as biodiesel feedstock. *Energy*, 33, 1646–1653.

293. Shah, S. N., Sharma, B., K., Moser, B. R. (2010). Preparation of biofuel using acetylatation of jojoba fatty alchols and assessment as a blend component in ultralow sulfur diesel fuel. *Energy Fuels*, 24, 3189–3194.

294. Arters, D. C. (2003). Biofuels and Alternative Fuels—Performance and the Role of Additives. Hart World Fuel Conference—Europe 2003, May 19–21, Brussels.

295. Mittelbach, M., Schober, S. (2003). The impact of antioxidants on biodiesel oxidation stability. *JAOCS*, 80(8), 817–823.

296. Lin, J., Hunt, R., Hemmens, D., Jacques, C., Kaczmarek, D., Hirsch, P. (2005). 7th Generation infineum cold flow additive for B100 FAME. *IBF7733*.

297. Mähling, Sondjada, R., Koschabek, R., Couet, J., May, M. (2011). Performance and compatibility of novel cold flow improvers for B100 and Bxx fuels. 8th International Colloquium on Fuels, January 19–20, Esslingen, Germany.

298. Wilson, D. (1997). Improving the Quality of RME by the Use of Performance Chemicals. 1st International Colloquium on Fuels, January 16–17, Esslingen, Germany.

299. Dinkov, R., Hristov, G., Stratiev, D., Aldayri, V.B. (2009). Effect of commercially available antioxidants over biodiesel/diesel blends stability. *Fuel*, 88, 732–737.

300. Karavalakis, G., Stournas, S. (2010). Impact of antioxidant additives on the oxidation stability of diesel/biodiesel blends. *Energy Fuels*, 24, 3682–3686.

301. Ingendoh, A. (2011). Protection of biodiesel against oxidation and decomposition. Fuels 8th International Colloquium, January 19–20, Esslingen, Germany.

302. Sarin, R., Arora, A. K., Ranjan, R., Gupta, A. A., Malhorta, R. K. (2007). Bio-diesel lubricity: correlation study with residual acidity. *Lubr. Sci.*, 19, 151–157.

303. Geller, D. P., Goodrum, J. W. (2004). Effects of specific fatty acid methyl esters on diesel fuel lubricity. *Fuel*, 83, 2351–2356.

304. Munoz, M., Moreno, F., Monné, C., Morea, J., Terradillos, J. (2011). Biodiesel improves lubricity of new low sulphur diesel fuels. *Renew. Energy*, 36, 2918–2924.

305. Fischer, J. (2011). FAME als Blendkomponente für Diesel und Heizöl, International Mineral Oil Technology Congress, April 5–6, Stuttgart.

306. Cammack, R., Frey, M., Robson, R. (2001). *Hydrogen as a Fuel—Learning from Nature.* Taylor & Francis, London.

307. Züttel, A., Borgschulte, A., Schlapbach, L., eds. (2008). *Hydrogen as a Future Energy Carrier.* Wiley-VCH Verlag, Weinheim.

308. Barclay, F. J. (2006). *Fuel Cells, Engines and Hydrogen—An Exergy Approach.* Wiley, Chichester.

309. Farrauto, R., Schäfer, A., Schwab, E., Urtel, H. (2011). Hydrocarbon reforming catalyst and new reactor designs for compact hydrogen generators. *Oil Gas Eur. Mag.*, 37 (1), 40–44.

310. Balat, M. (2009). Possible methods for hydrogen production. *Energy Sources, Part A*, 31, 39–50.

311. Gehrke, H., Ruthardt, K., Mathiak, J., Roosen, C. (2011). Hydrogen: a small molecule with large impact. *Oil Gas Eur. Mag.*, 37 (1), 29–33.

312. Vauk, D., Grover, B., Walter, S., Kuttner, H. (2008). Hydrogen choices. *Hydrocarb. Eng.*, 13 (7), 83–90.

313. Wang, L., Murata, K., Inaba, M. (2004). Conversion of liquid hydrocarbons into h_2 and co_2 by integration of reforming and the water–gas shift reaction on highly active multifunctional catalysts. *Ind. Eng. Chem. Res.*, 43, 3228–3232.

314. Broadhurst, P. V., Cotton, B. J., Vasudeva, A. (2004). Feed for thought. *Hydrocarb. Eng.*, 9 (3), 37–40.

315. Barba, J. J., Hemmings, J., Wheeler, F., Bailey, T. C., Horne, N. (1997–98). Advances in hydrogen production technology: the options available. *Hydrocarb. Eng.*, 3(1), 48–54.

316. Meyers R. A., ed. (2007). *Handbook of Petroleum Refining Processes*. McGraw-Hill, New York.

317. Dydkjaeer, I., Madsen, S. W. (1997–98) Advanced reforming technologies for hydrogen production. *Hydrocarb. Eng.*, 3(1), 56–65.

318. Antoniak, K., Gołębiowski, A., Narowski, R., Ryczkowski, J. (2010). Sulfur-resistant alkali metals promoted Co-Mo catalyst for water–gas shift conversion process in coal-derived syngas. 2nd International Symposium on Air Pollution Abatement Catalysis, September 8–11, Cracow.

319. van Selow, E. R., Cobden, P. D., Verbraeken, P. A., Hufton, J. R., van den Brink, R. W. (2009). Carbon capture by sorption-enhanced water-gas shift reaction process using hydrotalcite-based material. *Ind. Eng. Chem. Res.*, 48, 4184–4193.

320. Bohn, C. D., Müller, C. R., Cleeton, J. P., Hayhurst, A. N., Davidson, J. F., Scott, S. A., Dennis, J. S. (2008). Production of very pure hydrogen with simultaneous capture of carbon dioxide using the redox reactions of iron oxides in packed beds. *Ind. Eng. Chem. Res.*, 47, 7623–7630.

321. Daza, C., Kiennemann, A., Moreno, S., Molina, R. (2009). Stability of Ni–Ce catalysts supported over Al-PVA modified mineral clay in dry reforming of methane. *Energy Fuels*, 23, 3497–3509.

322. Omata, K., Kobayashi, Y., Yamada, M. (2009). Artifical neural network aided screening and optimization of additives to $Co/SrCO_3$ catalyst for dry reforming of methane under pressure. *Energy Fuels*, 23, 1931–1935.

323. Song, Y., Liu, H., Liu, S., He, D. (2009). Partial oxidation of methane to syngas over Ni/Al_2O_3 catalysts prepared by modified sol–gel method. *Energy Fuels*, 23, 1925–1930.

324. Effects of hydrothermal conditions of ZrO_2 on catalyst properties and catalytic performances of Ni/ZrO_2 in the partial oxidation of methane (2010). *Energy Fuels*, 24, 2817–2824.

325. Wang, Y., Wang, W., Hong, X., Li, B. (2009). Zirconia promoted metallic nickel catalysts for the partial oxidation of methane to synthesis gas. *Catal. Commun.*, 10, 940–944.

326. de Castro, J., Riviera-Tinoco, R., Bouallou, C. (2010). Hydrogen production from natural gas: auto-thermal reforming and CO_2 capture. *Chem. Eng. Trans.*, 21, 163–168.

327. Papavasilion, J., Avgouropoulos, G., Ionnides, T. (2010). Effect of dopants on the performance of CuO–CeO_2 catalysts in methanol steam reforming. *Appl. Catal. B: Enviromental*, 69 (3–4), 226–234.

328. Conant, T., Karim, A. M., Leberbier, V., Wang, Y., Girgsdies, F., Dchlogl, R., Datye, A. (2008). Stability of bimetallic Pd–Zn catalysts for the steam reforming of methanol. *J. Catal.*, 257 (1), 64–70.

329. Wang, W., Wang, Z., Ding, Y., Xi, J., Lu, G. (2002). Partial oxidation of ethanol to hydrogen over Ni–Fe catalysts. *Catal. Lett.* 81 (1–2) 63–68.

330. Morgenstern, D., A., Fornango, J. P. (2005). Low-temperature reforming of ethanol over copper-plated Raney nickel: a new route to sustainable hydrogen for transportation. *Energy Fuels*, 19 (4), 1708–1716.

331. Byrd, A. J. Pant K. K., Gupta, R. B. (2007). Hydrogen production from ethanol by reforming in supercritical water using Ru/Al_2O_3 catalyst. Energy Fuels, 21(6), 3541–3547.

332. Palma, V., Palo, E., Castaldo, F., Ciambelli, P., Iaquaniello, G. (2011). Catalytic activity of CeO_2 supported Pt–Ni and Pt–Co catalysts in the low temperature bio-ethanol steam reforming. *Chem. Eng. Trans.*, 25, 947–952.

333. Wang, H. (2007). Hydrogen production from a chemical cycle of H_2S splitting. *Int. J. Hydrogen Energy*, 32, 3907–3914.

334. Yu, G., Wang, H., Chuang, K.T. (2009). Upper bound for the efficiency of a novel chemical cycle of H_2S splitting for H_2 production. *Energy Fuels*, 23, 2184–2191.

335. Jacoby, M. (2009). Hydrogen from sun and water—Photocatalysis: three-component catalyst evolves hydrogen with exceptional efficiency. *Chem. Eng. News*, 87 (32), 7.

336. Ritter, S. K. (2010). Fuel From the Sun—Cobalt water-oxidation catalysts benefit from federal initiatives to harness solar power to make fuel. *Sci. Technol.*, 88 (26), 26–27.

337. Dubey, P. K., Sinha, A. S. K., Talapatra, S., Koratkar, N., Ajayan, P. M., Srivastava, O. N. (2010). Hydrogen generation by water electrolysis using carbon nanotube anode. *Int. J. Hydrogen Energy*, 35(9), 3945–3950.

338. Balat, M. (2007). Hydrogen in fueled systems and the significance of hydrogen in vehicular transportation. *Energy Sources, Part B*, 2, 49–61.

339. Das, L. M. (1990). Fuel induction technique for a hydrogen operated engine. *Int. J. Hydrogen Energy*, 15 (11) 833–842.

340. Eberle, U. (2011). Hydrogen for automotive applications and beyond. *Oil Gas Eur. Mag.*, 37 (1), 34–39.

341. Hackeney, J., Neufville, R., (2001). Life cycle model of alternative fuel vehicles: emission, energy, and cost trade-offs. *Transport. Res. Part A*, 35, 243–266.

342. Ogden, J. M. Williams, R. H. Larson, E. D. (2004). Societal lifecycle costs of cars with alternative fuels/engines. *Energy Pol.*, 23 (1), 7–27.

343. Eberle, U., von Helmolt, R. (2010). Sustainable transportation based on electric vehicle concepts: a brief overview. *Energy Environ. Sci.*, 3, 689–699.

344. von Helmolt, R., Eberle, U. (2007). Fuel cell vehicles: status 2007. *J. Power Sources*, 165, 833–843.

345. Eberle, U., Felderhoff, M., Schüth, F. (2009). Chemical and physical solutions for hydrogen storage. *Angew. Chem. Int. Ed.*, 48, 6608–6630.

346. Norbeck, J. M., Heffel, J. W., Durbin, T. D., Tabbara, B., Bowden, J. M., Montano, M.C. (1996). Hydrogen fuel for surface transportation. Society of Automotive Engineers, Troy, MI.

347. Winter, C.-J., Nitsch, J. (1988). *Hydrogen as an Energy Carrier—Technologies, Systems, Economy*. Springer-Verlag, Berlin.

CHAPTER 5

Fuel Additives

In modern automotive fuels, a combination of several chemical additives is used in order for the fuel to meet the desired performance level [1–3]. These chemical additives in small dosages combine to add or improve properties of virgin fuels that cannot be obtained through the refining processes. The most important are additives that improve the flow of gasoline and diesel oils [5]. Sometimes the additive is even used to realize better margins by diverting a value-added product to other applications [4–6]. For example, several additives are used to improve the yield of middle distillates obtained by cutting deeper into the bottom of the crude barrel [7]. Additives are also used in other petroleum products such as heating oil, aviation fuels, and lubricants in order to improve their performances.

There are six reasons for using additives in fuels [1–6]:

- To improve handling properties and stability of the fuel
- To improve combustion properties of the fuel
- To reduce emissions from fuel combustion
- To provide engine protection and cleanliness
- To increase in the economic use of the fuel
- To establish or enhance the brand image of the fuel

Conventionally, chemical compounds added in high concentrations (typically >1%) at the refinery are called blending components, and compounds added in lower concentrations (typically <1%) at the refineries are called refinery (functional) additives. The even lower concentrations of chemical compounds added at depots and terminals by companies are called performance additives [1–3]. Besides these additives, there are the additive mixtures sold for use to the consumer, called "after market products" or "secondary products." Blending components are mainly refined petroleum streams and oxygenates, whereas functional and performance additives are mainly mixtures of chemical compounds dissolved in solvents. The concentrations of additives in fuels are not regulated. Typical concentrations of selected groups of additives for engine gasolines are listed in Table 5.1. The additive concentrations are

Fuels and Fuel-Additives, First Edition. S. P. Srivastava and Jenő Hancsók.

TABLE 5.1 Motor engine gasoline additives and their functions

Additive	Functions	Treat Level (mg/kg)
Antiknock additives	Increase of octane number	10–1000
Combustion improvers	Improve combustion characteristics	5–50
Detergents and dispersants	Additives to clean and keep clean (injection and exhaust system)	20–1000
Octane requirement increase inhibitor	Inhibit octane requirement increase by removing deposits in combustion chamber	20–100
Corrosion inhibitors	Corrosion protection of fuel system	5–50
Antioxidants	Increase storage stability, prevent resin formation	10–50
Metal deactivators	Deactivation of metal surfaces that act as oxidation catalysts	5–20
Antiwear additives	Decrease wear (e.g., of fuel pump)	10–50
Friction modifiers	Fuel saving by decrease friction between moving parts	30–50
Dehazers (demulsifiers)	Inhibit haze formation (promote water coalescence)	3–50
De-icing additives	Inhibit ice formation in carburetor	5–30
Antistatic agents	Increase conductivity	2–10
Dyes	Differentiate fuels	2–20

highly proprietary, and the treat levels will vary to a large degree depending upon the product, the fuel distributor, the refinery and the type of additive used. Many manufacturers use additional additives to provide value additions.

Diesel fuels contain another set of additives [1,3,8,9]. Some of these appear similar to the additives, according to function, used in gasoline, but they may differ in chemical composition and structure. Various types of markers are also used in diesel fuel, and in kerosene, to indicate their origins. Descriptions of diesel fuel additives and their suggested approximate dosage are given in Table 5.2.

The additive concentration can be varied depending on the mixture's chemistry and the chemical compositions of the fuels. But the dosage should be determined experimentally, at the optimum concentration to meet fuel standards.

Classification of Additives by Functions

Antiknock agents Tetra-ethyl lead, methyl-cyclopentadienyl- manganese tricarbonyl, ferrocene, and iron pentacarbonyl

Cetane improvers Alkyl-nitrates and nitrites, peroxides, and nitro-and nitroso- compounds

Detergents/dispersants Long-chain alkyl-amines, alkanol-amines, amides, polybutene-amine, polyether-amine, alkenyl-succinimide, and Mannich-bases

Antioxidants and stabilizers Butylated-hydroxytoluene (BHT), 2,4-dimethyl-6-tert-butyl-phenol, 2,6-di-tert-butyl-phenol (2,6-dtbp), alkyl/aryl-phenylene-diamine, ethylene-diamine, and alkyl/aryl-aminophenol

TABLE 5.2 Additives of Diesel fuels and their functions

Additive	Functions	Treat Level (mg/kg)
Cetane improvers	Increase cetane number (easier cold start, lower emissions, noise, and fuel consumption, longer engine life time)	100–300
Combustion improvers	Lower emission (improve combustion of particle matters)	10–30
Deposit control additives	Additives to clean and keep clean, inhibit deposit formation; decrease fuel consumption and CO_2 emission	30–330
Antioxidants	Prevent resin and inhibit deposit formation, increase storage stability	5–30
Corrosion inhibitors	Corrosion protection of fuel system	10–20
Metal deactivators	Deactivation of metal surfaces that act as oxidation catalysts; increase storage stability, decrease the catalytic effect of especially the copper ions	5–20
Demulsifiers	Prevent, inhibit, and terminate the haze formation caused by water or insoluble compounds	10–20
Lubricity improvers	Increase lubricity in cases of low sulfur and decrease final boiling point of gasoils (fuel pump)	25 100
Cloud point depressants	Decrease initial temperature of paraffin crystallization	150–500
Pour point depressants	Decrease pour point	75–350
Flow improvers/wax crystal modifiers/wax dispersants	Ensure favorable cold flow properties	150–500
Wax-antisettling-additive	Inhibit paraffin settlement	100–200
Friction modifiers	Reduce friction (lower fuel consumption)	50–100
De-icing Additives	Inhibit ice crystal formation	2–10
Biocides	Suppress formation of microorganisms and bacteria to prevent quality degradation	1–10
Antifoam agents	Inhibit foam formation during filling	1–5
Antistatic agents	Increase conductivity	2–10
Deodorants	Suppress or neutralize odors	5–10
Dyes	Differentiation of fuels	5–10

Metal deactivators *N,N*- disalicylidine 1,2- propane-diamine

Corrosion inhibitors Alkyl-succinates, amines salts, dimer acids, and salts of carboxylic acids

Flow improvers and wax antisettling additives Various polymeric compounds containing nitrogen

Lubricity improvers Carboxyl-acids with 8–14 carbon numbers and their esters, and amides of fatty acids

Demulsifiers Long-chain alkyl phenols and alcohols, long-chain carboxylic acids, and long-chain amines

Antifoaming agents for diesel fuel Silicones

Biocides Succinimides and imidazolines

Drag reducers High molecular weight polymers

Fluidizer oils Light mineral oils, polyester, polybutene, polyalpha-olefin (PAO), and polyalkoxylates

Dyes Solvent Red 24, Solvent Red 26, Solvent Yellow 124, and Solvent Blue 35, etc.

Some of these specific additives are either used in the refinery fuel distribution system or in the vehicle system:

Additives for Gasoline Distribution Systems [1,3]

Antioxidants

Metal deactivators

Antistatic agents

Corrosion inhibitors

Sediment reduction agents

Dyes

Dehazers

Additives for gasoline vehicle system [1,3]

Antiknock additive (was tetra ethyl lead, which is now phased out)

Anti-valve seat recession additive (also phased out due to metallurgy change in the engines)

Carburetor detergents (gradually being phased out due to the introduction of injectors)

Deposit control additives

Deposit modifiers

Friction modifiers

Lubricity improvers

Additives for Diesel Distribution System [1,3,8]

Antifoam agents

Antistatic agents

Biocides

Corrosion inhibitors

Sediment reduction agents

Dyes

Demulsifiers

Flow improvers/wax crystal modifiers/wax dispersants

Metal deactivators

Markers to check origin

Stabilizers

Additives for Diesel Vehicle System [1,3,8]

Cetane improvers

Combustion improvers

Deposit control additives

Injector detergents

Lubricity improvers

Friction modifiers

Additives for gasoline and diesel distribution systems are used in refineries to meet minimum fuel specifications at the optimum cost without compromising on the yield of the products. For example, diesel fuels of acceptable flow properties can be obtained in the refineries by reducing the wax content of the fuel. Wax removal, however, is much more costly than the use of a flow improver that controls or modifies the wax crystals at lower temperatures and also does not reduce the yield. Additives like antioxidants, metal deactivators, and stabilizers help in maintaining the quality of the fuels during storage and transportation. Drag reducers help in pumping more fuels through a given pipeline. These additives are relatively high molecular weight polymers. They lose their activity due to the shear produced in the pipeline, and therefore they should be added at regular intervals during the pipe line transportation. Antistatic additives are used for safety reasons to disseminate static electricity generated during the product transfer and movement. Dyes and markers are used for product differentiation and for checking adulteration. Thus the additives are used in the distribution system would be varied according to the need and their treat level.

Additives for the vehicle fuel system are beneficial only when the fuel enters into the intake system of the engine. These additives are generally introduced into the fuel at the marketing terminal, but sometimes at the refinery when the quality demand is high. Such additives in combination with solvents and carrier fluid are also marketed as after market products in small cans, which can be added in a specified dose at the time of taking fuel from the gasoline/diesel pumps. Several independent companies market such products with additional claims like added fuel economy, lower maintenance costs, and longer engine life due to better engine cleanliness.

Fuel quality standards have undergone a ratcheting-up gradation with progressive improvements in engine design and more stringent environmental regulations. These changes in fuel quality have involved reductions in sulfur, aromatics, benzene, polyaromatics, olefins, and lead and improvements in octane numbers, cetane numbers, oxidation stability, and storage stability. In addition, the fuels are also treated to control deposit formation in the engines. All these changes, despite the heavy investments and process changes in the refineries, cannot be met by refiners without the use of fuel additives to perform the desirable functions.

Fuel additives have been used in the oil industry since 1925 or 1930. Initially, cetane improvers, octane number improvers (tetra-ethyl-lead), dyes, and antioxidants were used to raise fuel qualities. These were followed by the incorporation of metal deactivators, corrosion inhibitors, deposit modifiers, and anti-icing agents in 1940s. Diesel stabilizers and lubricity improvers were introduced in the 1950s. Deposit control additives, diesel flow improvers, and demulsifiers were developed in the 1960s and 1970s. In 1980s, diesel and gasoline detergent additives and drag reducers for pipelines were introduced. From 1990 onward, many of these additive chemistries have been improved. In the meantime, anti-valve seat recession additives and wax dispersants for diesel fuels have been developed.

Unlike lubricant additives, fuel additives are used only in small concentrations ranging from 2 to 1000 ppm (parts per millions, mg/kg). Dyes, drag reducers, and antifoam and metal deactivators are used in the lower range, while deposit control additives and flow improvers at the higher level. Other additives are used in the intermediate range of 100 to 400 ppm.

There are other minor additives used for specific purposes, such as the combustion catalyst (usually an organometallic compound that lowers the ignition point of fuel in the combustion chamber reducing the burning temperature from 425–650 °C) and burn rate modifier (increases the fuel burn time resulting in an approximate 30% increase of the available BTUs from the fuel).

5.1 CONSUMPTION OF ADDITIVES (DEMANDS)

There are few new data about the consumption of fuel additives in the free scientific literature. At the end of the 1990's, total consumption of fuel additives was around 600,000 tons. Gasoline additives accounted for more than 50%, diesel fuel additives more than 40%, and aviation additives around 5%. Among the additives, detergent dispersants have the biggest share, at over 50%. This is followed by flow improvers at around 13%, cetane improvers at around 8%, antioxidants at around 7%, octane number improvers around 6%, and lubricity improvers at around 5%. Anti-icing additives and corrosion inhibitors have 3%, while the remaining share is other additives.

Deposit control additives (detergent dispersants) are the single largest group of fuel additives in the world. In 2000 their global demand was estimated [5] at about 200,000 MT (metric tones). European fuel additives demand, as documented in ATC document 52, "Fuel Additives and the Environment," was estimated as listed below (for 1998) [1]:

Deposit control	70, 000
Cetane improver	20, 000
Injector cleaning	25, 600
Lubricity improver	5250
Flow improver	30,000
Antioxidants	2200

Antistatic	50
Corrosion inhibitor	1250
Dehazer	2550
Diesel stabilizer	200
Metal deactivator	200
Dyes, deodorants, etc.	400
Lead antiknock	Nil after year 2000
Anti-valve seat recession	Nil, used for older vehicles only

In effect, the OECD countries consumed 83.8 thousand tons of gasoline and 80.1 thousand tons of diesel fuel additives in 1998. This is equal to 0.066% (ca. 660 mg/kg) treat level for motor gasolines (127 million tons) and 0.0615% (ca. 615 mg/kg) treat level for diesel fuels (130 million tons), respectively [1].

The market value of the US special fuel additives was 625×10^6 USD in 1998, 795 million USD in 2003, and about 1.1 billion USD in 2008. This means about 4.9% (between 1998–2003), and about 6.4% (between 2003–2008) increases on a year-to-year basis [11].

By 2012, the value of US special additives rose to 1.6 billion USD. The estimated quantity of detergent-dispergant additives was 371 million pounds (168.3×10^3 t) in 2011, which was worth 440 million USD (Fredonia Group Inc., Cleveland, OH). Morcover 90% of this additives value was for gasolines in the United States.

The demand for diesel fuel cold flow improvers showed a fast rise too. Lubricity improver demand grew at an above average rate of 3.5% a year from 2007 to 2012, mainly due to the need to reduce sulfur levels in diesel fuels. In addition, the federal Renewable Fuel Standards (which were expanded in 2007) and state level renewable fuel requirements have been behind a rise in the use of biodiesel fuel. Biodiesel fuel has poorer cold flow properties than conventional diesel has and requires higher treat levels of cold flow improver additives. The worldwide cold flow improver market for middle distillate fuels was estimated at 84,000 metric tons, and worth about 125 million USD in 1999. Europe had the largest market, with 36,000 metric tons valued at 56 to 60 million USD of cold flow improver consumption in 1999. North America and the Far East represented about 25% and 11% of the worldwide market, respectively.

Overall, the fuel additive market is expected to grow in average 2% to 2.5% a year. These demand estimates are very rough and reflect only the importance of the fuel additives. The recent global recession has had a negative effect on demand. But the demand for fuel additives will align itself in the future with the production of higher quality fuels in which the use of specific additives is inevitable.

The EPA (of the United States) has detailed regulations (40 CFR 80 series, and others) for the application of fuels and additives in internal combustion engines. These regulations are based on the theory that the emission products of fuels or additives endanger the public health or impair the performance of motor vehicle emission control equipments.

5.2   ENGINE DEPOSITS AND THEIR CONTROL

During combustion all fuels and engine oils form some deposits in the engine. No degree of refinement and processing of the fuel and lube oil in the refinery can eliminate the formation of deposits in the engines. Deposits of soot, sludge, lacquer, and varnish are formed, either due to the incomplete combustion of fuel or due to the degradation of engine lubricating oil. The burning of fuel in the presence of air (oxygen and nitrogen) theoretically results to CO_2, H_2O, and SO_x. Some amount of NO_x will also be formed if the temperature is favorable (high enough). However, ideal combustion is not possible, and some products of decomposition are formed that are rich in carbon and yet contain hydrogen, oxygen, and sulfur too. These products are polar in nature and tend to get attracted to each other, forming bigger aggregates.

5.2.1   Deposits in Gasoline Engines

Deposits in both gasoline and diesel engines and in their fuel systems adversely affect the operation of the vehicle.

Deposits in Fuel System   For example, deposits on the body of the carburetor throttle and in the venturies increase the fuel to air ratio of the gas mixture entering into the combustion chamber of gasoline engines. As a result the amounts of unburned hydrocarbons and carbon monoxide increase in the exhaust gas. This leads to rough idling and stalling. The high fuel to air ratio also reduces the fuel efficiency of the vehicle. Additionally, heavy deposits on the engine intake valves restrict the flow of the gas mixture into the combustion chamber. This restriction starves the engine of air and fuel and results in loss of power.

In the modern automobiles the carburetor was first replaced by the port fuel injector, and then by the direct injection system during the last decade, which is a very delicate and precise fuel metering device. It is important to keep the metal surfaces of the injector clean to allow undisturbed and preset flows of air and fuel vapors. The clearance between the pintle and the orifice is of the order of magnitude 0.05 mm, and any restriction due to deposits can seriously affect the fuel spray pattern and air–fuel ratio in the cylinder.

Port fuel injectors in the engines monitor and deliver precise amounts of gasoline, and these must be kept clean to provide good engine performance. This area is especially prone to the formation of deposits, when the engine is swiched off. The remaining gasoline at this point is subjected to high temperatures and can undergo thermo-oxidative degradation, forming gum-like materials that can trap other contaminants forming deposit precursors.

Deposits at the intake valve can again change the fuel–air ratio and restrict the flow of air and fuel into the cylinder (Figures 5.1 and 5.2). This affects the fuel economy, emissions, and engine performance. Deposits on the valves also increase the possibility of valve failure due to burning and improper valve seating. In addition,

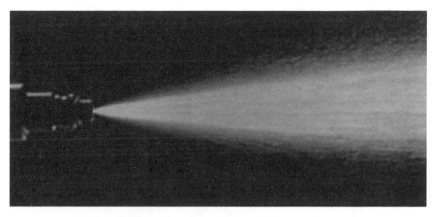

FIGURE 5.1   Good spray pattern with clean injectors

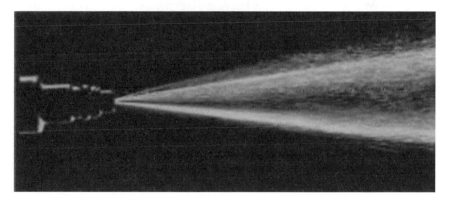

FIGURE 5.2   Deposits, leading to improper spary of fuels

if these deposits break off and enter into the combustion chamber, mechanical damage of the piston, piston rings, engine head, and the like, can take place.

Excessive deposit formation (Figure 5.3) can also take place in the intake port and on the intake valve. Several factors are responsible for these deposits. Contaminations of the air, engine blow-by, exhaust gas recirculation, lubricating oil (oxidized material and metal-containing additives), and oxidation products of gasoline, all contribute to formation of these deposits. The composition of gasoline such as olefins, aromatics, boiling range, oxygenates, and sulfur content plays role in intake valve deposits. Engine design also influences the deposit formation in the intake valve system, since the temperature of the valve varies with the operating cycle. These deposits can build up sufficiently to cause valve stem sticking. Sometimes deposits break off and can lodge between the valve and valve seat, causing valve leakage and valve burning. These combined effects result in power loss, loss of fuel economy, and increased CO and HC emissions.

FIGURE 5.3   Deposits on intake valves without detregent additive

Combustion Chamber Deposits   All gasoline engines are prone to develop deposits [12,13] in the combustion chamber due to the condensation of partially oxidized and thermally degraded fuel and lubricant components on cold surfaces. These can further degrade at the location of the condensation to form bigger solid deposits. Such deposits in combustion chamber (CCD) reduce the available space for combustion and the small crevices in the deposits increase the surface area of the chamber. These deposits lead to increased exhaust gas emission, carbon knock due to the mechanical interference between piston top and cylinder head, and an octane requirement increase (ORI) due to the higher compression ratio.

Combustion chamber deposits have also been found to increase NO_x emissions due to the higher (surface) temperature. However, this can lead to a marginal increase in fuel economy, due to the increased thermal efficiency brought about by the insulating effect of the deposits. The presence of these deposits in the combustion chamber often causes the following problems:

- Reduction in the volume of the combustion zone causing higher compression ratio
- Inhibited heat transfer between the combustion chamber and engine cooling system
- Reduction in the operating efficiency of the engine

The formation of deposits and their accumulation in the combustion chamber further contribute to the knocking of the engine, which can result in stress fatigue and wear in engine components such as in pistons, connecting rods bearings, and cam rods.

Because of all these problems associated with the formation and accumulation of deposits in the combustion chamber, fuel intake, and exhaust systems of an internal combustion engine, the application of appropriate detergent fuel additives is important, to inhibit the deposition and dissolve the existing deposits in the engine.

Detergents are conveniently incorporated into a multifunctional additive package containing several other functional chemical additives. Most premium gasolines marketed around the world contain multifunctional additive packages, including a deposit control detergent for smooth running of the engines. Higher dosages of these additives can clean up deposits (already existing) in the injector. Smaller dosages can keep clean the injector. These are differentiated as clean-up mode and keep-clean mode fuel detergent additives. While the keep-clean dosage of the detergent is incorporated into commercial fuel supplies, the clean-up mode dosage is generally marketed as after-market products. However, modern additive packages are able to provide both functions.

In recent years the tighter tolerances and design changes incorporated into modern engines to meet emissions and fuel economy performances have led to "carbon rap" and CCD flaking-induced driveability problems. These problems demand that the combustion chamber deposits be controlled in the engine from all sources, including additives. However, conventional detergent additives contribute to an increase in combustion chamber deposits.

The mineral oils used as carrier fluid also lead to formation of CCD. To control CCD, special detergent molecules in combination with synthetic carrier fluid have been found to be more effective in controlling combustion chamber deposits [14]. But their dosage has to be sufficiently high to incorporate this property into the fuel.

Combustion chamber deposits show only one advantage: better fuel economy due to the improved thermal efficiency derived from the insulating effect of the deposits. Nevertheless, deposits increase the octane number requirement of the engine [15,16], which contributes to loss of power and accceleration and also to higher NO_x emission. (The octane requirement of the engine increases mainly due to the formation of combustion chamber deposits that lead to the reduced combustion chamber volume and consequently compression ratio increases.)

In the last 10 to 15 years, direct injection spark ignition (DISI) or gasoline direct injection (GDI) engines were introduced as an alternative to conventional port fuel injection spark ignition (PFI SI) engines. Interest in these engines emerges from benefits in fuel efficiency and exhaust emissions. The direct injection of gasoline into the spark ignition engines allows the manufacturers to decrease engine fuel consumption significantly and, at the same time, to maintain the engine performance characteristic and level of gaseous emissions. The fuel–air mixture in such engine types is often lean and stratified (as opposed to stoichiometric and homogeneous in conventional PFI SI engines), resulting in improved fuel economy. Although there are many differences between the two engine technologies, the fundamental difference is in the fuel introduction strategy. In a traditional PFI SI engine, fuel is injected inside the intake ports, coming in direct contact with the intake valves, whereas in the DISI engine, fuel is directly injected into the combustion chamber.

But the direct injection spark ignition (DISI) or gasoline direct injection (GDI) engines have introduced some further deposit problems [6–8]. Therefore there is a need to control combustion chamber deposits or at least not to allow more CCD than it is formed from the base fuel and lubricant as well as from the additives. Most conventional deposit control additives do not control CCD in GDI engines; some are even reported to contribute to higher CCD.

Recent studies have also shown that DISI engines are prone to deposit buildup, and that in some cases these deposits are hard to remove using conventional deposit control fuel additives [17–19]. There is a concern that in DISI engines, performance and fuel economy benefits may diminish as deposits build up on various surfaces of these engines over a period of time. Therefore the development of effective fuel detergents or "deposit control" additives to prevent or reduce such deposits in DISI engines is particularly important. A new type of detergent-dispersant additive and combinations had to be developed [20].

At the end of the second millennium, restart problems of these gasoline engines were noticed in increasing numbers for short-distance travel especially at cold climate conditions. The main reason was found to be an insufficient solubility of the deposits in unburned fuels at short-term operation of the (cold) engine. Consequently the deposit just loosens on the wall of the piston, bursts, and then flakes off. The solid flakes block partially the inlet of the suction valves; further they inhibit the valves, perfect closing and opening, making the buildup of suitable compression difficult. Eventually these deposits inhibit the restart of the engine. To prevent this phenomenon, detergent-dispersant additives of excellent solubility have to be used [9]. If the gasoline does not contain this additive, then deposit control additives have to be added to the fuel.

5.2.2 Deposit Control Additives (Detergent Dispersants)

Detergent-dispersant additives are blended in fuels to prevent the formation of different deposits and to clean the fuel system and combustion chamber for different engines. Their importance is reflected by the fact that their share is about 40–50% of all additives used [21]. Deposits can form both from fuel and from the lubrication oil.

Mechanism of Action of Deposit Control (Detergent-Dispersant) Additives   All detergent and dispersant (DD) additives have a polar hydrophilic and a nonpolar hydrophobic group in their structure (Figure 5.4). In some molecules the polar group is directly attached to the long hydrocarbon chain and in some attached through another linkage molecule, such as in polyisobutylene-sucinimide (an amine group linked through the maleic-anhydride radical; see later discussion).

The different mechanisms of DD additives are represented in Figure 5.5 [11]. They form a protective film on metal surfaces, besides their dispersant, detergent, solubility, and neutralization of acids activities [22–24].

The nonpolar group dissolves the additives in the fuel, and the polar group attracts the deposit particles, which also have some polarity. All deposit control additives act through this mechanism. The polar group attaches to the deposit particles and removes them from the metal surface through the detergent action. Besides ensuring the good solubility of the additives in the fuel, the carrier oil helps in removing the deposit particles from the metal surfaces.

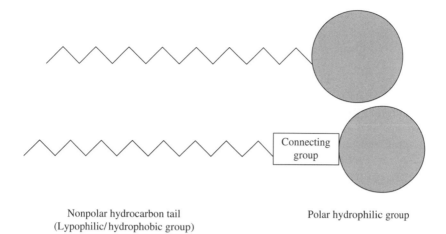

FIGURE 5.4   Typical deposit control, detergent/dispersant molecule

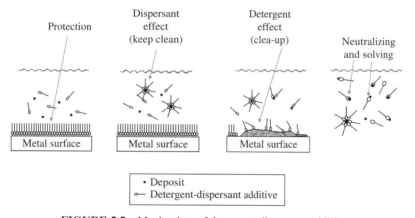

FIGURE 5.5   Mechanism of detergent-dispersant additives

Classification of Deposit Control Gasoline Fuel Additives   The deposit control additives that have been in use since the 1950s can be classified into six generations according to the chemistry of the different additives:

Year	Generation	Structure	Threat Level (mg/kg)
1955–	First generation	Fatty-acid-amides	100
1960–	Second generation	Polybutene-succinimides	400–500
1970–	Third generation	Polyolefin-amines	400–500
1980–	Fourth generation	Polyether-amines	350–400
1990–	Fifth generation	PIB-Mannich-bases	300–350
2000–	Sixth generation	Succinimide-, Mannich-base derivatives	350–500

These products are often used in combination with either a mineral oil based carrier fluid or a polyether based synthetic fluid for more effective cleaning of IVD and CCD.

The DD additives of gasoline fuel have also been classified. The following Generations I through V classification is based on the performance characteristics of the additives as compared to the chemistry basis described above [25]:

Generation I These gasoline additives provide both "keep-clean" and "clean-up" performances in carbureted and fuel-injected vehicles. The keep-clean mode usually requires a lower dosage of detergent, at an amount that is not capable of cleaning any existing deposits in the engine. The clean-up mode utilizes a higher dosage of detergent but is used only in older engines with deposits.

Generation II These additives provide Generation I and a higher intake valve deposit (IVD) "keep-clean" performance.

Generation III These additives provide Generation II plus the IVD "clean-up" performance. This performance is measured by the same tests as IVD, but the procedure is modified to first generate deposits and then the engine is run in the clean-up mode. Most Generation I, II, and III performance additives also leave some combustion chamber deposits. Generation III additives may sometimes modify CCD to avoid an octane rating increase (ORI) in the engine. However, Generation III additives do not have the capability to remove CCD.

Generation IV These detergent additives have Generation III plus CCD "keep-clean" capabilities. Unlike Generation I, II, and III additives, these do not leave combustion chamber deposits, but do not clean already formed CCD. Thus these additives are most suitable in new engines.

Generation V These additives provide Generation IV plus CCD "clean-up" performance. The most modern category additives are capable of removing combustion chamber deposits, remove engine rap, and reduce the octane rating increase (ORI) of an engine. Often field tests are carried out to evaluate CCD clean-up performance.

This is an open-ended classification and new generations may be added in the future [25].

Detergent-Dispersant Effective Compounds All fuel deposit control detergent additives have a polar and a long nonpolar hydrophobic group. The hydrophobic hydrocarbon group has an average molecular weight ($\bar{M}_n$) between 750 and 2500 (a number of about 1000 is preferable), which activates the solubility of the additive in fuel.

The most important classes of chemical compounds used as fuel detergents are the following [26–28]:

Fatty-acid-amides
Ester-amines
Aliphatic hydrocarbon-substituted phenols

Poly(oxyalkylene)-amide

Poly(oxyalkylene)-hydroxyaromatic

Hydroxy-, nitro-, amino-, and aminomethyl- substituted aromatic esters of poly-alkyl-phenoxy-alkanols

Polyalkyl-phenoxy-amino-alkane and a poly(oxyalkylene)-amine

Polyamine-containing poly(oxyalkylene) aromatic esters

Aminocarbamates of polyalkyl-phenoxyalkanols

Polyisobutene-amino-alcohols

Alkalene-succinimides

Polyisobutenyl (PIB)-succinimide and/or -amid and/or -ester

Polyolefin (PIB)–amines

Polyether-amines or poly(oxyalkylene)-amines, amides

Poly(oxyalkylene)-amine-alcohol-amides

Mannich-bases

There is an exhaustive patent literature dealing with deposit control additives alone. In the following pages we discuss some select DD additives.

Fatty Acid Amide   Carburetor deposit control additives based on fatty acid amide chemistry were first developed in the 1950s and were used in a 50-ppm dosage. The earliest gasoline carburetor detergents were based on the reaction products of a fatty acid ester, such as coconut oil, and a mono- or di-hydroxy-hydrocarbyl-amine, such as diethanolamine, or dimethyl-amino-propyl-amine (US Pat. No. 4,729,769). The fatty amines of these additives were capable of preventing the formation of deposits in the intake system but could not remove existing deposits.

Polyolefin (PIB) -Amines and Their Derivatives   Polyolefin (mainly polyisobutylene: PIB) amines have been used extensively as gasoline detergent dispersants to control deposits in the engine fuel system. These are mostly produced by hydroformylation (an oxo reaction with carbon monoxide and hydrogen) of the reactive polyisobutylene (double bound in the α-position≥ 70–80%), of an average molecular weight at about 1000, to yield polyisobutyl-alcohol—and a subsequent reductive amination with ammonia and hydrogen under pressure in the presence of catalyst(s) to yield polyisobutylamine [28–30].

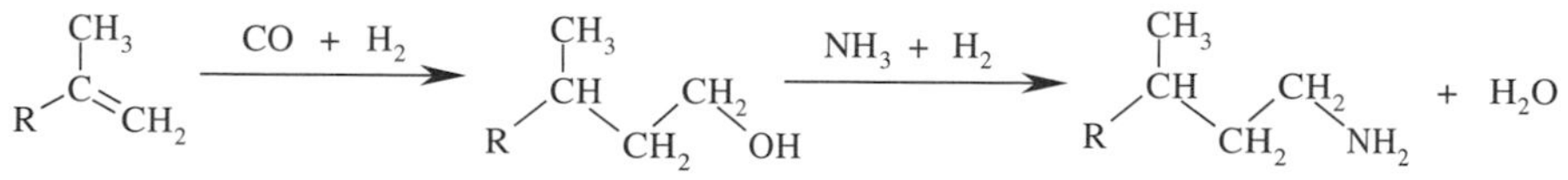

Polybutene-amines are also prepared by chlorination of polyisobutylene of about 1000 molecular weight, and a subsequent reaction with mono- or diamines or oligoamines, such as diethylene-triamine or triethylene-tetramine, and then alkanol-amines, such as amino-ethyl-ethanolamine. The disadvantage of the chlorinated intermediate product is that the final product has some chlorine.

The preparation of olefin-terminated polyisobutenes are known to be by cationic polymerization of isobutene or isobutene-containing hydrocarbon streams in the presence of boron trifluoride complex catalysts (European Pat. Nos. A 807 641 and WO 99/31151). The polyisobutenes thus obtained have a high content of olefinically unsaturated terminal groups. However, the polyisobutenamines prepared by functionalizing these polyisobutenes have unsatisfactory viscosity at low temperatures, with undesirable side effects in engines such as sticky valves. This problem can be corrected by adding large amounts of nondetergent carrier oils. The viscosity of conventional polyisobutenamine is due to the wide molecular weight distribution and higher number of polyisobutenyl radicals, with an exceeding, molecular weight of $\bar{M}_n = 1500$.

It has been found that polyisobutenes having a high content of more than 80 mol% of olefinic terminal groups and that a polydispersity of less than 1.4 can be prepared by cationic polymerization of isobutene in the presence of an initiator and a Lewis acid. Due to the high proportion of reactive olefinic terminal groups, the polyisobutenes obtained in this manner can be converted into polyisobuten-amines by the known processes. In an Opel Kadett engine, 200 ppm of the product, when added into the gasoline, provided adequate intake valve cleanliness in an CEC F-05-A-93 test. The CEC F-16-T96 procedure was followed to determine valve stickiness. The result was 0.1% of polyisobutenamine and 30 to 150 ppm of poly-1-butene oxide as the carrier fluid proved useful in the fuel [32].

Based on general opinion, the optimal polyisobuthylene-amine additives have a lower than or equal to polyisobuthenyl root polydispersity of 1.4 (Mn/Mw: molecular ratio of number and mass mean).

The PIB amines containing also propylene oxides have been found to be excellent DD additives (EP-A696572). The most efficient DD additives are the hydroxyalkyl substituted amines, which are the reaction products of a polyolefin epoxide and a nitrogen-containing compound such as ammonia, a mono-, or polyamine [33–35].

Alkalene (Polyalkylene) Succinic(Anhydride)Acid Derivatives   The application of succinic anhydride derivatives as an ashless dispersant in engine oils was demonstrated already in the 1960s. Since then, several thousand patents have been issued that deal with the synthesis and application of succinimides in both fuels and lubricants. In fuels these succinimides are used as deposit controls in the fuel intake system and in the combustion chamber. In lubricants they are mainly used to keep the oxidation products, contaminants, and soots dispersed in the oil phase. Over the years different variations of succinimides have been developed. In detergents the polar group is directly linked to the hydrocarbon chain, and in dispersants based on succinimides; the polar group is attached to the long hydrocarbon chain via an intermediate linkage. Polyisobutylene succinimide of following structure is an example of this type of disperssant:

R is a long alkyl chain, typically polyisobutylene (PIB) of about 1000 molecular weight connected with a succinate linkage group, which in turn is linked with the polar amine group ($x=1-4$).

In a typical polyisobutylene succinimides, PIB is the long nonpolar hydrocarbon group, succinic anhydride is the linkage, and polyamine is the polar group. The following types of succinimides dispersants have been reported [36,37]:

- Polyisobutylene-mono-succinimides (see above)
- Polyisobutylene-bis-succinimide

R_1 and R_2 : PIB ($\overline{M}_n$= 800–1200); m = 2–4

- PIB-succinic acid-esters (structure of succinic esters of polyols is provided below)

R: PIB ($\overline{M}_n$ = 900–1000)
R′: hydrocarbonyl group (C: 8–12)

- PIB-succinic acid-amide-ester-imide

$x \geq 30-300$, $n \geq 1$, W: alkyl-, polyalkyl-, polyol group

- Derivatives of di-PIB-succinicacid (anhydride)

+ amines, polyamines, alcohols, amino-alcohols

- Derivatives of PIB-di-succinicacid-anhydride

+ amines, polyamines, alcohols, amino-alcohols

The derivatives of succinic anhydrides are usually produced in two steps. The first step is the preparation of polyisobutylene-succinic anhydride (PIBSA) by -ene reaction between polyisobutylene and maleic anhydride. The reaction can be thermally or catalytically initiated. In the second step, the anhydride is used to acylate amines, amino alcohols, alcohols, and so forth. The properties of the final products depend on the structure and purity of the intermediate. Production of PIB succinimides requires a polyisobutylene (PIB) of appropriate molecular weight, maleic anhydride, and polyamines [38,39].

PIBs are produced by the polymerization of C_4-olefins containing isobutene, with boron trifluoride used as a catalyst. The polymerization of isobutene is sensitive to certain impurities like 1-butene and di-isobutylene. The higher butane content and higher temperature give rise to a low molecular weight product. The polymerization process is exothermic, and low temperatures are required to control the reaction.

Conventional C_4 raffinate from steam cracking contains about 40% isobutylene, about 40% n-butene, and about 20% butane. Polymerization of this mixture with $AlCl_3$ yields products where the double bond is located at the middle of the molecule. Such polymers have less reactivity and are not sufficiently suitable for dispersant manufacturing. The product also contains chlorine and is not desirable.

A process using a BF_3 catalyst and concentrated isobutylene yields good quality PIB where the double bond is located toward the end of the chain. The reaction with BF_3 proceeds through a protonation of isobutylene, leading to the formation of a carbonium ion [38]. The chain growth mechanism further leads to the formation of a cationic polymer that, on removal of the proton from the methyl group, forms the α-olefin polymer (PIB) in a yield of about 85%. This is the so-called reactive PIB. Conventional polyisobutylene contains only about 10% α-olefin, 42% β-olefin isomers, and 15% tetra substituted olefin. The molecular weight, the percentage of the terminal double bond in isobutylene, and the concentration of the active material determine the molecular structure of the PIB and its reactivity. The highest reactivity is obtained when the double bond is at the end of the chain, since this is the point of least hindrance. The higher is the terminal double-bond content of the PIB, the faster is the reaction between PIB and maleic anhydride [39]. Thus, by varying the molecular weight of PIB, the terminal double-bond content, the maleic anhydride–PIB ratio, and the reaction conditions, various PIB succinimides of different properties can be obtained. Tetra-substituted olefins, $>C = C<$ do not react with maleic anhydride. β- olefin polymers obtained by the polymerization of butene-2 react with maleic anhydride at much higher temperatures.

Mono-PIBSA is obtained by the reaction of PIB with maleic anhydride in a molar ratio of 1:1. When this ratio is 1:2, PIB-di-SA is obtained [40,41]. PIB-di-succinicanhydride can be formed easily by the α-olefin polymer. The β-olefin isomer can also form bis-succinicanhydride if chlorine is used in the reaction:

PIB-di-succinicanhydride

The reaction of PIB with maleic anhydride can be carried out thermally [42] by heating the two at 150–250 °C for 2 to 10 hours. Alternatively, polyisobutylene can be first chlorinated and then reacted [43] with maleic anhydride. The reaction can also be carried out in a single step [44] by passing chlorine gas into the mixture of PIB and maleic anhydride. The thermal process has the advantage that the reaction product does not contain chlorine, which is not desirable from the environmental considerations. There are several options to control the chlorine content in the final product [45]. The thermal process proceeds via a diene formation, while the chlorine-assisted process proceeds via a Diels–Alder reaction.

Succinic anhydride derivatives from polyisobutylene of about 1000 molecular weight and maleic anhydride can be prepared in the presence of a free radical initiator (e.g., di-tertiary-butyl-peroxide; $T = 120$–180 °C). In this case even two MA molecules can react with one PIB molecule depending on the PIB/MA molar ratio [39]. The PIB:SA ratio can be defined as a function of the PIB:MA molar ratio and the process parameters. Based on this, if two PIB molecules react with 1 MA molecule, then mainly Di-PIB-succinic-anhydride molecules form, while in opponent molar ratio PIB-di-succinicanhydride forms:

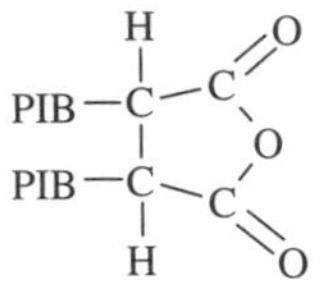

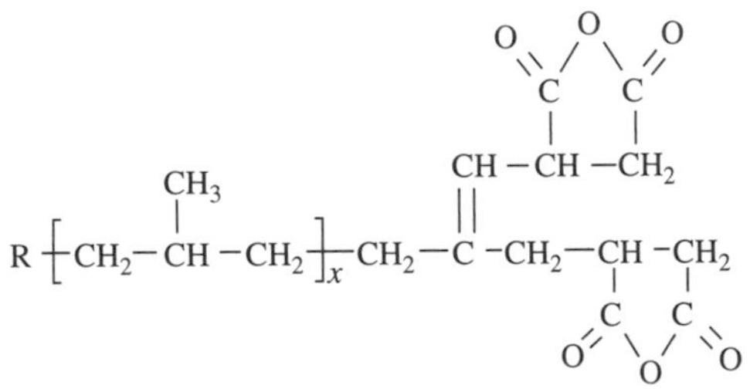

Of course, it is only a theory; in practice any kind of MSA:PIB ratio may be formed that can be determined by the following equation [46]:

$$CR = \frac{8.91 \times 10^{-4} \times TAN \times \overline{M}_{nPIB}}{A_s - 8.79 \times 10^{-2} \times TAN}$$

where CR is the coupling ratio, A_s is the active material content of the sample (%), and TAN is the total acid number (mg KOH/g).

The second step of the production of DD additives is the reaction of different PIBSA, PIBDISA, and DIPIBSA intermediates with polyamines or alcohols or aminoalcohols, and the like [47–49]. Two moles of PIB-mono-succinicacidhydride react with 1 mole of polyamines to yield bis-succinimide (see above). DiPIBSA on acylation with polyamine yields di-PIB-succinimide:

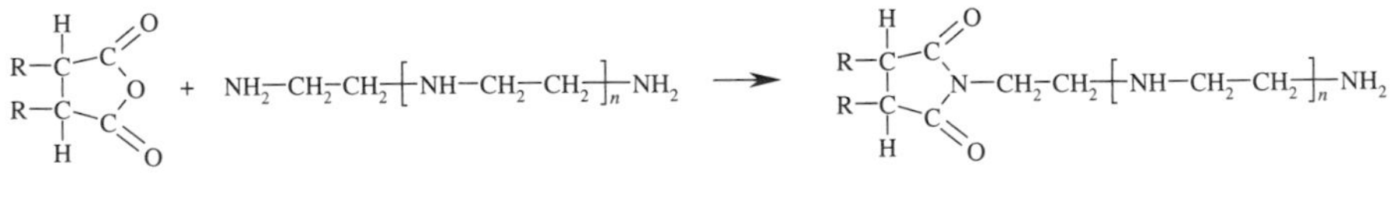

The sucinic anhydride group(s) attached to the PIB group(s) possesses polyfunctionality, and a different mole ratio of polyamine or polyhydric alcohols can be attached to produce a wide variety of deposit control additives. These succinimides and/or succinate esters and/or succinicamides can be tailored to different molecular weight and different molecule structure [37,46].

It has been found that certain ethers of polyalkyl or polyalkenyl N-hydroxyalkyl succinimides [48,49] provide excellent control of engine deposits, especially intake valve deposits.

Mannich Bases   The reaction of ammonia or amine, formaldehyde, and phenol is called the Mannich reaction. This reaction can be used for the manufacture of Mannich base DD additives (e.g., EP 1,123,814):

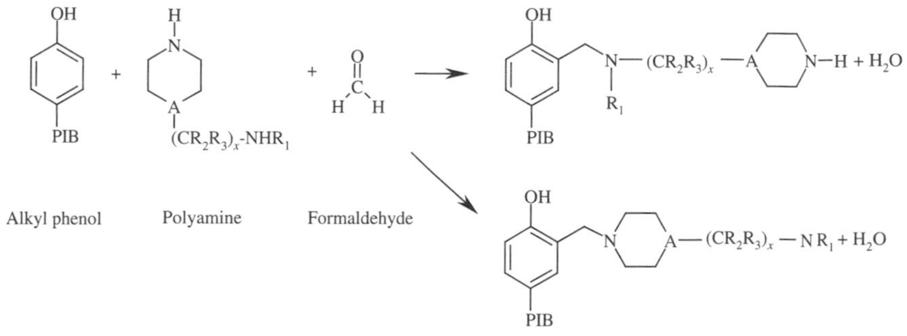

Mannich base DD additives

where,

A = carbon or nitrogen

R_1, R_2, R_3 = hydrogen or low alkyl group (C_{1-6})

$-CR_2R_3$ = units consists of R_2 and R_3

x = positive integer from 1 to 6

Aromatic hydroxyl compounds such as phenol can be alkylated with polyolefins using acid catalysts for the preparation of polyalkenylphenols [50,51]. Generally, polyisobutylene is the preferred polyolefin for preparing fuel detergent. This Friedel–Crafts alkylation reaction leads to a mixture of mono-, di-, and polyalkylation products. Such mixtures are unsuitable for many industrial applications. Polyisobutenes having a high content of β-olefin terminal units (more than 45 mol%), that is, a low content of α-olefin terminal units (e.g., 65% or less), can generally be alkylated with good results in the presence of a Lewis acid alkylation catalyst. This effect occurs, when using polyisobutenes having beta double bonds and to a lesser extent when the double bonds are situated further toward the interior (gamma-positioned bonds, etc.). Polyisobutenes with an average molecular weight $\bar{M}_n$ number of about 1000 containing at least 70 mol% of alpha and/or beta double bonds are preferably used for the alkylation, since these contain at least 70 mol% of terminal methyl-vinylidene groups $[-C(-CH_3)=CH_2]$ (α-olefin) and/or dimethyl-vinyl groups $[-CH=C(CH_3)_2]$ (β-olefin). Polyisobutenylphenols are thus prepared by alkylating an aromatic hydroxy compound with monoethylenically unsaturated and homopolymeric polyisobutenes in the presence of a Lewis acid alkylation catalyst.

A Mannich reaction product can be produced from a polyisobutylene-substituted phenol (in which at least 70% of the terminal olefinic double bonds is of the vinylidene type), an aldehyde, and ethylenediamine (EDA) (US Pat. No. 5,876,468, issued in 1999; Moreton). This compound is a more effective detergent in hydrocarbon fuels than the Mannich compounds prepared from 3-(dimethylamino) propylamine, diethylenetriamine, and triethylenetetramine. The phenol can also be alkylated with a polyoxyalkyl group, which on reaction with an aldehyde and amine would yield the Mannich base. This process, however, produces undesirable bis-products along with the desirable mono-polymeric amine:

$$\text{Polyoxyalkyl}(POA) + \text{phenol} \rightarrow POA - \text{phenol}$$

$$POA - \text{phenol} + \text{formaldehyde} + \text{amine} \rightarrow POA\,\text{phenol amine}\,(\text{Mannich base})$$
$$+ \text{bis} - POA\,\text{phenol amine}$$

At the Mannich condensation conditions, some portion of the reactants, generally the amine, remains unreacted. The reaction products of the Mannich processes also commonly contain small amounts of insoluble by-products of the Mannich condensation reaction that are high molecular weight products (resin) of formaldehyde and polyamines. The amine and amine by-products cause haze formation during storage, and in diesel fuel formulations, rapid buildup of piston ring groove deposits and skirt varnish. These drawbacks are overcome by adding long-chain carboxylic acids like oleic acid during the reaction to reduce the amount of solids forming from the Mannich reaction. This is considered to render the particulate polyamine-formaldehyde condensation product soluble through formation of an amide-type linkage [52–54].

Another solution of these problem is the transamination. Thus eliminating the need of using a fatty acid (US Pat. No. 4,334,085, issued Jun. 6, 1982, Basalay and Udelhofen). Transamination has been defined as the reaction of a Mannich adduct based on a single-nitrogen amine with a polyamine to exchange the polyamine for the single-nitrogen amine. The TEPA replaced 80–95% of the diethylamine in the Mannich. The effectiveness of the Mannich condensation products is improved by the addition of a carrier fluid, including liquid polyalkylenes, and polyoxyalkylenes [55–57], in reducing intake valve deposits and/or intake valve sticking (US Pat. No. 5,634,951, Jun. 3, 1997).

However, the Mannich base may also contain a small quantities of low molecular weight amine and an amine-formaldehyde intermediate that provide [52] anticorrosion properties.

It has been found that by using a di-substituted hydroxyaromatic [58] compound that has only one site for the Mannich reaction to occur (i.e., only one ortho- or para-position being unsubstituted), in combination with a secondary amine having only one hydrogen capable of entering into the Mannich reaction, products are obtained in higher yields and are more effective at controlling intake valve deposits compared to Mannich base products derived from a hydroxyaromatic compound having two or three reactive sites or Mannich base products derived from primary amines or amines having more than one active hydrogen.

The Mannich product is obtained by reacting a di-substituted hydroxyaromatic compound in which the hydrocarbyl substituent is polypropylene, polybutylene, or an ethylene α-olefin copolymer having a number average molecular weight in the range of about 500 to about 3000 and a polydispersity in the range 1–4, one or more secondary amines, and at least one aldehyde. Dibutyl amine as the secondary amine, and formaldehyde as the aldehyde (molar ratio of the above-substituted hydroxyaromatic compound to dibutyl amine to formaldehyde of 1.2:0.9), has been most advantageous in controlling IVD. Carrier oils such as poly (oxyalkylene) compounds further enhance the effectiveness of these Mannich condensation products in reducing intake valve deposits and/or intake valve sticking. Mannich base detergents provide combustion chamber deposit control in a cost-effective manner [57].

A Mannich reaction product (high molecular weight alkyl-substituted phenol, amine, and aldehydes) blended with a polyoxyalkylene compound and poly-alpha-olefin had improved performance and better compatibility (less insoluble material, haze, and flocs) [55–58]. Improved anticorrosion properties are provided as well by this fuel additive composition.

Polyetheramines Polyetheramines contain primary amino groups attached to the end of a polyether backbone. The polyether backbone is normally based on butylene oxide, propylene oxide (PO), ethylene oxide (EO), or mixed BO/PO/EO. Polyetheramines are a large group of compounds consisting of monoamines, diamines, and triamines based on the polyether core structure. Secondary, hindered, and polytetramethylene glycol (PTMEG) based polyetheramines have also been reported. The wide range of molecular weight, amine functionality, repeating unit type, and distribution can provide flexibility in the molecule to impart specific properties and solubility in water or hydrocarbons.

The general structures of the polyether-amines and polyether-polyamines are the following [59,60]:

$$\text{Alkoxy} -\!\!\left[\, CH_2 - \underset{\underset{C_2H_5}{|}}{CH} - O \,\right]_z\!\!- CH_2 - \underset{\underset{C_2H_5}{|}}{CH} - NH_2$$

Butylene-oxide based polyether-amine ($\overline{M}_n \approx 1800$)

$$\text{Alkoxy} -\!\!\left[\, CH_2 - \underset{\underset{C_2H_5}{|}}{CH} - O \,\right]_z\!\!- CH_2 - \underset{\underset{C_2H_5}{|}}{CH} - NH - CH_2 - CH_2 - CH_2 - N \underset{\diagdown CH_3}{\overset{\diagup CH_3}{}}$$

Butylene-oxide based polyether-polyamine ($\overline{M}_n \approx 1800$)

Hydrocarbyl-substituted poly-(oxyalkylene) polyamines may be prepared by the reaction of a suitable hydrocarbyl-terminated polyether alcohol with a halogenating agent such as HCl, thionyl chloride, or epichlorohydrin to form a polyether-chloride, followed by a reaction with a polyamine. The polyether chloride may be reacted with ammonia or dimethylamine to form the corresponding polyether amine or polyether dimethylamine (US Pat. No. 4,247,301;1981 Honnen). Primary amines can be prepared by the amination of butylene oxide (BO) capped alcohols. This method results in primary amines with the terminal end group.

Another possible synthesis of ether amines is by the cyanobutylation reaction of an alcohol having 3 to 22 carbon atoms with 2-pentenenitrile to form a branched alkyl-ether-nitrile. The alkyl-ether-nitrites on hydrogenation form alkylether amines [61].

Other derivates are the ether amine compounds. When they are reacted with formaldehyde and hydrogen, methylated tertiary amines are obtained [62,63].

Polyalkene-alcohol polyether-amines are suitable for use as carrier oils, detergents, and dispersing agents in fuel and lubricant compositions, and they are an improvement over polyisobutene-amines [64]. Although polyether-amines exhibit good performance, their usability is subject to certain restrictions, since these are highly polar and their solubility in fuel would be limited. It is therefore, necessary to use long-chain alkanols as the initiator molecules for the preparation of the polyether-amines.

DD derivates of polyether-amines are the etheramine alkoxylates. These are produced by reaction with a polyether-amine and alkylene oxide (ethylene-oxide, propylene-oxide, a butylene-oxide, or a mixture) in the presence of methanol, ethanol, propanol, or their mixtures to form etheramine-alkoxylate [61].

Application of Some DD additives

Port Fuel Deposit Control Additives   Port fuel injector deposit control additives were developed in the United States in 1985 and were used in about 100 ppm to overcome a drivability problem caused by injector fouling. Deposits on the intake valves also create drivability problems, and in both the United States and Europe, fuel detergents were required to keep the valves clean. Improved additives like PIB-amine or PIB-succinamide in combination with carrier fluids were required to keep intake valves clean.

The additives based on PIB-succinamide, PIB-amine, polyisobutene-amino alcohol, and/or polyether amines can prevent and eliminate deposits in both keep-clean and clean-up modes. This is because they have excellent heat stability in the zones of relatively high temperatures, such as in the intake valves [21–23].

DD Additives for DISI Engines   Prior to DISI, the main deposit control additive focus was on intake valve deposits (IVD) in the traditional port fuel injection (PFI) engines. There is thus a need for fuel additives to meet DISI engine requirements in addition to their good performances in PFI engines. The fuel additives in modern engines have to be reformulated and tested in DISI engines to meet deposit control requirements in these engines [65].

The polyether amines, also known as poly(oxyalkylene)-amines, are used to control fuel injector deposits in direct injection spark ignition gasoline engines [66]. It has been reported that the combination of aromatic esters of polyalkyl-phenoxy-alkanols with poly(oxyalkylene)-amines provides a fuel additive [67,68] that controls engine deposits in direct injection spark ignition gasoline engines and have been correlated to the octane requirement increase (ORI). The wider use of direct injection spark ignition (DISI) vehicles has led to an increase in the range of performance requirements for gasoline additives.

Carrier Oils for DD Additives   Selection of the carrier oil used in formulating a fuel deposit control additive is very important, because it improves the performance of the detergent additives. The carrier oil acts as a solvent for the fuel detergent and helps in removing the deposits from the engine parts. Carrier oils can be a mineral oil, a synthetic oil, or a semisynthetic oil. Examples of suitable synthetic carrier oils are polyolefins, (poly)esters, (poly)alkoxylates, and in particular, aliphatic polyethers, aliphatic polyether-amines, alkyl-phenol-polyethers, and alkylphenol-initiated polyether-amines. Polar carrier oils are more effective but may create a solubility problem in the fuel. Therefore a careful balance is required in formulating a fuel additive package containing detergent and carrier oil.

Adducts of butylenes-oxide with alcohols such as tridecanol etherified with butylene oxide have excellent solubility in fuels but are comparatively costly products. Generally, fully synthetic additive packages have better keep-clean properties than mineral oil based. Fully synthetic additives also have lower viscosities, especially at lower temperatures, than mineral oil based formulations.

Carboxylic acid alkoxylates are also good carrier oils and can be prepared [69] by reacting a carboxylic acid (coconut fatty acid, tall oil fatty acid, tallow fatty acid, oleic acid, or soya fatty acid) with a lower molecular weight alkylene oxides (ethylene oxide, propylene oxide, butylene oxide, or their mixtures) in the presence of an alkaline solution. When these alkoxylates are combined with a nitrogen-containing fuel detergent, they significantly improve the intake valve detergency of the nitrogen-containing fuel detergents. Carrier oils based on the condensation of propylene oxide have been used in combination with various detergent molecules to control engine deposits (European Pats., No. A-0 704 519 and No. A-0 374 461). A combination of a detergent additive with a synthetic carrier based on polyethers to control intake valve deposits in the engines [US Pat., No. 6,840,970, Jan. 11, 2005] has been reported.

DD Partpackages   In the last 10 to 15 years, co-utilization of at least two DD additives of different structures has become more useful [21–23]. The different functions of DD parts can be utilized for higher severity. Consequently the additive manufacturers now produce so-called partpackages of combined DD effects.

The Mannich detergent combined with a fluidizer (polyether-amine or a polyether, or a mixture of both) and a succinimide detergent is very effective in reducing intake valve deposits in gasoline fueled engines, especially when the weight ratio of the detergents to fluidizer is 1:1 [70].

A mixture of polyolefin-amine and polyetheramine is useful as a fuel additive since these two fuel detergents do not contribute to an increase in combustion chamber deposits in port fuel injected internal combustion engines [71].

A gasoline detergent additive consisting of a mixture of a polyether-amine, an aliphatic hydrocarbon-substituted amine, and a Mannich reaction product has been found to improve fuel economy and reduce fuel consumption, engine wear, and emissions [72,73].

5.2.3 Deposits and Their Control in Diesel Engines

Diesel engines use either indirect injection (IDI) or direct injection (DI) of fuel. In the past IDI was used for small diesel engines. However, most modern engines have switched to direct injection, which is more fuel efficient. The fuel is injected into the hot compressed air, where it is auto-ignited. For good combustion and low emissions, atomization of the air-fuel mixture and control of its ratio are important. For smooth combustion, initially a small quantity of fuel is injected to initiate the combustion in modern diesel engines. This is then followed by the progressive injection of fuel into the burning mixture. This approach requires very precise injector designs. A well-dispersed and atomized spray pattern is required for good air-fuel mixing to ensure smooth combustion. Accordingly, clean fuel injectors are critical for efficient diesel engine operation. The fuel injector is a high precision component designed to meter diesel fuel to a high degree of accuracy. The tip of the injector is in direct contact with the combustion zone and any deposits at that place can alter the performance of the injector significantly.

Super clean injectors have become an important requirement in modern diesel engines, as high-pressure injection systems are increasingly being used in both light-duty and heavy-duty engines. The performance of a diesel engine in terms of power, fuel consumption, and emission depends mainly on the cleanliness of the injector, since excessive deposits disrupt the injector spray pattern, and hinder the air–fuel mixing. Both fuels and lubricants can form deposits in the nozzle area of the fuel injector, which is exposed to high cylinder temperatures. The amount of deposit formed depends on the fuel composition, engine design, lubricant composition, and operating parameters [21,23]. Today the high levels of deposits that can build up in the injectors of diesel engines are mainly due to the presence of unstable biodiesel blendstock, or even trace metal concentrations of fuel [22].

As mentioned above, injector coking deposits form on the tips of fuel injectors and within the fuel spray holes. The internal diesel injector deposits (often abbreviated to IDID) are formed within the injector body, which can be at the armature group, on the piston and nozzle needle, and inside the nozzle body. These deposits slow the response of fuel injector and cause the moving internal parts to stick. The result is loss of control of injector event timing and of the quantity of fuel delivered per injection. The vehicle then experiences rough engine running, as well as unwanted variations in power and loss of fuel economy, among other drivability problems [74,75].

Basically two types of internal injector deposits can form: waxy or soap-like deposits (mainly in US engines) and carbonaceous or lacquer-like deposits (in European engines).

A number of theories have been suggested for IDID buildup. Some focus on mechanical systems: higher temperature and pressure in common rail system, tighter tolerances in injectors, multiple injection events per cycle, fuel (hydrocracked and/or reduced aromatic and/or reduced polar species content and/or inorganic ions, Na, Ca, and/or biodiesel); others on fuel additives: corrosion inhibitors, lubricity additives and/or deposit control additives, cold flow improver/wax antisettling agents. However, none of these have been proved [23,75].

Deposits can also build up in the tubes and runners that are part of the exhaust gas recirculation (EGR) system. The collection of these deposits can reduce EGR flow, resulting in knocking ("hard operation") of the engine and an increase in NO_x emissions in diesel engines.

Soot is the most harmful contaminant in modern diesel engines. Soot is generated during the combustion process and can be introduced into the oil during the scraping of oil-containing soot from the ring-liner area. Even in modern diesel engines meeting the latest emission regulations, the soot level may reach 5% or more before the oil is changed. When long drain oils and emission strategies like exhaust gas recirculation (EGR) are used, the soot level gets higher.

In direct injection diesel engines, the time for fuel evaporation and air-fuel mixing is considerably short. Therefore combustion of local fuel-rich mixtures may be incomplete, resulting in the soot formation in DI diesel engines. This condition can be ressened if the fuel contains oxygen so that incomplete combustion due to the shortage of oxygen is avoided [76]. Oxygen-containing fuels like methanol, dimethyl-ether ($H_3C–O–CH_3$), and ethanol are a source of extra oxygen for the fuel.

Soot formation can be completely eliminated if the fuel oxygen content is higher than 30%. Dimethyl ether contains 34% oxygen and is ideally suited to suppress soot formation in diesel engines. However, oxygenated fuels also cause energy loss, and their calorific value is lower than that of conventional fuels.

Deposit control additives of diesel fuels also have a polar molecule part and a nonpolar hydrocarbon chain. The deposit precursors attach to this molecule and remain bound with the micelle. Under normal conditions, the deposit control molecule forms a thin film on the metal parts of the fuel system. This film is the first barrier against the formation of deposits and can also remove any deposits from the metal surface slowly by detergent-cleaning action.

Diesel fuel DD additives are generally polyalkenyl-succinic-acid derivatives, Mannich bases, and some ashless polymeric compounds [21,23,77]. These are similar to the detergent dispersants used in gasoline, and they are typically used in 100- to 500-ppm (mg/kg) dosages in additive packages for diesel fuels, which contain other components like antioxidants, a metal deactivator, a corrosion inhibitor, an antifoam additive, and a cetane improver, as is especially the case in multifunctional fuel additive packages.

Improved DD efficiency has been reported (Peugeot XUD9 test) for mixtures of di-polyisobutylene-succinimides and polyisobutylene-di-succinincimides and derivates of polyisobutylene-disuccinic-anhydrides, as well as their good compatibility with flow improvers (MDFI) and paraffin dispergators ("Aral test" at −13 °C for 16 hour) [78].

Recently, a type of modified additive structure of polyisobutylene-succinimid compounds was reported that incorporated a fatty-acid-methyl-ester molecule [79,80]:

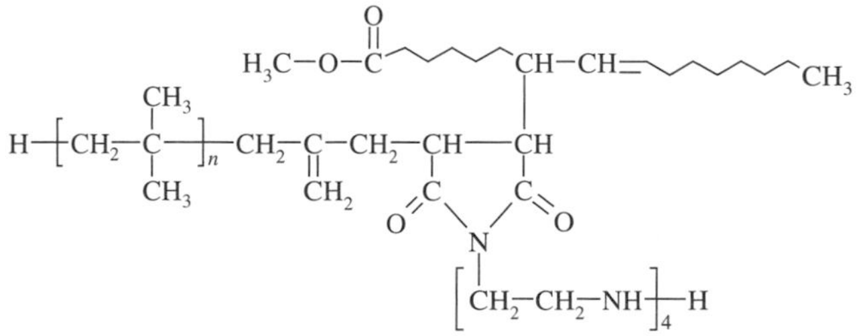

Besides the high DD efficiency of these additive molecules, whose structures were confirmed by IR, ^{13}C, and 1H-NMR, they have excellent anticorrosion and antiwear effects. So far there are no known negative effects owing to the three molecules being incorporated into one structure. Tests have further shown that this additive is very effective for diesel fuels that are practically sulfur free (≤ 10 mg/kg).

The newest experimental results have proved that co-utilization of the conventional succinimide type deposit control detergents (DCA) and monocarboxyl acids prevents the formation of internal diesel injector deposits (Peugeot DW10 test). Tests in North America of internal diesel deposits of the 6.8-liters common rail engine (tier off-road emission standard) have, however, detected Na salts. These are the Na salts of dodecyl-succininc-acid, and during the experiments significant performance loss was observed. It was nevertheless demonstrated that the lubricity-improved monocarboxyl acid does not form internal diesel injector deposits [81,82].

The newest additive packages thus offer a solution for both internal and external nozzle coking problems for both the old and latest technology diesel engines. The new diesel fuel deposit control additives proved capable of totally eliminating injector fouling in the new CEC F-98-08 DW10 test and completely preventing flow loss in the older CEC F-23-01 Peugeot XUD-9 test method. The performance improver package of these additives inhibited the formation of deposits and cleaned up all deposits. There was no performance loss. The anticorrosion, antifoaming, and the cetane number improver proved excellent in pure fossil or biodiesel-containing diesel fuels [82].

5.2.4   Detergent Additives and Exhaust Emissions

Fuel packages contain about 40–50% DD additives; consequently it is useful to know their effects on the quality and quantity of the exhaust gas emission [21,23]. That is to say, while detergent additives are effective in reducing the formation of deposits in internal combustion engines, it is difficult to evaluate the advantage to exhaust emission, as there are many factors that can affect engine emissions. In other words, detergent additives do not reduce directly the exhaust gas emissions from vehicles, but rather reduce the deposits that affect engine performance negatively. Therefore fuel additives reduce emissions lost due to deposit formation in the engine by keeping the fuel system clean [83].

Spark-Ignition Engines   There are three types of deposits in spark-ignition engines: deposits in the injector systems (DI), intake valve deposits (IVD), and combustion chamber deposits (CCD). The studies of IVD and PFID have been inconclusive concerning emissions. Some studies have shown that reducing IVD by means of detergents in fuels can reduce HC and CO emissions compared to untreated fuel. Even so, a negative effect of detergents on CO and NO_x emissions, probably due to the increase of CCD, has been observed. It has been reported in a US study that engine knock increased with the increased use of detergent additives [84].

A study performed in Venezuela showed that the use of detergent additives led to a significant reduction of CO (6–10%) and HC (13–24%) emissions from test trucks [85]. However, a NO_x increase (2–3%) was also observed. Fuel economy was improved by 0.5–1%. Similarly, an evaluation by the European Commission, ACEA, and EUROPIA [86] has estimated an average emission reduction (due to prevention of deterioration) of CO by 10–12%, HC by 3.0–15%, NO_x by 5 to –5%, and fuel economy by 2–4%.

Combustion chamber deposits typically increase the charge temperature and flame propagation rate and reduce the heat loss to the coolant. This is reflected as an increase in the octane requirement, increased NO_x emissions, and a reduction in maximum power but an improvement in fuel economy. Thus the effects of CCDs are quite variable in different engines and can go in different directions [87]: For example, tailpipe CO change –50 to 30%, HC change –30 to 20%, NO_x increase 0 to 50%, CO_2 reduction 2 to 10%, fuel economy improvement 2 to 10%, and ORI increase 1 to 10.

Diesel Engines   The formation mechanisms and the effects of deposits in diesel engines are different than those in gasoline engines. Injector fouling is the important exhaust emissions concern, since the injector in the diesel engine is the most critical design component. Deposits in the injection nozzles due to fuel coking affect the injection rate, spray pattern, and the amount of fuel injected. The obstruction to the injector flow can result in increased exhaust emissions. Detergent additives can reduce the deposits in the injector nozzles, but it is also difficult to estimate the benefits to the exhaust emission due to lack of data. Detergent additives alone do not improve the emission performance of diesel engines but reduce the risk of increased emissions due to deposit formation. Thus additives do indirectly help reduce emissions to some extent.

An evaluation of the existing data was conducted in 1995 jointly by the European Commission, ACEA, and EUROPIA [86]. The resulting document provides estimates of average emission reduction (due to prevention of deterioration) with the use of detergent additive. In light-duty diesel engines the emission reduction of CO was 8–12%, HC by 15–29%, NO_x by 1–2%, and PM by 22–10%. A similar trend was also observed in heavy-duty diesel engines (CO reduction by 10–14%, HC by 14–15%, NO_x by 2%, and PM by 10–15%).

CRC (Project No. E-6) also investigated the effects of combustion chamber deposits on vehicle emissions and fuel economy. A field test program on 28 test vehicles driven for 15,000 miles was performed with two different additive concentrations. Emission tests were carried out to investigate the influence of CCD on emissions and fuel economy; IVD and CCD were also rated for each engine. The study concluded that the additive treated fuels led to a statistically significant increase of NO_x emissions and fuel economy. The results also indicate that the presence of CCD causes a decrease in tailpipe and engine-out HC emissions. This is due to the thermal insulation provided by CCD, facilitating higher combustion temperatures through reduced energy loss to the engine coolant.

5.2.5   Tests for DD Additives in Engines

Verifications of the efficiency of DD fuel additives is done partly by engine tests and partly by fleet tests. Investigations must be able to defect the difference between the fuels and the additives. In this section the most important criteria used in engine tests are summarized.

TABLE 5.3 Test methods for the evaluation of deposits

Property	Criteria	Test Method
Carburettor cleanliness	Merit	CEC F -03-T
Fuel injector cleanliness	Percent flow loss	ASTM D 5598
Intake valve sticking	Pass/fail	CEC F- 16-T
Intake valve cleanliness I	Merit	CEC F-04-A
Intake valve deposits, Opel Kadett engine test	mg/valve	CEC-F-02-T-79
Intake valve cleanliness II		
Method 1 (4 valves average)	Average mg/valve	CEC F-05-A-93 or
Method 2 (BMW engine)		ASTM D 5500
Method 3 (Ford engine)		ASTM D 6201
Combustion chamber deposits		
Method 1	% of base fuel	ASTM D 6201
Method 2	mg/engine	CEC F-20-98
Method 3	% mass at 450 °C	TGA- FLTM BZ 154-01

Tests for Gasoline Engine   Four major automobile manufacturers (BMW, General Motors, Honda, and Toyota) have developed a TOP TIER standard for detergent-containing gasoline. This document describes the deposit control performance of unleaded gasoline at the retail level, with the purpose of minimizing deposits on fuel injectors, intake valves, and combustion chambers. The standard requires gasoline not to contain metallic additives, especially methylcyclopentadienyl manganese tricarbonyl (MMT). An additive must be certified to have met the minimum deposit control requirements established by the US EPA in 40 CFR Part 80. Also an additive has to be registered with the EPA in accordance with 40 CFR Part 79.

Listed in Table 5.3 are the test methods utilized for the evaluation of deposits in the carburetor, intake valve, injector, and combustion chamber of the engines.

Intake Valve Keep Clean Initial Performance   According to ASTM D 6201, Standard Test Method for Dynamometer Evaluation of Unleaded Spark-lgnition Engine Fuel, the intake valve deposit (IVD) must be clean for top performance. Tests demonstrating the base fuel minimum deposit level and the performance of additives must be conducted using the same engine block and cylinder head. All results must be derived from operationally valid tests in accordance with the test validation criteria of ASTM D 6201. The IVD results must be reported for individual valves and as an average of all valves.

Fuel Injector Fouling Initial Performance   Fuel injector fouling must be measured using the TOP TIER fuel injector fouling vehicle test available from General Motors. GM will run the test on a first-come-first-served basis and must make the method available to those who wish to run the test on their own.

TABLE 5.4 Diesel fuel tests in industrial engines

Test	Investigated Function	Reference
Cetane number	Ignition characteristics	CEC-PF-022
Peugeot XUD9 injector cleanliness	Detergent-dispersant injector deposits	CEC-PF-023
Peugeot DW10 Euro V injector[a]	Detergent-dispersant injector fouling	CEC F-098-08
RVI MIDS emission test	Detergent dispersant	R-49-88/77/CEE
Mercedes OM 336LA Euro 1, emission test	Detergent dispersant	R-49-88/77/CEE
Mercedes OM 336LA Euro 2, emission test	Detergent dispersant	R-49-88/77/CEE
Diesel fuel lubricity	Lubricity	CEC-PF-006
Diesel pump lubricity	Lubricity	CEC-PF-032
	Detergent dispersant	CEC-IF-035

[a] Four cylinder; 2.0 liter; direct injection; turbo charged common rail, max. 1600 bar.

Intake Valve Sticking Initial Performance   The sticking tendency of intake valves must be determined using either the 1.9 L Volkswagen engine (Wasserboxer according to CEC F-16-T-96) or the 5.0 L 1990-95 General Motors V-8 engine (SWRI IVS test). Two options are available for demonstrating intake valve sticking. A test must demonstrate no stuck valves during a cold start. A stuck valve is defined as one in which the cylinder compression is less than 80% of the normal average cylinder compression pressure.

Combustion Chamber Deposit Initial Performance   Combustion chamber deposits (CCD) must be collected and weighed along with IVD using ASTM D 6201, Standard Test Method. However, ASTM D 6201 does not contain a procedure for collecting and measuring CCD. Adapting a scrape and weigh procedure developed by CARB is recommended. Results for individual cylinders and an average must be reported.

Performance Engine Test for Diesel Fuel   The industrial engine tests of diesel fuel are summarized in Table 5.4 and some additional special tests in Table 5.5. These tables contain the ignition characteristics and lubricity of diesel fuels as well as the DD effects.

Diesel engine injector's cleanliness is evaluated by measuring the percentage of air flow loss according to the CEC (PF-023) TBA test procedure. This test is also recommended in WWFC- 2006 for categories 3 and 4 diesel fuels (limit max. 85% air flow loss). Injector cleanliness performance evaluation is also carried out by ASTM D 5598 test in a 2.2-liter four cylinder Daimler Chrysler engine test. This test is also required by the US EPA for certifying the additive performance for injector cleanliness.

TABLE 5.5 Extended diesel fuels tests—special engine tests

Test	Investigated Function	Comment
VW Passat, injector cleanliness test	Detergent dispersant	Modern 1.9 liters, IDI
Mercedes OM 604, injector cleanliness	Detergent dispersant	Modern 2.2 liters, IDI
Mercedes OM366LA—Euro 1	Detergent dispersant	6 liters turbocharged truck engine
PV1 MIDS 06-20-45 long-term/sideeffect test	Lubricity	10 liters turbocharged truck engine

A new injector fouling test for modern direct injection diesel engines (CEC F-098-08) was approved in March 2008. This test directly measures engine power, which is a function of the level of injector fouling. This test is based on the Peugeot DW10 engine, where the standard production injectors have been replaced with more advanced Euro V type injectors, as these are more sensitive to deposit formation. The DW10 test uses the Peugeot 4-cylinder, 2.0-liter, direct-injection, turbocharged, common rail engine with a maximum injector pressure of 1600 bar. It is designed to simulate injector fouling with a highly sensitive Euro 5 Siemens injector systems and measures power loss after 32 hours.

Compared to the well-known CEC F-23-01-XUD9 test (using an older indirect-injection engine), the new procedure is more intensive. This DW10 test shows greater sensitivity for specific deposit control detergents compared with the CEC F-23-01-XUD 9 method.

In Europe, injector cleanliness is evaluated by a Peugeot XUD-9A/L test that is run for 10 hours. This test (CEC F-23.A-00) is an intensive test for evaluating the deposit control ability of additives intended to prevent injector fouling and is also a requirement of the World Wide Fuel Charter—2006.

Remarks

1. Combustion chamber deposits can be measured by M102E, M111, VW Polo, and Kadett tests.
2. Table 5.4 can be supplemented by special dynamometer tests.
3. The M111 test can be performed by engines using two fuels.

5.2.6   Advantages of Using DD Additives in Fuels

Detergent-dispersant additives in fuels assure a number of advantages for the customers:

- Clean fuel supply system
- Uniform fuel injection (no deposits inhibiting free flow)
- Efficient combustion

- Pressure increase with a lower gradient, lower noise
- Higher performance
- Optimal driving characteristics
- Lower fuel consumption and upkeep costs (5–10%, due increase of the maintenance interval)
- Lower exhaust gas
- Lower harmful material in exhaust gas

Beyond these developments, the utilization of additives has stimulated research and development of new kinds of engines constructions in which the injector deposits would activate fuel consumption, and be particularly sensitive toward deposits because of their "high swirl" and fast combustion [60].

5.3 ANTIKNOCK ADDITIVES (OCTANE NUMBER IMPROVERS)

5.3.1 "Knocking"

Gasoline engine knocking is a combustion phenomenon that takes place when the air-fuel mixture in the gasoline engine does not burn smoothly or evenly. Knocking in gasoline engines is an undesired acoustic occurrence when the air and fuel mixture ignites before the upper point of piston in spark ignition vehicles. As the fuel in a gasoline engine is ignited by the spark plug, the flame tends to move away from the spark plug, and the heat and pressure generated by the first portion travels to and compresses the last portion of the fuel. Chemically, combustion is a multistage oxidation process where the organic hydroperoxides are generated initially. These decompose to form free radicals that start branch chain reactions. These chain reactions start local auto ignitions/detonations, ahead of the flame front in the unburned hydrocarbon air mixture due to the high temperature and pressure generated in the last portion of the fuel. This violent detonation creates pressure fluctuations and noise. This noise is called knock, and if it continues, it can damage engine parts like the valves and the piston. Some heat is also lost, and the fuel efficiency of the engine is decreased. This problem is more pronounced in high-compression gasoline engines.

5.3.2 Octane Number

The common type of octane rating is the research octane number (RON). RON is determined by a test engine of variable compression ratios at controlled conditions, and the results are compared to those for mixtures of isooctane and n-heptane (the isooctane content of mixture in v/v%). There is another type of octane rating, called the motor octane number (MON), or the aviation lean octane rating, which is a better measure of how the fuel behaves under a load; it gives information about the combustion/compression tolerance properties of the mixture in the actual engine.

TABLE 5.6 RON and MON of different chemicals components

Compound	RON	MON
Methane	135	122
n-Butane	91	92
n-Pentane	62	62
n-Hexane	25	26
n-Heptane	0	0
n-Octane	−10	−14
Isooctane	100	100
Benzene	101	99
Toluene	111	95
Xylene	117	110
Methanol	133	99
Ethanol	130	96
Isopropanol	118	98
n-Butanol	92	71
t-Butanol	105	95
MTBE	118	100
ETBE	118	102
TAME	111	98

MON is determined at 900-rpm engine speed, instead of the 600 rpm for RON. MON testing uses a same test engine as that for RON testing, but the fuel mixture is preheated, the engine speed is higher, and the ignition timing is varied to further stress the fuel's knock resistance. Depending on the composition of the fuel, the MON of a modern gasoline would be about 8 to 10 points lower than the RON. However, there is no direct linkage between RON and MON [88].

An average of RON and MON is called the antiknock index (AKI). Sometimes it is also called as the road octane number or the pump octane number.

The RON and MON of n-heptane and isooctane are 0 and 100, by definition. Table 5.6. lists the approximate octane ratings of various chemical compounds.

5.3.3   Octane Number Improver Additives

An antiknock compound is added in gasoline to reduce engine knocking by increasing the fuel's octane rating.

Lead-Tetramethyl and Lead-Tetraethyl   The discovery that lead-containing compounds have an antiknock effect led to the widespread use of these additive to develop engines of higher specific power and a higher compression ratio. Lead-tetraethyl was discovered by a German chemist in 1854, but remained commercially unused for many years. In 1921, tetra-ethyl-lead was found to be an effective antiknock agent by the General Motors Corporation research group, and they patented the use of TEL as an antiknocking agent and called it "ethyl." By 1923, leaded gasoline was being sold in the United States. In 1924, Standard Oil of New Jersey (ESSO/EXXON) and General Motors founded the Ethyl Gasoline Corporation to produce and market TEL.

The use of TEL then continued in the petroleum industry for several decades. The mechanism of organo lead compounds as an antiknock agent is to decompose to lead oxide during the combustion process. Then lead oxides deactivate the free radicals that are responsible for the chain branching auto ignition reactions. Thus auto ignition is shifted to a higher temperature and pressure regime.

When lead compounds were used as antiknock compounds, the lead oxides tend to generate deposits in the combustion chamber. This problem was solved by using scavengers like ethylene dichloride (8–18%) and -dibromide (ca. 18%), which convert nonvolatile lead oxides into volatile lead halides [89].

Tetra-ethyl-lead is produced by reacting chloroethane with a sodium–lead alloy.

$$4NaPb + 4CH_3CH_2Cl \rightarrow (CH_3CH_2)_4Pb + 4NaCl + 3Pb$$

In place of sodium, lithium is used too, but the process is not used commercially. TEL has four weak C–Pb bonds that break at the elevated temperatures encountered in internal combustion engines, resulting in lead and lead oxides and ethyl-free radicals. The lead and lead oxides scavenge free radicals formed in the combustion reactions. This prevents the ignition of unburned fuel during the engine's power stroke. Lead alone is the reactive antiknock agent, and TEL serves as a gasoline-soluble lead carrier.

Combustion of the TEL produces carbon dioxide, water, and lead:

$$(CH_3CH_2)_4Pb + 13O_2 \rightarrow 8CO_2 + 10H_2O + Pb$$

Lead can be oxidized further to yield a lead (II) oxide:

$$2Pb + O_2 \rightarrow 2PbO$$

The use of antiknock additives permits greater efficiency and higher power output because of the higher compression ratios they produce. The use of TEL in aviation gasoline has enabled the development of large supercharged engines. In the aviation industry, TEL is still in use to formulate aviation gasoline, although efforts are being made to find its substitute.

The use of TEL in gasoline was started in the United States, whereas in Europe, alcohol was still being used. The advantages of leaded gasoline come from its higher energy content and better storage stability, and this eventually led to a universal switch to leaded fuel. One of the greatest advantages of TEL over other antiknock agents is its low concentrations requirement. Typically 1 part of TEL is required to treat 1260 parts of gasoline. TEL biocidal properties also helped prevent fuel degradation due to bacterial growth.

In most Western countries TEL went out of use in the late twentieth century because of concerns over lead pollution of air and soil and the accumulative neurotoxicity of

FIGURE 5.6   Structure of methyl-cyclo-pentadienyl-manganese-tricarbonyl (MMT)

lead, since it accumulates in living organisms and in humans too. The use of TEL as a fuel additive also has detrimental effect on catalytic converters [90,91].

Decrease in the use of TEL has come with the use of other antiknock additives of less toxicity (MMT, MTBE) and the application of high octane blending compounds in increasing amounts (reformates), [92,93]. TEL was phased out in 1996 in the United States and in 2000 in the European Union. This was followed by several non-EU member countries [94,95].

The vehicles produced before TEL's phase-out required modifications to run on unleaded gasoline, since lead provided antiwear protection to the valve seats. Additives were developed to provide antiwear properties in fuel to compensate for the absence of lead, for example, K-naphtenate [96,97]. Subsequently, metallurgy was changed by the installation of hardened exhaust valves and value seats that did not require additional antiwear/lubricity additives.

However, TEL remains an ingredient of 100LL octane aviation fuel (avgas) for piston-engine aircraft and until recently in formula racing cars. The EPA and others are working on a lead-free commercially viable replacement for avgas. The current formulation of 100LL (low-lead) aviation gasoline contains much less lead than in previous fuels.

Several alternative antiknock agents were proposed and developed as substituent of lead-containing fuels, for example: methyl-cyclo-pentadienyl-manganese-tricarbonyl (MMT) [98,99], ferrocene [100], iron pentacarbonyl [101], and dialkyl carbonates [102,103].

Methylcyclopentadienyl-Manganese-Tricarbonyl   Methylcyclopentadienyl-manganese-tricarbonyl (MMT, Figure 5.6) was discovered by the Ethyl Corporation (now Afton), in 1954, for improving the octane number of gasoline. The manganese atom in MMT is bonded to three carbonyl groups as well as to the methylcyclopentadienyl ring. These hydrophobic organic ligands make MMT highly lipophilic, which may increase bioaccumulation. A synergistic effect between MMT and paraffinic fuels has been reported.

Methylcyclopentadienyl-manganese-tricarbonyl (MMT) has been used to boost octane of gasoline for many years in the United States, Canada, Australia, and some other countries in Asia, Africa and some Eastern European countries. MMT is suspected to be a strong neurotoxin and respiratory toxin. Additionally, MMT impairs the effectiveness of automobile emission control and increases pollution of vehicles.

Various OEM also do not recommend the use of MMT (see World Wide Fuel Charter—2006 September). The European Directive of 2009/30/EC also permits the use of MMT to a limited extent, pending risk assessment. In Europe the presence of the metallic additive methylcyclopentadienyl-manganese-tricarbonyl (MMT) in fuel is limited to 6 mg of manganese per liter since January 2011. The limit will be 2 mg of manganese per liter in January 2014.

Ferrocene   *Ferrocene* is an organo metallic compound with the formula $Fe(C_5H_5)_2$. It is a typical metallocene consisting of two cyclopentadienyl rings bounded on opposite sides of a central iron atom. Such organometallic compounds are also known as sandwich compounds [100]. The ferrocene was tried as an antiknock additive in many countries. Its iron content in form of iron-oxide contaminates the spark plug, the exhaust gas treating catalysts, and the exhaust system. The deposits can cause the spark plug to misfire.

Ashless Antiknock Additives   Ashless antiknock additives consist of mainly carbon, hydrogen, oxygen, and nitrogen. Some examples of nitrogen-containing additives are 1,3-diimino-2-hydroxy-propane, and methyl-aniline, and an example of an oxygen-containing additive is dimethyl-carbonates.

Beside the application of the aforementioned metal-containing and ashless additives, the octane number deficiency caused by the phase-out of lead-tetraethyl and -tetramethyl was initially compensated with the increased amount of blending of aromatic hydrocarbons (e.g., benzene and toluene) of high octane numbers (101–111) too[104]. Because of their carcinogenetic and toxic properties, they were gradually changed to oxygen-containing compounds of high octane numbers (MTBE, ETBE, TAME) [105,106]. Bioethanol of a high octane number has been used in increasing amount over the last five years [107,108]. Nowadays this has helped solve the octane number deficiency. (The advantageous properties of ethanol in fuel were already known in the year 1921, and the ethanol was utilized in Europe at this time.) Oxygenates not only have high octane numbers but the oxygen atoms in the molecules promote the cleaner burning, resulting in lower CO and hydrocarbon emissions. The blending volume of oxygenates is in the range of 1–15%, accordingly they are discussed among the motor gasoline-blending components in Chapters 3 and 4. Research and development activities on antiknock additives are still continuing.

5.4   CETANE NUMBER IMPROVER

Cetane is an unbranched open chain alkane molecule with 16 carbon atoms. This molecule has been assigned a cetane number of 100, while alpha-methyl naphthalene has been assigned a cetane number of 0. The cetane number is the index number of the autoignition property of the diesel fuel and diesel fuel components. All other hydrocarbons in diesel fuel are indexed to cetane as to how well they ignite under compression. Since there are a large number of chemical compounds in diesel fuel, each having a different cetane rating, the overall cetane number of the diesel is the average cetane rating of all the components. Accordingly, the cetane number measures how quickly the fuel starts to burn (auto-ignites) in diesel engines. This is the time period between the start of injection and start of combustion (ignition) of the fuel. In certain diesel engines, the higher cetane fuels have shorter ignition delay periods than lower cetane fuels. There is a direct relationship between the cetane number of a diesel fuel and its ignition point. However, it is not absolutely correct from the point of view of measurement technique. The measurement of the cetane number is based on the measure of the ignition delay of the motor fuel, which is the delay of the injection and the start of the ignition/burning. Consequently the later the start of the burning/ignition of the compressed motor fuel is, the lower the cetane number is. Thus there is a decrease in the efficiency of the ignition and the peak of the pressure in the cylinder. For the abovementioned reasons, there is a decrease in the efficiency of the engine and an increase of emissions. Therefore a major factor in diesel fuel quality is the cetane number. Commercial petroleum derived diesel fuels generally have a natural cetane number of about 40–55. Alcohols have much lower cetane values and require the addition of a cetane improver for smooth engine operation [109,110].

Cetane improvers alter combustion in the engine. They ensure early and uniform ignition of the fuel. These also prevent premature combustion and an excessive pressure increase in the combustion cycle. The combustion is smoother, the efficiency of burning is better, as well as the operation of the engine. The effects of cetane number improvers, in the necessary amounts and cetane number boosting effects, depend on the ratios of high and low cetane components in the base diesel fuel. Typically alkyl nitrate additive treatments can increase cetane by about 3 to 7 numbers (1:1000 treatment ratios, 0.1%). With the high natural cetane premium base fuels (containing a high percentage of paraffins) and a 1:500 (0.2%) treatment ratio, cetane can be increased up to a maximum of 8 to 10 numbers [109–111].

Generally, diesel engines run well with a cetane number of 40–55. Fuels with higher cetane numbers that have shorter ignition delays provide more time for the fuel combustion process to be completed. Hence higher speed diesels engines operate more effectively with higher cetane number fuels.

The current standard for diesel fuel sold in the European Union is covered by EN590:2009+A1:2010, with a minimum cetane number of 51. The premium diesel fuel has a cetane number as high as 60. In North America, most states adopt ASTM D975 as their diesel fuel standard, and the minimum cetane number is set at 40, with typical values in the 42–45 range. The premium diesel fuel's cetane number is much higher, depending on the distributor.

Measuring an engine's cetane number is costly, therefore a cetane index was introduced to determine the ignition of gas oils. This is calculated using the easy measurable properties of the distillate and density of fuel.

The cetane index (CI) can be calculated by the following expression:

$$CI = 454.74 - 1641.4169\rho + 774.74\rho^2 - 0.554t + 97.803(\log t)^2$$

where

ρ = density of fuel at 15 °C, g/cm^3
t = middle average boiling point, °C

The cetane index are only used for the relatively light distillate diesel oils. For heavy (residual) fuel oil, two other scales are used CCAI and CII:

Calculated carbon aromaticity index (CCAI)

$$CCAI = D - 140.7\log(\log(V + 0.85)) - 80.6 - 483.5\log\left(\frac{t + 273}{323}\right)$$

where

D = density on 15 °C, kg/m^3
V = kinematic viscosity, cSt (mm^2/s)
t = temperature of viscosity measurement, °C

Calculated ignition index (CII)

$$CII = (270.795 + 0.1038T) - 0.254565D + 23.708\log\log(V + 0.7)$$

where

D = density on 15 °C, kg/m^3
V = kinematic viscosity, cSt (mm^2/s)
t = temperature of viscosity measurement, °C

5.4.1   Cetane Number Improver Additives

Several different additives have been tried to increase the cetane number of diesel fuel. These include peroxides, nitrites, nitrates, nitroso-carbamates, tetra-azoles, and thio-aldehydes. Alkyl nitrates such as amyl-nitrate, hexyl-nitrate, and mixed octyl-nitrates have been used commercially with good results. These chemical cetane improvers readily decompose to form free radicals, which inturn promotes the rate of initiation. This increased rate of chain initiation leads to improved ignition of diesel fuel [109,112–114]. Alkyl nitrates have proved most important in commercial use as cetane improvers. 2-Ethylhexyl nitrate (EHN) has been used as a commercial cetane improver for a number of years and today is the predominant cetane improving additive [115]:

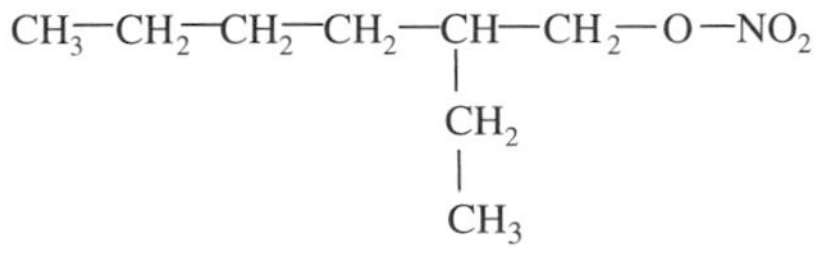

$$CH_3-CH_2-CH_2-CH_2-\underset{\underset{\displaystyle CH_3}{\overset{\displaystyle |}{\underset{\displaystyle |}{CH_2}}}}{CH}-CH_2-O-NO_2$$

Most cetane improvers contain alkyl nitrates that readily break down to provide additional oxygen for better combustion. These also break down and oxidize fuel in storage. During long-term storage these additives can generate organic particulates (compound fragments), water, and sludge—all of which can degrade fuel quality.

The alkyl nitrates are prepared by nitrating the corresponding alcohol. The alcohols are nitrated by adding them to a mixture of nitric acid and acetic anhydrate at $-10\,°C$ to $0\,°C$ [116].

Di-tertiary-butyl peroxide (DTBP) was first recognized as an effective cetane improver in the 1940s [117]. However, due to its high cost, DTBP has not had widespread use compared to 2-ethylhexyl-nitrate. It's low popularity did not change even after the ARCO Chemical Company reported a new technology to reduce the cost of DTBP to a level comparable to that of 2-EHN.

Dialkyl-peroxides can be synthesized by the reaction of an alcohol and/or an olefin (1-butyl alcohol and/or isobutylene) with an organic hydroperoxide (1-butyl hydroperoxide), using an acidic catalyst. The DTBP has a potential advantage over alkyl nitrates in reducing NO_x emissions, since it does not contain nitrogen [118].

A thermally or oxidatively unstable cetane improvement additive can degrade fuel quality and could lead to poor engine performance. Thus, for commercial acceptance the cetane improver additives must be stable, thermally and oxidatively, under actual storage conditions. DTBP has been reported to be more thermally stable as compared to alkyl nitrates in low sulfur diesel fuels [119,120].

A fuel treated with DTBP does not show significant loss in the cetane number after heating for 100 hours at $92\,°C$. The half-life for DTBP at $70\,°C$ in diesel fuel is in excess of 10,000 hours. Even at $100\,°C$, the half-life of DTBP in diesel fuel is over 300 hours. Although the thermal decomposition rate of the peroxide is much faster than the nitrate, the peroxide additive is quite stable under typical fuel system temperatures. The stability of DTBP was demonstrated by the standard accelerated oxidative stability test ASTM D274, and the long-term storage stability test ASTM D 4625 did not show any gum formation in most of the diesel fuels containing DTBP. Some inherently unstable fuels, however, show degradation that can be controlled by the addition of a small amount of antioxidant.

The cetane response of DTBP was inversely related to the aromatic content of the fuel. Thus low aromatic fuels respond better to the additive. Fuels with a high aromatic content or highly aromatic blend stocks, like light cycle oil (LCO) or light cycle gas oil (LCGO), respond poorly to cetane improvers. A highly aromatic (87%) LCGO blend stock did not respond at all to either of the cetane improvers, DTBP or EHN. This low response for the aromatics may be attributed to the higher activation energy required for nitrate- or peroxide-free radicals to react with an aromatic fragment compared to the aliphatic hydrocarbon fragment of the fuel.

These products should be handled with caution because of their potential explosive nature. Cetane improvers have been reported to improve fuel economy [121], engine durability [122], and emissions [123] in heavy-duty diesel engines.

It is also reported that higher cetane numbers improve cold start properties and in normal climatic conditions decrease emissions.

5.4.2   Cetane Number Measurement

Cooperative Fuel Research Test   The cetane number is determined by a Cooperative Fuel Research (CFR) engine test (ASTM D-613, ISO 5165) under standard conditions. The compression ratio (and therefore the peak pressure within the cylinder) of the engine is gradually increased until the time between fuel injection and ignition is 2.407 ms. The resulting cetane number is then calculated by determining which mixture of cetane (hexadecane) and isocetane (2,2,4,4,6,8,8-heptamethylnonane) will result in the same ignition delay (the cetane number is the n-cetane content of this mixture in v/v%).

Ignition Quality Tester   Another reliable and more precise method of measuring the cetane number of diesel fuel is the ignition quality tester (IQT). This instrument makes the measurement simpler than the CFR engine test. The fuel is injected into a constant volume combustion chamber in which the temperature is approximately 575 °C. The fuel combustion and the high rate of pressure change within the chamber define the start of combustion. The ignition delay of the fuel can then be calculated as the time difference between the start of fuel injection and the start of combustion. The fuel's derived cetane number can be calculated using an empirical inverse relationship to ignition delay (ASTM D-6890).

5.4.3   Cetane Index

The cetane index (CI) is calculated based on the density and distillation range of the fuel. The ′4-point method ASTM D 4737 is based on density, 10 v/v%, 50 v/v%, and 90 v/v% recovery temperatures. The 2-point method is defined in ASTM D 976, and uses just density and the 50 v/v% recovery temperature. This 2-point method tends to overestimate the cetane index and is not generally recommended. Cetane index calculations cannot account for cetane improver additives and therefore do not measure total cetane number for additive-treated diesel fuels. The CFR diesel engine test is primarily related to the actual cetane number, and the cetane index is only an estimation of the base (nonadditive treated) cetane number.

5.5   FUEL ANTIOXIDANTS (STABILIZERS)

The main objective in adding antioxidants to fuels is to inhibit the formation of sludge and deposits and the darkening of color [124,125]. The causes of these unwanted effects are several. In this section some of these causes are reviewed.

Fuels are fast-moving petroleum products, but they are sometimes stored for one to three years or more and are subject to deterioration. In military establishments fuels are stored for longer durations due to various operational and logistics needs. It is essential that the fuel keeps its quality over this period [125–128]. The reduction of sulfur in gasoline and diesel fuels has greatly improved stability, but has also created an additional problem resulting from the removal of the natural antioxidant.

The increasing demand for fuel products has led to the blending of cracked stocks by refineries. The cracked stocks have lower stability than the straight run stocks. The reason of this is that cracked products contain high amounts of olefins, aromatics, sulfur and nitrogen compounds that quickly degrade. For example, gasoline produced with catalytic cracking contains olefins and diolefins that tend to form gums on storage [129]. This gum can deposit in the automotive fuel system, especially in carburetor and intake system and affect vehicle performance.

The oxidation liability of biodiesels due to olefin double bonds is well known. Their use, though these fuels are already blended, is only possible with use of additives [130,131]. The presence of dissolved metals, even in trace amounts, accelerates the degradation [124]. Importantly, the peroxides formed during the oxidation reactions can damage the material of the seals (an elastomer).

5.5.1   Increasing Storage Stability

The fuels produced through hydrotreating processes have improved stability due to the saturation of olefinic and aromatic molecules [132]. However, the cost of stability improvement by additive treatment [133] is generally lower than that of hydro processing. Often a middle path is followed where the combination of two gives the cost-effective solution.

5.5.2   Oxidation of Fuels

The oxidation process taking place in fuels is a chain reaction resulting in insoluble "gum" as the end product. This gum can form in storage tanks, refinery pipelines, and the fuel-injection system of gasoline engines. It can clog fuel pipelines, carburetors, intake manifolds, and intake ports. Fuel stability is affected by several factors. Most prominent are the olefins and acid–base reactions in fuels.

During autoxidation, fuels slowly absorb oxygen from the atmosphere in the initial phase, called an induction period, and form hydroperoxides. These hydroperoxides form free radicals and initiate a chain reaction that generates the oxidation products and ultimately forms gum [133,134].

It is known that hydrocarbons can also generate free radicals in the presence of light or heat energy:

$$RH \, (\text{hydrocarbon molecule}) + \upsilon \rightarrow R* + H* \; (\text{free radicals})$$

$$R* + O_2 \rightarrow ROO* \, (\text{peroxide})$$

$$ROO* + RH \rightarrow ROOH + R'$$

This chain can further branch off, producing more free radicals:

$$ROOH \rightarrow RO* + O*H$$

$$RO* + RH \rightarrow R* + ROH$$

$$RH + O*H \rightarrow R* + H_2O$$

Termination of this chain occurs when the supply of air/oxygen is cut off or the reactive fragments are destroyed due to the formation of following types of stable molecules:

$$R* + R* \rightarrow RR$$

$$2ROO* \rightarrow ROOR + O_2$$

There are a series of following complex reaction taking place in the presence of an antioxidant:

$$ROO* + AH(antioxidant) \rightarrow ROOH + A*$$

$$A* + O_2 \rightarrow AOO* \text{ Peroxy radical}$$

$$AOO* + RH \rightarrow AOOH + R*$$

$$AOO* + AH(antioxidant) \rightarrow AOOH + A*$$

$$AOO* + AOO* \rightarrow \text{dimerized radical}$$

The sludge can also form from the reaction of nitrogen-containing compounds (e.g., indols) and phenalenes that originate in the unsaturated compounds. The dark color of diesel fuels is caused by "rubber-like" compounds that are soluble in oils, while the sludge is caused by the slower peroxidation reactions.

5.5.3 Chemical Mechanism of Antioxidants

Antioxidants are organic chemical compounds that decrease the oxidation and gum-forming tendencies of fuels due to autoxidation process and controlling the acid–base reactions [124].

Antioxidants donate their hydrogen atom to the peroxy radical, reducing its activity and thus the chains are terminated [124,135,136,137,138].

5.5.4   Types of Antioxidants

The hindered phenols and the phenylene diamine are the two most common antioxidants used today to control the gum formation in fuels. These are usually used in 5 to 100 ppm dosage.

Steric Hindered Phenols   2,6-Di-tertiary butyl-para-cresol and 2,6-di-tertiary butyl phenol and their mixtures are known to be good fuel antioxidants [139,140]. Hindered phenols are more effective when the olefin content in the fuels is lower. Di-tertiary butyl-para-cresol is a common fuel additive, known as DBPC or BHT:

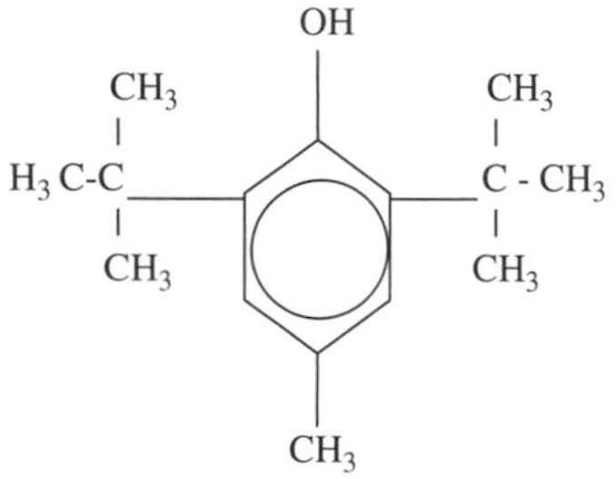

Mixtures of liquid alkyl phenols have been reported to provide good oxidation protection in both fuels and lubricants. The combination of alkyl phenols shows particularly good balancing antioxidant properties in the middle distillate: 1–8% 2-tertiary butyl phenol; 4–12% 2,6-di-tert-butyl phenol; 4–12% 2-tri-tertiary butyl phenol; 65–80% 2,6-di-tertiary butyl-4-n-butyl phenol [141].

This mixture can be produced by an aldehyde catalyzed alkylation reaction using n-butanol and tertiary butylated phenol in the presence of a strong base consisting of NaOH/ KOH. Butylated hydroxy toluene (BHT) or 2,6-di-tert-butyl-para-cresol (DBPC) is produced by the reaction of isobutylene with para-cresol in the presence of an acidic catalyst. Isobutylene first reacts at the ortho to hydroxy group, 2-position, followed by reaction of isobutylene group at the 6-position.

The reaction is carried out with several acidic catalysts, including a solid catalyst like the ion exchange resins, silica-alumina, and γ-alumina. A highly active solid catalyst has been reported that has good selectivity and provides good yield [135,142].

DBPC or BHT in pure form is a white crystalline material that is soluble in oil. However, if the phenol and para-cresol mixture is alkylated with isobutylene, a mixture of 2,6,-di-tert-butyl phenol and cresols is obtained, which may be liquid depending on the ratio of the two phenols. Many commercial products are available at a relatively low cost in liquid form, with different ratios of these butylated phenols and cresols. Such products are being used in many industrial oil applications [143].

Di-alkyl-para-Phenylenediamines   and   Alkylated   Di-phenyl   Amines
Phenylenediamines like *N,N′*-di-secondary butyl-p-phenylenediamine and mixtures of aromatic diamines are more effective in gasolines containing higher amounts of cracked products:

R—HN—⟨benzene⟩—NH—R'

Aromatic amines are a radical chain destroyer. The chain carrying the peroxy radical is scavenged by the phenol or amine due to the hydrogen atom donation. The resulting radicals are resonance stabilized and are eventually destroyed by another peroxy radical.

Diphenyl amines or substituted diphenyl amine are prepared [143] by reacting phenol with an aniline compound in the presence of a hydrogen transfer catalyst. Phenol is first converted to cyclohexanone by the accepting protons. Aniline is then added dropwise in the presence of alkali metals or alkaline earth metal hydroxides or carbonate as the co-catalyst and the noble metals of group VIII as the hydrogen transfer catalyst. Di-phenylamine can be alkylated with di-isobutylene (US Pat. No. 5,520,848) or nonane, or with another alkyl group using a Friedel–Craft reaction and a Lewis acid catalyst [144]:

Strucure of diphenyl amine antioxidant

The hindered amines also react with free radicals to form stable intermediates that do not support chain reaction. These alkylated di-phenylamines are effective antioxidants because the alkyl R*, alkoxy RO*, and alkyl peroxy ROO* free radicals can extract the hydrogen atoms from these molecules and form stable RH, ROH, and ROOH molecules. Di-phenylamine is converted to an aminyl radical, which further reacts with the peroxy radicals to yield a nitroxyl radical. This chain continues to end up with the formation of 1,4-benzoquinone and alkylated nitrosobenzene [145].

As compared to DBPC/BHT, the alkylated di-phenylamines are able to quench two times more free radicals under the oxidation conditions of moderate temperature of about 120 °C. At higher temperatures the oxidation mechanism with alkyl di-phenylamine will proceed via a slightly different mechanism [146]. The nitroxyl radical further reacts with a secondary alkyl radical to yield an N-sec-alkoxy diphenyl intermediate that re-arranges itself to generate alkylated di-phenylamine. Thus in one inhibition, cycle the alkylated di-phenyl amine traps one alkyl and one alkyl peroxy radical, but on a subsequent regeneration, it can trap all 12 free radicals per molecule of antioxidant. Alkylated di-phenylamines are therefore very efficient antioxidants at both moderate and high temperatures.

Synergistic effects have been observed when these amines are used in combination with hindered phenols such as DBPC. Di-phenylamine reacts faster with peroxy radicals than hindered phenol and generates an aminyl radical. The hindered phenol then donates hydrogen [145] to the aminyl radical, and di-phenylamine is regenerated. Thus the hindered phenol will be first consumed during the oxidation process. Di-phenylamine will then begin to be consumed by the free radicals (this explains the synergism of the combination of these two antioxidants in lubricating oil).

In fuels, the smaller molecule of alkylated phenylene diamines is used in small dosages as compared with lubricants, since fuels are not subjected to high temperatures for long durations [146]. The following compound is especially effective in aviation fuels:

N-N'-di-butyl-para-phenylene diamine

Aviation fuel uses the following approved antioxidants [147–149]: 2,6-di-tertiary butyl phenol (DBP); 2,6-di-tertiary butyl-4-methyl phenol (DBPC or BHC); 2,4-dimethyl-6-tertiary butyl phenol; 75% 2,6-di-tertiary butyl phenol and 25% mixed tertiary and tritertiary-butyl phenols; 75% di- and tri-isopropyl phenyl and 25% di- and tri-tertiary butyl phenols; 72% 2,4-di-methyl-6-tertiary butyl phenols and 28% monomethyl and di-methyl-tertiary butyl phenols; N,N'-di-isopropyl para-phenylene diamine; N,N'-di-secondary butyl-para-phenylene diamine.

A strong basic additive based on an organic amine can prevent such a reaction and control the stability of fuels. N,N'-Di-methylcyclohexyl amine in the dosage of 25 to 200 ppm can react with the weakly acidic compounds present in the fuel to form a fuel soluble product.

N,N- di-methylcyclohexyl amine

It has been reported that tertiary amines are more effective in fuels than primary or secondary amines [150]. Straight-chain primary amines are not regarded as a good fuel stabilizer since these are not good peroxide-free radicals' quenchers as compared to the secondary or tertiary amines. However, special primary amines having branched alkyl chains with the primary amino group attached to a tertiary carbon have been reported [151] to deliver excellent oil solubility, thermal/oxidative stability, and strong basicity and to boost fuel stability. The stabilizing properties of tertiary alkyl primary amines are the result of radical scavenging, hydroperoxide decomposition, acid scavenging and their ability to disperse gums and particulates. In the distillate fuels, these amines also have the capability of complexing with trace metal ions and to act as a metal deactivator. Amines therefore have multifunctional applications [152].

Fuels containing lower sulfur and a lower aromatic content generate higher levels of peroxides on aging [153]. In such fuels, especially gasoline, the phenylene diamine type of antioxidants are more effective than hindered phenols due to their superior peroxide neutralizing capabilities. These amines are not very effective in diesel fuels because they lead to the formation of higher levels of sediments [154].

The responses of the various stabilizers depend on their concentration, the composition of the fuel, and storage conditions. It has been found that sediments obtained by accelerated aging and ambient storage have different properties, and this finding explains why there is no correlation with the total gum of the fuels under different test conditions [155].

Color Stabilizer   Diesel fuels get darkened in storage due to several changes taking place in the fuel. Minor components of the fuel such as pyrroles, phenols, and thiophenols oxidize during storage to form quinoid structures, which condense with one another or with other active hydrogen compounds to produce highly colored bodies [156,157]. These colored substances may also increase in molecular weight until they separate out as insoluble sludge. Antioxidants are effective in inhibiting gum formation; however, these are less effective in preserving the color of diesel fuel. Nitrogen compounds such as tertiary amines, imidazolines, and tertiary alkyl primary amines having the structural unit ($C–C–NH_2$) have been shown to be effective in preserving color during storage. In US Pat. No. 3, 149, 933, Sept. 22, 1964, alkyl amino phenols were used as stabilizers for liquid fuels.

5.6   METAL DEACTIVATORS/PASSIVATORS

Trace amounts of metals like copper or their soluble compounds accelerate the oxidation of fuels by catalyzing the reaction, thereby forming gums and deposits at a faster rate. Copper and its alloys are extensively used in the chemical processing plants and also in the distribution and automotive fuel systems. Refinery streams contain several acidic components, like phenols, mercaptans, and naphthenic acids, that react with copper to form soluble copper compounds. Copper ions exist in both monovalent (cuprous, Cu^{+1}) and divalent (cupric, Cu^{+2}) forms. The divalent copper ion is more common and is not a pro-oxidant, but monovalent copper ion is highly reactive and a pro-oxidant [158].

Whenever gasolines are treated with phenolic antioxidants, the adverse gum formation effect has been noticed. This is because the phenolic compounds reduces the divalent copper into a monovalent copper, which catalyses the oxidation process. When the antioxidant is depleted in the reduction process, the monovalent copper accelerates the formation of gums. Thus inhibited gasolines show lower induction time as compared to the uninhibited product. This problem has been resolved by using a metal deactivator compound, to passivate the copper metal. Generally, *N,N'''*-di-salicylidine-1,2-propane diamine is used as a metal deactivator. This compound chelates with the dissolved copper

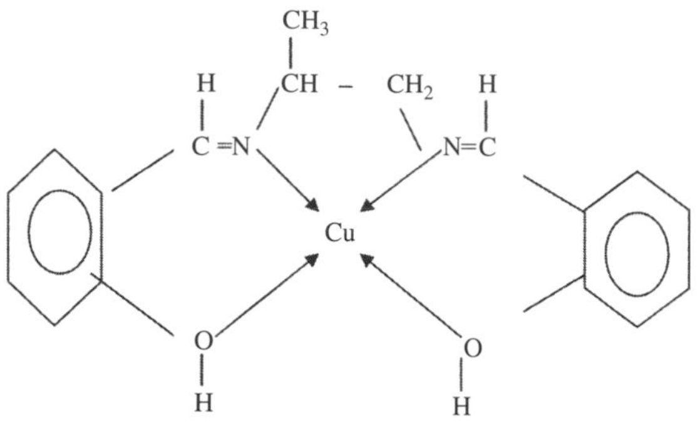

FIGURE 5.7   Copper adduct with *N,N'*-di-salicylidene-1,2-propane diamine

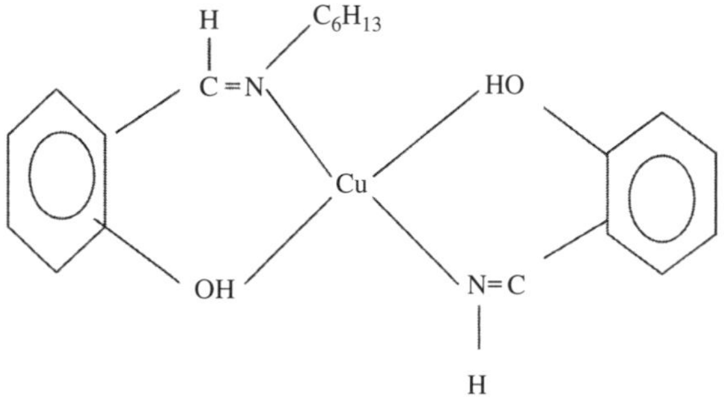

FIGURE 5.8   Copper adduct with N-salicylidene-hexane amine

compounds that are not pro-oxidant. The metal deactivator compound also migrates to various other metallic parts and protects the gasoline from further dissolution of copper [158]. In fuels, the salicylidene type of MDA is effective only in 1 to 5 ppm.

Diesel fuel contains higher levels of metal like Fe, Cu, Ni, Co, and Mn as compared to gasoline. These also increase when the crude oil contains higher amounts of metals. In diesel gas oils, tertiary alkyl primary amines are also claimed [159] to reduce sediment formations on metal surfaces. These tertiary-alkyl-amines, like the t–$C_{12\text{-}18}$ amine, have no α-hydrogen attached to nitrogen and therefore have better oxidation stability. These molecules inhibit the reaction responsible for sludge formation and also disperse gums and sediments.

Aviation turbine fuels also use the following metal deactivator in combination with an antioxidant:

N,N'-Di-salicylidene-1,2-propanediamine (Figure 5.7)
N,N'-Di-salicylidene-1,2-cyclohexanediamine (Figure 5.8)

A recent review [158] describes the development and applications of metal deactivators in the petroleum industry.

5.7   CORROSION INHIBITORS

Trace contaminations of commercial gasolines with water cannot be avoided. Sometimes the moisture may be picked up from the atmosphere. This moisture along with air (oxygen) can attack iron and other metal in storage tanks, pipelines, tankers, and fuel tanks of automobiles, leading to severe corrosion problems. In addition, rust particles can clog fuel filters and carburetor/injectors orifices and adversely affect engine performance.

The extent of rusting depends on the temperature, humidity, exposure to the environment, and their duration. Corrosion is the outcome of the reaction of acidic compounds on metals. Organic acids are formed by the oxidation of fuels and lubricants. Sulfur in engine oils, and in the fuels on combustion, produces sulfurous/sulfuric anhydrides and sulfuric acid. Oxides of nitrogen are also formed at elevated temperatures. These factors lead to rusting and corrosion of metal surfaces [160].

5.7.1   Mechanism of Rusting/Corrosion

The development of rust on a metal surface is an electrochemical phenomenon that occurs due to the reaction of water and oxygen with the metal [161]. This happens in the following three steps (Figure 5.9):

1. Oxidation of metal
2. Reduction reaction
3. Charge transfer from anode to cathode

If any of these three steps are prevented, corrosion will not take place. Metals are thermodynamically unstable in water, and different reactions take place under acidic and alkaline conditions [161]. These can be represented by the following two reactions:

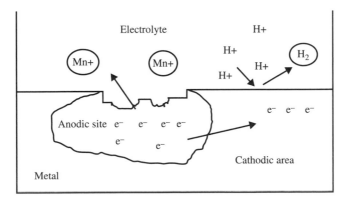

FIGURE 5.9   Electrochemical corrosion of metals

- Reaction at pH 7 or higher:

$$M + nH_2O \rightarrow M^{n+} + nOH + n/2H_2$$

- Reaction at pH 7 or lower (i.e., under acidic conditions) as takes place with hydrogen ion:

$$M + nH^+ \rightarrow M^{n+} + n/2H_2$$

However, most solutions have dissolved oxygen in them, and this oxygen also reacts with metals under acidic and alkaline or neutral conditions. The following reactions take place with dissolved oxygen:

- Reaction at pH 7 or higher:

$$M + n/2H_2O + n/4O_2 \rightarrow M^{n+} + nOH^-$$

- Reaction at pH 7 or lower, takes place with hydrogen ion and oxygen:

$$M + nH^+ + n/4O_2 \rightarrow M^{n+} + n/2H_2O$$

In the solutions, the metal (M) passes from a metallic state to an ionic state of valance n, and hydrogen is evolved. In both processes a change in charge is involved, that is, M to M^{n+} and nH^+ to $n/2H_2$. In this change a transfer of electron takes place from M to H^+. In all these reactions with oxygen and water, electrons are consumed; therefore the rate of corrosion will be higher than in the previous reactions [162].

In the case of iron, dissolved oxygen is more important than the hydrogen ion reaction when the pH is greater than 4. With iron, and aerated water, a black layer of Fe_3O_4 and $Fe(OH)_2$ is formed. In the presence of dissolved oxygen (mineral oils can dissolve 8–9 vol% of air), a layer of FeOOH and red Fe_2O_3 is formed.

According to the definition given by National Association of Corrosion Engineers (NACE), a corrosion inhibitor is a substance that retards corrosion when added to an environment in a small concentration. These inhibitors control corrosion by adsorption at the metal surface, or they induce the formation of a thick corrosion product, or they change the characteristics of the environment resulting in less aggressive effect. Corrosion inhibitors have to form a type of bond with the metal surface within a range of pH and create a layer so that the corroding ions do not penetrate it [161].

Rust inhibitors [162] therefore prevent the adsorption of oxygen or water molecules on the metal surface. In a nonaqueous system where water and oxygen are only in trace amounts, the chances of rusting are minimized, and it is easier to control rusting in such a system by the use of an organic rust inhibitor with a polar group. In contrast, rust prevention is difficult in aqueous systems where large amounts of water and oxygen are present and more aggressive chemicals are required to protect the metal surfaces from rusting. In such a system a combination of inorganic compounds that passivate the metal surface and organic compounds forming an adsorption layer are employed. However, the solubility of organic compounds in water is limited, and the adsorbed layer can be displaced by more polar molecules of water present in the system.

Most of the inorganic corrosion inhibitors [163] are a conjugated base of weak inorganic acid, and therefore act as a Lewis base. The metal, in contrast, acts as a Lewis acid. The attraction between the two forms an insoluble layer on the metal surface. The effectiveness and stability of this layer depends on the chemical nature of the inorganic compounds and the interaction of the base fluid with this layer. The adsorption of an organic compound on the metal surface is controlled by the activation energy provided by the electrochemical corrosion process.

5.7.2 Anticorrosion Compounds

The most successful materials that have been used as additives against corrosion are high molecular weight carboxylic, sulphonic, or phosphoric acids; salts of these acids; and products of neutralization of the acids with organic bases such as amines [164,165,166]. When dissolved in the hydrocarbon, these materials have the property of forming an adsorbed film on the metal in contact with the liquid. This film is hydrophobic and displaces the water from the metal surface. The polar end of the inhibitor molecule becomes attached to the metal surface, while the other end of the molecule is dissolved in the fuel. If sufficient inhibitor is present, a compact barrier layer of its molecules on the metal prevents a penetration by water. Corrosion is then prevented by this impenetrable monomolecular film [162].

Partially esterified alkylsuccinic acid or succinicanhydride amides are also effective rust (corrosion) inhibitors. Although amine derivatives of phosphates and sulphonates have been predominantly used, new compounds/formulations have also appeared. Some examples are long chain of unsaturated hydrocarbon containing 15 to 20 carbon atoms condensed with naphthalene before sulphonation, a substituted succinic acid and an aliphatic tertiary amine, and the magnesium salt of a substituted succinic acid. Some new anticorrosion agents incorporate amines into the mixture.

To prevent corrosion, polar compounds like the above-mentioned carboxylic acid (dodecenyl-succinic acid), amides, and their salts are used in small quantities, ranging from 5 to 20 ppm. These products are polar in nature and get attached to the metal surface, forming a protective layer that prevents the attack of water on metals (Figure 5.10). Half esters of dodecenyl-succinic acid are useful rust inhibitors too.

A detergent and anticorrosive additive, containing amide or imide groups has been [167] obtained by mixing 60–90 wt% of a polyalkenyl, a carboxylic acid, a diacid, and an anhydride, with an average molecular mass 200–3000, 0.1–10% of the carboxylic acid compound, a monoacid and an anhydride, containing 1–6 carbon atoms per chain, and 10–30% of a primary polyamine. Early studies have showed these synergic components to be effective corrosion inhibitors and lubricity improvers [168].

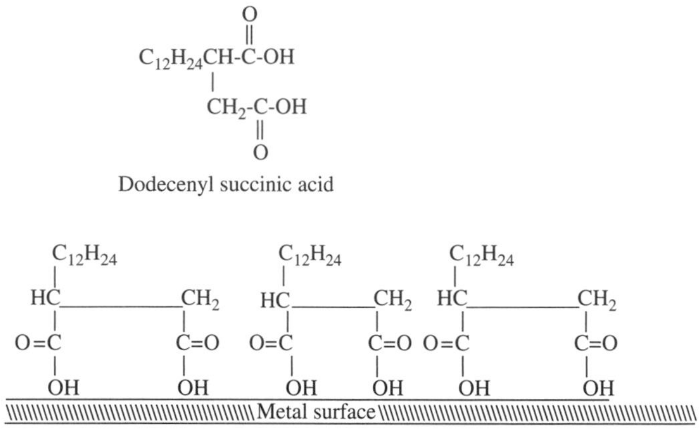

$$
\begin{array}{c}
O \\
\parallel \\
C_{12}H_{24}CH\text{-}C\text{-}OH \\
\mid \\
CH_2\text{-}C\text{-}OH \\
\parallel \\
O
\end{array}
$$

Dodecenyl succinic acid

$$
\begin{array}{ccc}
C_{12}H_{24} & C_{12}H_{24} & C_{12}H_{24} \\
\mid & \mid & \mid \\
HC\underline{\quad\quad}CH_2 & HC\underline{\quad\quad}CH_2 & HC\underline{\quad\quad}CH_2 \\
\mid \quad\quad \mid & \mid \quad\quad \mid & \mid \quad\quad \mid \\
O{=}C \quad C{=}O & O{=}C \quad C{=}O & O{=}C \quad C{=}O \\
\mid \quad\quad \mid & \mid \quad\quad \mid & \mid \quad\quad \mid \\
OH \quad\quad OH & OH \quad\quad OH & OH \quad\quad OH
\end{array}
$$
\\Metal surface\\\

FIGURE 5.10   Metal surface protected by a corrosion inhibitor (dodecenyl-succinic-acid)

5.8 ANTISTATIC AGENTS

In conventional diesel fuel, the high concentrations of sulfur-containing molecules (>500 mg/kg) have been sufficient to give significant intrinsic conductivity, so that static electric charging has not been a problem. However, as sulfur levels in diesel are reduced, the risk of static charging during pumping has increased significantly. Ultra-low sulfur diesel (50 mg/kg sulfur, and lower) with a low aromatic content has poor lubricity and electrical conductivity. Highly refined petroleum products, like jet fuels and ultra-low sulfur diesel fuels, have very low electrical conductivity since the conducing ionic species are removed during the intensive refining processes [169]. GTL (gas to liquid) fuels do not contain any conducing ionic species [170].

Under turbulent flow the generated static electricity cannot be dissipated by fuels having low electrical conductivity. When a fuel of low electrical conductivity flows through a pipe, a separation of electrical charges can occur and the static charge is built up in the liquid. This charge separation can lead to high voltages with the possibility of spark discharges that can ignite the fuel vapor. The charge generation is increased by high pumping rates or by the fuel vapor and also by contact with the equipment having a high surface area, such as water separators and fuel filters. The filter separators and splash loading can also increase the charge considerably.

Several reports of road tanker explosions in Europe followed the introduction of ultra-low sulfur diesel (ULSD), despite the use of grounding leads. Truck explosions in Europe were also reported in several publications around the same time. The cause was traced to a static charge induced spark ignition of the fuel vapor during the fuel transfer operations. It is therefore necessary to treat such diesel fuel with an additive to restore electrical conductivity. This is especially important for aviation fuel.

If fuel conductivity is increased by an additive, the charge buildup in the liquid can go back to the container wall. These additives exist in ionic form in the fuel solution. Following antistatic additives have been used [171]:

1. Chromium salt of an alkylsalicylic acid (e.g., chromium di-alkylsalicylate)
2. Amine salt of carboxylic acid and polycarboxylic acids
3. Calcium di(2-ethyl hexyl/sulpho succinate) (e.g., manganese dodecenylsuccinate)
4. Quaternary ammonium compounds
5. An organic polymer that functions as stabilizing agent

The dosages were 1–10 mg/kg. In jet fuels the antistatic additive must meet high safety needs. The minimum conductivity requirement for jet fuel is generally specified as 50 pS/m.

The antistatic properties in liquid fuels containing less than 500 mg/kg sulfur can be improved by incorporating 0.001–1 mg/kg of hydrocarbyl monoamine or N-hydrocarbyl substituted poly(alkyleneamine), and 20–500 mg/kg of fatty acid containing 8–24 carbon atoms or its ester with an alcohol or polyols of about 8 carbon atoms [172].

5.9 LUBRICITY IMPROVERS

Lubricity improvers and friction modifiers both work through the action of film formation on the metal surfaces. However, lubricity improvers are meant for protecting the fuel pump from wear through the same mechanism of surface adsorption, since the friction modifiers reduce the friction between moving engine parts. Lubricity improvers are generally surface active compounds and get concentrated at the surfaces of separation, forming extremely thin adsorption layers. These thin layers are capable of producing marked changes in molecular nature and surface characteristics. This leads to a change in the kinetics of the processes involved in the transfer of substances across surfaces of separation and in the second place to the changes in the condition of molecular interaction between the two contacting surfaces. These include conditions of cohesion, adhesion, friction, and molecular interaction. Thus the addition of small quantity of a surface active ingredient can generate many changes and control many technological processes [173].

The activities of most fuel additives such as detergents, dispersants, demulsifiers/dehazer, lubricity improvers, and friction modifiers depend on this surface phenomenon. All surface compounds have a polar and a nonpolar group, and whichever predominates makes the compound either oil soluble or water soluble. All lubricity additives have a small polar group that gets attached to the metal surface and a long hydrocarbon nonpolar group. While these lubricity additives get adsorbed at the metal surfaces, they also act as antiwear additives for the fuel pumps and injectors at low loads.

The diesel fuel pump is an important part of the engine. It is lubricated by the fuel and no external lubricant is used. It is therefore important that the fuel have adequate viscosity and lubricity properties to protect the pump from wear. The lubricating

TABLE 5.7 Influence of sulfur content on the lubricity of diesel fuel

Diesel Fuel Sulfur Content (wt%)	Wear Scar Diameter (mm)	
	Load 500 g	Load 1000 g
0.0	1.00	1.10
0.05	0.98	1.12
0.1	0.88	0.1.05
0.2	0.66	0.78
0.5	0.68	0.77
1.0	0.65	0.72

properties in the diesel fuel are due to the heavier hydrocarbons present, some of which are sulfur and polar compounds.

Sulfur has been progressively reduced in the diesel due to environmental considerations. The removal of sulfur compounds by hydrotreatment of the distillate fuels, in combination with increasing injection pressures (ca. 2000–2500 bar) in fuel systems in modern engines, has caused concern over lack of fuel lubricity. During hydroreatment, aromatics and olefins are saturated, and sulfur is converted to hydrogen sulfide. Oxygen- and nitrogen-containing compounds also loose their polarity. While these changes are good for emission reduction, they result in poor lubricity in low-sulfur (below 500 mg/kg, especially below 50 and 10 mg/kg) fuels.

Hydrodesulphurization decreases the lubricity reserve of gas oil by 80–200 µm (HFRR) and the load-carrying capacity by 1500 N in the HiTOM (high-temperature oscillating machine). A high-frequency reciprocating rig (HFRR) procedure has been developed to assess the lubricity of diesel fuel and also of aviation turbine fuel. The HFRR specification limits wear scar diameter of max. 400–460 µm.

Model sulfur compounds do not restore the lubricity of a hydrotreated diesel fuel. However, dibenzothiophene can increase the load-carrying capacity. Therefore the removal of sulphurized compounds is not the only reason for the loss of lubrication of ultra-low-sulfur diesel fuels. Some fuels containing higher levels of sulfur have also shown poor lubricity. In a refinery, the diesel fuel becomes a blend of several streams and components, and these have different levels of lubricity.

Two streams of total cycle oil (TCO) from two Indian refineries have been separated into saturates, aromatics, and polars by a combination of analytical techniques. The measurement of lubricity of the separated streams by HFRR showed that the saturates have poor lubricity while the aromatics and polars have adequate lubricity.

The saturated fraction of TCO contains mainly paraffins, isoparaffins, and alkyl naphthalenes. The incorporation of the separated polars and aromatics back into the saturates improved the lubricity of the saturates [174]. Diesel fuels containing different sulfur levels have been evaluated [175] using a ball on cylinder lubricity tester (BOCLE) at two different loads and 240 rpm, 250 °C, for 30 minutes. The data are provided in Table 5.7.

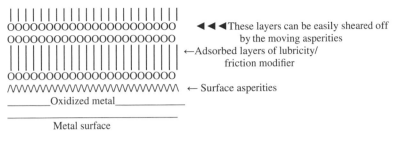

(O: polar head to metal surface)

FIGURE 5.11   Adsorption of lubricity additive/friction modifier

These data clearly show that as sulfur levels are reduced, the wear scar diameter increases. Inadequate lubricity can lead to problems of excessive wear of fuel-lubricated components such as fuel pumps, fuel injectors, and increased emissions.

These lubricity improvers, oiliness additives or film strength improvers, are fats, fatty esters, amides, fatty oils, and esters. These compounds have slight polarity and get preferentially adsorbed at the metal surface by physical forces and reduce friction. However, as the load between the surfaces increases, the adsorbed additive layer is squeezed out, returning to metal to metal contact and increased friction. Clearly, these additives need to function where the temperatures and loads are high enough, to prevent the removal of adsorbed film formed under low to moderate stress conditions (Figure 5.11). Under pure hydrodynamic lubrication, the viscosity of the fuel also plays a big role in imparting lubricity.

Low-sulfur- and nitrogen-containing motor fuels have low lubricity due to the lack of lubricity components. Adequate amounts of lubricity additive in about 25 to 500 ppm, are necessary, and these should be long-chain ashless polar compounds to form an adequate protective layer on the metal surface. Biodiesel has been reported to have good lubricity because it is an ester. The addition of 5% biodiesel to a low-lubricity diesel fuel has shown considerable improvement in lubricity [176] as assessed by the HFRR (WSD 200 μm).

Fuel compositions have been described to meet these requirements. Such additives contain nitrogen, aromatic rings and have relatively high molecular weights [177]. There are very large numbers of lubricity additives described in the patent literature. Some important patents disclosures are reviewed in the following pages.

US Pat. No. 3,273,981 describes a lubricity additive that was prepared by mixing a polybasic acid, or a polybasic acid ester, with C1–C5 monohydric alcohols with a partial ester of a polyhydric alcohol and a fatty acid, such as glycerol monooleate, sorbitan monooleate, or pentaerythritol monooleate. This mixture finds application in jet fuels.

US Pat. Nos. 2,252,889; 4,185,594; 4,208,190; 4,204,481; and 4,428,182 describes antiwear additives for diesel fuels consisting of fatty acid esters, unsaturated dimerized fatty acids, primary aliphatic amines, fatty acid amides of diethanolamine, and long-chain aliphatic monocarboxylic acids.

US Pat. No. 3,287,273 describes lubricity additives that are reaction products of a dicarboxylic acid and an oil-insoluble glycol (alkane diols or oxa-alkane diols). The acid is a dimer of unsaturated fatty acids such as linoleic or oleic acid. Examples of lubricity additives are certain carboxylic acids or fatty acids, alkenylsuccinic esters, bis(hydroxyalkyl) fatty amines, hydroxyacetoamides, and castor oil.

EU Patent Application 798,364, Oct. 1, 1997, describes an additive comprising a salt of a carboxylic acid and an aliphatic amine, or an amide obtained by dehydration-condensation. The additive reduces the amount of deposits and improves the lubricity of the fuel. It is also said to impart antiwear property to diesel fuel of low sulfur content.

Carboxylic acid or its derivative substituted by at least one hydroxyl group, an ester, and an alkanolamine has been used as fuel lubricity additive [178].

Fuel additives acting both as detergent and as lubricity additive are rare. A fuel composition comprising a middle distillate fuel having a sulfur content of 0.2% by weight or less, oleyl diethanolamide and a mixture of cold flow improvers, and ashless dispersants provides adequate lubricity to the fuel [179].

The use of a poly(hydroxy-carboxylic acid) amide or ester derivative of an amine, an aminoalcohol or a polyol linked to the poly(hydroxycarboxylic acid) via an amide or ester linkage, as a fuel additive acting as a detergent as well as a lubricity additive in fuel compositions has been described [180].

A polyol ester fuel additive exhibits improved lubricity and friction and wear performance. The ester [181] has between about 1% and about 35% unconverted hydroxyl groups and is characterized as having a hydroxyl number from about 5 to about 180. The ester may also be synthesized from a polyol and a polybasic acid. The ester comprises the reaction product of an alcohol having the general formula $R(OH)_n$, where R is an aliphatic group, cycloaliphatic group, or a combination thereof having from about 2 to 20 carbon atoms, and n is at least 2 where the aliphatic group is branched or linear, and at least one branched or linear acid [181]. The suggested additive concentrations are 10 to 10,000 ppm.

Blends of diethanolamine derivatives and biodiesel have been used for improving lubricity in low-sulfur fuels [182].

A lubricity additive based on the reaction products of an alkylated polyamine and urea or isocyanate, or the salt adducts of these reaction products, has been described [183] as suitable for diesel fuels.

US Pat. No. 4,609,376 (Craig et al.) describes an antiwear and lubricity improver additive based on an ester of a monocarboxylic or polycarboxylic acid and a polyhydric alcohol, where the ester contains at least two free hydroxyl groups.

EU Publication No. 0,798,364 describes diesel fuel additives comprising a salt of a carboxylic acid and an aliphatic amine, or an amide obtained by dehydration-condensation between a carboxylic acid and an aliphatic amine.

US Pat. No. 5,833,722, filed Aug. 9, 1996 (Davies et al.) describes a nitrogen-containing compound such as the salt of an amine and carboxylic acid, and an ester of a polyhydric alcohol and a carboxylic acid to enhance lubricity of the fuel.

US Pat. No. 6,328,771 describes fuel lubricity enhancing salt compositions prepared by the reaction of carboxylic acids with a heterocyclic aromatic amine.

It has long been observed that winter grades of diesel fuel, which are of lower viscosity and contain lower concentrations of waxy fractions than summer grade fuels, have lower load-bearing capacity, and consequently have lower lubricity.

The lubricity of fuels, oxygenated fuels, and their mixtures, particularly diesel or aviation fuels having reduced sulfur and/or aromatic content, is improved by the addition of a product that can be obtained by reacting pentaerythritol/thiodiethylene glycol, 1,4-butanediol/1,4-propanediol/diethylene glycol/triethylene glycol/diethanolamine/ or glycerol with sunflower oil or coconut fat and with methyl 3-(3',5'-di-tert-butyl-4'-hydroxyphenyl) propionate. These products also improve corrosion inhibition [184].

A fuel composition [185] having a sulfur content of not more than 50 ppm by weight and comprising at least 50 ppm of compounds like 1-amino naphthalene, 1,8-diamino naphthalene, or 5-aminoindole, 2-(2-aminophenyl) indole, and 8-aminoquinoline has been described. These compounds are capable of improving the antiwear and lubricity properties of a low-sulfur fuel.

Alkyl aromatics (alkyl group have 14 to 36 carbon atoms) with a hydroxyl group or an amide group also show lubricity properties [186].

About 200 mg/kg of carboxylic acid-, ester-, and amide-based additives can increase the lubricity in different fuels to meet the lubricity requirement of modern diesel fuels.

Certain amine salts of substituted aromatic carboxylic acids show lubricity performance [187].

Fuel oil compositions containing specific mixtures of esters of unsaturated monocarboxylic acids show improved lubricity properties [188].

An additive composition containing an ashless dispersant (acylated nitrogen compound) and a carboxylic acid (5–20 carbon atoms), or an ester of the carboxylic acid and an alcohol, provides an improvement in the lubricity of fuel oils [189] and exhibits improved solubility in the fuel oil [190].

5.10 FRICTION MODIFIERS

In an engine, about 18% of the fuel's heat value, which is the amount of heat released in the combustion of the fuel, is lost through internal friction in engine components, through the bearings, valve train, pistons, rings, water and oil pumps, and the like.

Only about 25% of the fuel's heat value is converted to useful work at the crankshaft. Friction occurring at the piston rings and parts of the valve train accounts for over 50% of the heat value loss. A lubricity-improving fuel additive, such as a friction modifier, capable of reducing friction at these engine components by one-third preserves an additional 3% of the fuel's heat value for useful work at the crankshaft. Therefore there has been a continual search for friction modifiers that improve the delivery of the friction modifier to strategic areas of the engine and hence improve the fuel economy of engines.

The major fuel-related deposit problem areas for PFI and DIG engines are injectors, intake valves, and the combustion chamber. Additionally, engine friction between the piston and cylinder, the valve train, and the fuel pump increase fuel consumption. In DIG engine technology, there is a friction-related durability issue with the high-pressure pump (up to 1500 psi—about 100-bar pumping capacity), which break down due to the inherently low lubricity of gasolines. So there is a desire in the petroleum industry to produce a fuel suitable for use in both PFI and DIG engines that can address the engine deposit and frictional requirements.

Refining of by deep hydro treatment removes the natural lubricity components of the fuel, such as certain aromatics, carboxylic acids, sulfur and nitrogen compounds, and esters. Unfortunately, commercial gasoline detergents and dispersants show very little friction-reducing capability unless very high concentrations are added to the fuel. These high detergent concentrations often reach levels where no-harm effects such as CCD become unacceptable. It has been suggested that separate friction modifiers can be added to gasoline to increase fuel economy by reducing engine friction. Fuel friction modifiers would also serve to protect high-pressure fuel pumps and injectors, such as those found in DIG engines, from wear caused by fuels of low lubricity. Worldwide regulations calling for a steep reduction in fuel sulfur levels increases this wear problem even further.

In selecting suitable components for a combined detergent-friction modifier additive package, it is important to ensure a balance of detergent and friction modification properties. Ideally, the friction modifier should not adversely affect the deposit control function of the detergent. In addition, the additive package should not adversely affect engine performance. For example, the additive package should not promote valve sticking or cause other performance-related problems. To be suitable for commercial use, the friction modifier additive also must pass all no-harm testing required for gasoline performance additives. This is often the biggest hurdle for commercial acceptance. The no-harm testing involves following assessment:

- Compatibility with gasoline and other additives present at a range of temperatures
- No increase in IVD and CCD
- No valve sticking at low temperatures
- No corrosion in the fuel system, cylinders, and crankcase

A friction modifier may be added to the gasoline as the lone additive or in combination with a detergent dispersant package that is fully formulated for fuel compatibility. In addition, a need may exist for a detergent-friction modifier additive concentrate for gasoline that provides fuel economy enhancement, deposit control, and friction reduction together. This modifier should be stable over the temperature range at which the concentrate may be stored, and not adversely affect the performance and properties of the finished gasoline or engine, and in particular, not lead to increased intake valve deposit problems. Compounds such as n-butylamine oleate, in particular when used in combination with a detergent, lead to increases in value deposits. The use of structural branching in the polyalkylene backbone of the fatty acid moiety of a branched saturated carboxylic acid salt of an alkylated amine has been found to increase the miscibility of the additive in fuel. However, solubilizing agents, like hydrocarbon solvents consisting of alcohols or organic acids, may also be included to improve miscibility.

The most common ashless friction modifiers are the fatty acid ester and the fatty acid amide. Among the esters, glycerol monooleate (GMO) has become quite popular, which is a partial ester of glycerol and oleic acid. The most effective molecule is the one where the terminal hydroxyl of the glycerol is esterified with oleic acid. When the β-OH is esterified, the product is not as effective as the friction modifier. Specific manufacturing techniques are applied to maximize the yield of α-ester. Sorbitan mono oleate (SMO) can also be prepared in the similar manner.

$$CH_2O-\overset{\displaystyle O}{\overset{\displaystyle \|}{C}}-R$$
$$|$$
$$CHOH$$
$$|$$
$$CH_2OH$$

Glycerol mono oleate

$$H_2N-\overset{\displaystyle O}{\overset{\displaystyle \|}{C}}-R$$

Oleyl amide

where R-COOH is oleic acid.

The oleic amide is prepared from the reaction of C_{18} straight-chain fatty acid, oleic acid, and ammonia. Mono-, di-, and tri-ethanol amines can also be reacted with oleic acid to yield corresponding ethanolamine oleates, which have good surface active properties and are useful in many lubricant formulations. All these compounds function by adsorbing at the metal surface and the adsorbed layers can shear off easily with the movement of asperities. These adsorbed layers will then quickly revert back to their original position, after the removal of stress.

A friction modifier is prepared [191] by combining a saturated carboxylic acid and an alkylated amine, such as n-butylamine isostearate for internal combustion engines.

Moreover, the use of the friction modifier in combination with a detergent package permits increased fuel efficiency without increasing the incidence of intake value deposits in combustion engines. The compositions [196] based on a mixture of a fatty acid; a fatty acid amide; a fatty acid ester; an amide, imide or ester derived from a hydrocarbyl substituted succinic acid or anhydrides; and an alkoxylated amine—are useful in enhancing the lubricity characteristics of fuels such as diesel fuel and gasoline. The compositions are also useful in reducing intake valve deposits and improving fuel economy.

Conventional port-fuel injection (PFI) engines form a homogeneous pre-mixture of gasoline and air by injecting gasoline into the intake port, whereas direct injection gasoline (DIG) engines inject gasoline directly into the combustion chamber like a diesel engine so that it becomes possible to form a stratified fuel mixture which contains greater than the stoichiometric amount of fuel in the neighborhood of the spark plug but highly lean in the entire combustion chamber. Due to the formation of such a stratified fuel mixture, combustion with the overall highly lean mixture can be achieved, leading to an improvement in fuel consumption over that of PFI engines, and approaching that of diesel engines.

It has been found that low sulfur liquid fuels can be incorporated with a reaction product of boric acid [192,193] having a particle size of 0.5 to 1 μm. The resulting products have improved frictional properties. WO 01/72930 A2 describes a mechanistic proposal for delivery of a fuel-born friction modifier to the upper cylinder wall and into the oil sump, resulting in upper cylinder, rings and valves lubrication. The friction modifier is packaged with fuel detergent dispersants such as polyetheramines (PEAs), polyisobutene amines (PIBAs), Mannich bases, and succinimides.

US Pat. Nos. 4,185,594, 4,208,190; 4,204,481; 4,428,182 describe some fuel friction modifiers for diesel fuels. These patents include fatty acid esters, unsaturated dimerized fatty acids, primary aliphatic amines, fatty acid amides of diethanolamine, and long-chain aliphatic monocarboxylic acids.

US Pat. Nos. 4,427,562; 4,729,769 describe a lubricant and fuel friction modifier prepared by reacting primary alkoxyalkylamines with carboxylic acids or by aminolysis of a formate ester.

EU Publication No. EP 947576, Oct. 6, 1999 (Fuentes-Afflick et al.) describes fuel compositions that include aliphatic hydrocarbyl substituted amines and/or polyetheramines and esters of carboxylic acids and polyhydric alcohols to improve fuel economy.

US Pat. No. 5,858,029 describes friction-reducing additives for fuels and lubricants involving the reaction products of primary etheramines with hydrocarboxylic acids to give hydroxyamides that exhibit friction reduction in fuels and lubricants.

US Pat. No. 4,617,026 describes the use of a monocarboxylic acid of an ester of a trihydric alcohol containing glycerol monooleate as fuels and lubricant friction modifier.

US Pat. Nos. 4,280,916; 4,512,903; 4,789,493; 4,808,196; and 4,867,752 describes fatty acid amides as friction modifiers.

EU Publication No. 0,869,163, A1 describes a method for reducing engine friction by use of ethoxylated amines.

A fuel additive consisting of a mixture of lithium didodecylbenzenesulfonate, t-butyl perbenzoate, and MEK peroxide has been described [194]. The additive is provided in a solvent such as diphenyl (also known as 1,1-biphenyl, bibenzene, phenylbenzene, and lemonene), which has a pleasant odor, to make it a more aesthetically pleasing product for the consumer. The additive can also be formulated into the fuel without the use of a solvent. This additive system has been demonstrated to improve fuel efficiency and also the performance of engines.

A combination of an aliphatic hydrocarbyl-substituted amine, or a poly(oxyalkylene) amine, and an ester of a carboxylic acid and a polyhydric alcohol has significantly reduced fuel consumption [195] in an internal combustion engine.

Another effective additive concentrate [197] for use in fuels, especially in gasoline internal combustion engines, consists of 0.2–10 wt% n-butylamine oleate as the ashless friction modifier, tall oil fatty acid, and a deposit inhibitor, along with a carrier fluid. The total treat level of this additive is 2000 ppm in the fuel. In this concentrate of 200 to 800 ppm of a deposit inhibitor (polyisobutylene amine, polyisobutylene succinimide, or polyether amine) is present.

5.11   DEHAZER AND DEMULSIFIERS

As we mentioned earlier in this chapter, fuels are often contaminated with trace amounts of water during storage or during transportation. The dispersion of water in the fuel gives a hazy appearance to the fuel. The dispersed or dissolved water can cause corrosion or carburetor icing in gasoline engines (these engines are now obsolete). The hazy appearance of fuel can be cleared up by using surface active compound to break the emulsion or dispersion. Such surface active compounds have both hydrophilic and hydrophobic properties in the same molecule to act at the interface of water and the hydrocarbons. Demulsification is the reverse of emulsification; the forces that are responsible for stabilizing the emulsion need to be overcome to bring about demulsification. All emulsions are thermodynamically unstable, although they may remain in emulsion form for a very long time because of the barrier preventing the aggregation of the dispersed droplets. Emulsion systems therefore exist in metastable form. Coagulation is the formation of larger aggregates through the process of particle cohesion leading to phase separation and emulsion breakage. The droplets merge to form bigger unstable drops, called coalescence, which determines the stability or instability of emulsions. The initiation of a demulsification process begins with Brownian diffusion and differential settling [198] (due to density differences), followed by flocculation, coalescence, and phase separation. The droplet sizes and their distributions, as well as the densities and viscosities of the internal and external phases in the emulsion, are important in deciding an emulsion's stability.

The smaller the dispersed droplet size, higher is the emulsion's stability. The greater the differences in the densities of the internal and external phases, the higher are the chances of quicker phase separation. The higher the viscosity of the external phase, the slower will be the phase separation process. These surfactants act by one of the following mechanisms [199]:

1. The demulsifier molecule adsorbs onto the dispersed water molecule at the oil–water interface and displaces the polar molecule responsible for the emulsification/dispersion of the water molecule. This increases the stearic forces with in the surface film, and the local surface tension increases, leading to a thinning of the film. This action promotes the coalescence of water droplets by molecular attraction, and water separates out from the fuel.
2. The water molecules get chemically attracted to the hydrophilic group of the surfactant and subsequently coalesce into bigger droplets that separate out from the fuel due to gravity differences.

Currently, from the nonionnic surfactants, the ethylene oxide condensates and polyoxy alkylenepolyol are the favored demulsifiers for fuels, since it is possible to design the molecules of appropriate molecular weight and solubility to control the demulsification processes. Ethylene oxide/propylene oxide/butylenes oxide can be reacted with an alkyl phenol, sorbitol, glycerol, or glycol to obtain these products. The ratio of the alkylene oxide in the molecule decides the solubility of product in water or the hydrocarbon. Among the anionic surfactants, the alkyl- and dialkyl-sulphosuccinates are important.

A careful selection of surface active compound is necessary to provide adequate water separation. It is convenient to incorporate such additives (in concentration 5–20 mg/kg) in a multifunctional package containing other performance additives. The effectiveness of demulsifiers in fuels can be evaluated, for example, by the ASTM D 1094 test method.

5.12 COMBUSTION IMPROVERS

The combustion of the fuels can be enhanced by the use of a catalyst. These are generally fuel-soluble organometallic compounds or complexes of iron, such as ferrocene, substituted ferrocenes, iron naphthenate, iron succinate, stoichiometric or over-based iron soaps (carboxylate or sulfonate), iron picrate, or iron carboxylate and iron beta-diketonate complexes. Iron carboxylate, such as iron tris(2-ethylhexanoate) has been preferred as a cost-effective source of the fuel-soluble iron. A wide range of "substituted ferrocenes" are known [200] that can be used as combustion improver catalysts.

Suitable iron complexes are methylcyclopentadienyl-iron, bis-methylcyclopentadienyl-iron, bis-ethylcyclopentadienyl-iron, and bis-1,2-di-methylcyclopentadienyl-iron. Also useful are the iron complexes of aromatic Mannich bases, prepared by reaction of an amine with an aldehyde or ketone followed by nucleophilic attack on an active hydrogen-containing compound such as the product of the reaction of two equivalents of (tetrapropenyl) phenol, two of formaldehyde, and one of ethylenediamine, and hydroxyaromatic oximes such as (polyisobutenyl) salicylaldoxime. Yet another complex is prepared by the reaction of polyisobutenyl phenol, formaldehyde, and hydroxylamine.

The synergistic combination of iron and an alkaline earth metal improves the regeneration of the diesel particulate trap, and there is a lower average backpressure across the trap. Thus the additive reduces overall emission. When a fuel containing a combustion improver is burned, any trapped material is oxidized at a reduced ignition temperature unlike fuel burned without this additive.

Copper compounds have been suggested as combustion improvers for diesel fuels. Diesel particulates consist of inorganic ash (due to engine wear particles and combustion products of lubricant oil additives), sulfuric acid (from sulfur in the fuel and lubricants), and hydrocarbons from incomplete fuel combustion. The hydrocarbons are typically further divided into SOF (solvent organic fraction; i.e., material extractable in dichloromethane) and hydrocarbonaceous soot.

Diesel fuels and diesel engines are especially prone to the emission of high levels of small size soot particulate matter in the exhaust gas when the engine is highly loaded, worn, or badly maintained. Particulate matter is also emitted from diesel engine exhausts when operated at partial load, and these emissions are normally invisible to the naked eye. The emission of black smoke and soot from diesel engines is hazardous to human health. Emissions of particulate matter of less than 10 micrometers of principle dimension ("PM10 matter") are claimed to cause a large number of respiratory diseases. It is suspected that these smaller particles penetrate deep into the lungs and adhere to mucocilliary tissue. While the mucocilliary system is thought to have evolved to cope with airborne dusts, pollens, and the like, it does not cope well with smaller particles, especially those less than 2.5 µm in aerodynamic diameter.

5.12.1 Conventional Approaches

Many modern engine designs use exhaust gas recirculation (EGR). The exhaust gas is recycled in a controlled way to the intake valve of a diesel engine. This does contribute to the reduction of oxides of nitrogen (NO_x), but there are two significant drawbacks associated with the use of EGR. First, particulates production is increased, and, second, soot particles in the exhaust gas get recirculated in the engine. Thus, in addition to any encountered emissions problems, engines running with EGR for prolonged periods of time can become choked with carbon particulates in areas such as the exhaust gas recycle lines, control valves, inlet ports and valves, and the piston top ring lands. Even the piston rings can become choked with carbon in the ring grooves. Carbon and other particles can also get into the engine lubricant, causing its deterioration.

As an alternative to reducing the production of particulates and other pollutants, several postcombustion treatments have been proposed. These include De-NO_x catalysts, hydrocarbon oxidation catalysts, and the use of particulate filters, especially diesel particulate filters (DPF), capable of oxidizing the collected material. The use of a filter is desirable in the light of recent evidence that suggests the total mass of particulates emitted to be less important than the number of ultrafine particles (as stated above, usually regarded as having a diameter of 2.5 µm or less).

A problem associated with the use of particulate filter traps is that of trap blockage, which causes an increase in exhaust backpressure and a loss of engine efficiency and/ or "chimney fires" resulting from sudden and intense burnoff of soot from highly loaded traps. Catalytic devices have been used to aid in trap oxidation. NO_2 is known to be a powerful oxidant. Using a by-pass system, it is possible to produce high concentrations of NO_2 in the exhaust gas when regeneration is required. However, these devices require a low-sulfur fuel (<50 ppm) to avoid increased sulfate emissions when using this method. Also low-speed engine operation can cause carbonaceous deposits to form on the active parts of the diesel engine oxidation catalyst and so inhibit the effectiveness of the catalyst until a sufficiently high gas temperature is available to regenerate the catalyst's active surface.

Diesel filters that feature "washcoats" comprised of metal ions capable of catalyzing soot oxidation are known and described in SAE papers (A. Mayer et al., SAE 960138; R. W. McCabe, and R. M. Sinkevitch, SAE 870009; B. Engler et al., SAE 860007). These do improve regeneration but can block the active sites by a coating of soot.

A number of fuel additives have further been proposed in to solve this problem (Miyamoto et al., SAE 881224; Martin et al., *I. Mech. E.*, November 1990; Lepperhoff et al., SAE 950369; Rao et al., SAE 940458; Ise et al., SAE 860292; Daly et al., SAE 930131). These additives reduce the soot ignition temperature to facilitate trap regeneration (i.e., decrease in backpressure), which occurs at high frequency during normal driving.

Iron-based additives are known for use in the regeneration of particulate filters and are described in WO-A-92/20762. The use of alkali and alkaline earth metal-based additives for the regeneration of diesel particulate filters is described in WO-A-96/34074 and WO-A-96/34075.

It has been found that a mixture of iron-containing and alkaline earth group metal-containing fuel-soluble or fuel-dispersible species are good combustion improvers. The alkaline earth group metal-containing species consists of strontium or calcium and the ratio of iron to alkaline earth group metal is from 10:1 to 5:4 [201]. This mixture acts synergistically to improve the regeneration of diesel particulate filter when added to the fuel prior to combustion. The additive combination enhances the oxidation of the trapped hydrocarbonaceous soot. The concentration of the metal-containing species in the fuel is about 10 to 30 ppm. The strontium compound was prepared by the reaction of $Sr(OH)_2 \cdot 8H_2O$ with poly(butenyl) succinic anhydride prepared, as described in WO-A-96/34075.

Processes for preparing ferrocene di-carboxylic acid di-esters have been described [202]. There are also many patents dealing with process for producing ferrous picrate fuel additives. These include US Pat. Nos. 2,506,539; 3,282,858; 4,073,626; 4,099,930; 4,129,421; 4,265,639; 4,424,063; 5,087,268; 5,359,103; 5,720,783; and 5,925,153.

Ferrous picrate as combustion improver has been prepared [203] by dissolving picric acid in a mixture of an aromatic hydrocarbon and an aliphatic alcohol in the presence of a trace amount of water and a metallic iron to form a ferrous picrate solution. This ferrous picrate solution is then blended with an organic solvent to form the ferrous picrate having less than 0.25% water [204–206].

Lanthanide compounds, particularly organometallic compounds of cerium [207], are known to be useful additives in fuel because they aid combustion. It is believed that these compounds adsorb onto the asphaltenes that are always present in fuel oil. During the combustion process, metal oxides are formed, and because of the catalytic effect of rare earth oxides on the combustion of asphaltenes, they reduce the quantity of solid unburned components released during combustion. Hence organometallic lanthanide additives in fuel have an effect of improving combustion and reducing harmful emissions.

> French Pat. No. 2,172,797 describes the organic acid salts prepared from cerium that are useful as combustion improvers. The use of organic acid salts of rare earth compounds is necessary since these compounds were found to be soluble in fuels.
> US Pat. No. 4,264,335 describes the use of cerium 2-ethylhexanoate for suppressing the octane requirement of a gasoline-fired internal combustion engine. Cerium 2-ethylhexanoate was found to be more soluble in gasoline than cerium octanoate.
> US Pat. No. 5,240,896 describes the use of a ceramic material containing a rare earth oxide. The ceramic material is insoluble in fuel. It is claimed that combustion of the liquid fuel is accelerated upon contact with the solid ceramic.
> EU Pat. No. 0485551 describes a device that conveys dry particles of a rare earth oxide directly to the combustion chamber of an internal combustion engine via the air intake.

The combustion carbonaceous products have reduced acidity if the hydrocarbonaceous fuels and additive compositions consist [208] of a fuel-soluble manganese carbonyl compound and a fuel-soluble alkali or alkaline earth metal-containing neutral or basic detergent salt. These compositions in addition to these components contain fuel-soluble ashless dispersants, a fuel-soluble demulsifying agent, an aliphatic or cycloaliphatic amine, and a metal deactivator. The deposition of sludge on critical engine or burner parts or surfaces is reduced and the fuels have improved stability and emulsibility characteristics. And the fuel compositions can result in decreased fuel consumption in diesel engines.

A review [209] of studies conducted in practical combustion systems such as oil-fired domestic and utility boilers, gas turbines, and diesel engines has demonstrated that metallic fuel additives can be effective in reducing soot emissions. Manganese, iron, and barium are the metals most often reported to be highly effective, although problems with metal oxide deposits on combustor surfaces sometimes prohibit their use. An evaluation of laboratory burner flame experiments reveals three distinct mechanisms by which the various metallic additives function to remove soot. Several mathematical models of soot reduction through additive use are discussed. Iron additives and their combustion products are relatively nontoxic, whereas the popular manganese additives and their oxides are fairly hazardous. Additives are only recommended for short-term use; combustor design modifications is the economically preferred long-term solution.

The composition containing a polyunsaturated aliphatic or alicyclic compound with 3–6 double bonds along with a derivative of di-hydrobenzo-gamma-pyrane as pseudo-catalyst has been claimed to posses combustion improver properties [210]. The composition may also include surfactants such as polyethylene glycol and/or one or more polyoxyalkene derivatives of polypropylene glycol, polyethylene glycol, or sorbitol dissolved in an ester of fatty acid.

An improved fuel additive comprises [211] a mixture of nitroparaffins (nitromethane, nitroethane, and nitropropane), and a combination of modified commercially available ester oil and toluene. The ratio of ester oil and toluene to nitroparaffins is preferably less than 20 vol%, with nitroparaffins comprising the balance of the additive. A method of preparing and using the additive formulation is also provided. A fatty acid ester of sorbitan [212] has been described which reduces deposit, corrosion and facilitate more efficient combustion.

5.12.2 Unconventional Approaches

Magnetic fields [213] have been applied to improve the combustion of fuels. An interesting patent claims that magnetic fields improves combustion efficiency by energizing the fuel molecules, which results in more complete combustion of the fuel in a combustion chamber. A series of magnets were arranged around the periphery of an activation chamber, and a fuel conduit was passed through the magnetic field region defined between the magnets. The fuel conduit was passed through the device many times, repeatedly exposing the fuel to magnetic fields within the device and thus increasing the amount of energy imparted to the fuel before combustion. A concentrating rod was set up to attracts the magnetic fields toward the center of the chamber to focus the magnetic fields through the fuel conduit.

Ion-exchange resin to which a fuel additive is chemically bound has been described to release the additive gradually [214]. The system is efficient as compared to the after market fuel additive bottles sold in the market.

> US Pat. No. 4,639,255 describes various solid fuel additive compositions. A pelleted additive formulation was developed using pellets containing about 25–75% by weight of additive. The additive is contained in a fuel-dispersible compacting agent, such as paraffin wax, and the additive pellets are sealed on their surface. Detergent additives are selected from polyisobutene amines and functional derivatives. A typical additive mixture of this novel composition comprises [215], for example,
>
> 1. 60–80% by weight of at least one polyisobutene amine having a number average molecular weight Mn of about 800 to 1500 g/mol
> 2. 20–40% by weight of at least one synthetic carrier liquid, such as an alkylphenol-initiated polyether, or mineral oil-based carrier oil

This is mixed with half the amount of molten commercial hard paraffin (melting point 55 °C) while stirring. After homogenization, the mixture is allowed to cool to room temperature. The paraffin on solidification encapsulates the additive. The solid fuel additive thus prepared is soluble in conventional fuels.

Drivers are resorting also to pills/tablets, powders, ion-exchange resins, inorganic salts, liquid additives, and magnets to align fuel molecules, and mechanical devices to increase the fuel economy of vehicles. These products are backed up by manufacturers' promises of increased fuel efficiency of the order of 10–30%. While the EPA evaluates and tests products, government agencies and the organized oil industry do not endorse any gasoline-saving products for cars. The author's experience of testing several such fuel-saving devices from all over the world also indicates negative feedback and exaggerated claims by the different manufacturers of these products. The best way to save fuel in vehicles is to drive efficiently, stay within the speed limits, and avoid sudden braking. Timely maintenance of the vehicle, proper tire pressure, changes of oil and filters on schedule are other parameters to improve fuel efficiency. A poorly tuned engine may increase fuel consumption by as much as 10–20%, depending on a car's condition. Good energy-efficient lubricating engines, gears, and transmission oils are now available and provide energy efficiency in addition to long engine life. Certified ashless deposit control additives in gasoline and diesel in appropriate doses clean the fuel system and keep the engine clean for optimum performance.

5.13 FLOW IMPROVERS AND PARAFFIN DISPERSANTS OF FUELS

The quality of diesel fuel has undergone tremendous change over the last decade. Aromatics, olefins, and sulfur contents have been drastically reduced. The final boiling point has been reduced too. Diesel fuel has become lighter. These changes have reduced paraffin chain length, and the reduced aromatics have changed paraffin's solubility, making fuels more paraffinic in nature. These changes obviously necessitate corresponding adjustments in the flow improver structures.

Polymers play important role in modifying the properties of both fuels and lubricants. Polymeric molecules are versatile as their molecular weights can be adjusted and functional groups attached at different positions to get the desired characteristics in petroleum products. A polymer is defined as a chemical compound in which large numbers of similar or identical atoms or groups of atoms are united by covalent bond. A polymer may be united with a low molecular weight substance or with another polymer by chemical linkage. The resulting compound is known as conjugated polymer. In practice, there is no significant difference between a high polymer and a macromolecular substance.

Large number of molecules having at least a double bond can be conveniently polymerized to yield a larger molecule called a polymer. Monomer compounds like ethylene, propylene, butylenes, styrene, isoprene, butadiene, acrylates, methacrylate, acrylonitril, vinylacetate, vinylchloride, and tetrafluoroethylene, can be conveniently polymerized.

Condensation polymers are also formed by the chemical reaction of two compounds, such as phenol/urea with formaldehyde, to yield a high molecular weight resin or polymer. The molecular forces between the two molecules, other than the primary valance bond, have great influence on the properties and behavior of polymers. Hydrogen bonding and Van der Waals forces are the two such forces. The strength of the hydrogen bond increases with the increase in electronegativity of the atom linked with the hydrogen atom. These linked atoms can be fluorine, oxygen, or nitrogen.

The hydrogen bond is mainly ionic or electrostatic in nature. Van der Waals forces are called secondary forces and are combination of several forces that are not as strong as the hydrogen bond. An X-ray examination of polyethylene indicates that the molecules have a shape and structure similar to that of long-chain paraffin hydrocarbons. Its properties are also similar to paraffin wax in many ways. The waxy nature and low softening temperature is due to the weak secondary forces and absence of hydrogen bonding. If the hydrogen atoms of the methylene groups are replaced by a hydrocarbon group as in polyisobutylene or polystyrene, the packing of the chain becomes difficult.

Polyisobutylene at room temperature is like rubber, and on stretching, it becomes crystalline, since the small methyl groups can pack closely together. However, the large phenyl group in polystyrene is irregularly spaced along the chains and prevents the molecules from forming a regularly packed crystalline arrangement. Consequently the paraffinic structure in polystyrene is completely lost, and its properties become similar to glass. The presence of a polar group in the paraffin chain increases the attraction among the chains, since the dipole forces are greater than the Van der Waals forces and operate over a larger distance. This dipole–dipole interaction influences the chain packing as well. As the intermolecular forces, due to the presence of polar groups, increase, the polymers become progressively harder, higher melting, insoluble, and brittle. The incorporation of amine, amide, imides, or hydroxyl groups in the hydrocarbon chain increases the intermolecular forces, because of the formation of stronger hydrogen bonds. Thus the properties of polymers can be modified substantially by the introduction of polarity by substitution or by copolymerization.

Within a group of polymers, these properties further depend on the chain length, structure, and average molecular weight. Fuel additives have to be oil soluble or dispersible. Polymers of very high molecular weight become oil insoluble. Generally, polymers used in fuels are of lower molecular weight as compared to the polymeric additives used in lubricants. The structure of polymers and their behavior in solution are important and govern the end properties. All polymer single crystals have the same general appearance. These are composed of thin, flat platelets of about 100Å thick and often several microns in lateral dimensions. Electron diffraction measurements indicate that the polymer chains are oriented normal to the plane of the lamellae. Since the molecules in the polymers are at least 100Å thick, the only possible explanation is that the chains are folded [216]. For example, in ethylene the molecules can fold in such a manner that only about five chain carbon atoms are involved in the fold. Polymers that have bulky substituents, closely spaced along the chain, often take a helical configuration in the crystalline phase. This helical configuration permits the substituents to pack closely without distortion of the chain bond. Polyethylene, polypropylene, polyisobutylene, poly(1-butene), and polymethyl methacrylate form helical structures. The size of the side groups in these polymers determines the number of units per turn. The helix in polypropylene, poly(1-butene), and polystyrene have three units per turn. Polymethyl methacrylate forms a helix with five units in two turns, while polyisobutylene forms a helix with eight units in five turns. In solutions, these high molecular weight polymers are found as randomly coiled chain molecules. These properties of polymers make them versatile, and their applications in fuel and lubricants are derived from these characteristics.

The flow improvers belong to several classes of polymers and copolymers with different degrees of polymerization and molecular weights. The effectiveness of the flow improvers and pour point depressants is a function of the nature of the paraffins present in the middle distillate, their dosage, molecular weights, and the structure of the polymer employed.

5.13.1   Characteristics of Middle Distillate Fuel at Low Temperatures

Cloud point and pour points are the two measurements used to characterize middle distillates at lower temperature. Cloud point (ASTM D 2500) is the temperature at which the paraffinic molecules present in the fuels begin to appear when the fuel is cooled at a specified rate, and fuel becomes hazy in appearance. In contrast, pour point (ASTM D 97) is the lowest temperature expressed as a multiple of $3\,°C$ at which the oil is observed to flow.

EN 116, a test method used for determining the filterability of diesel fuel, has also been used to characterize the low temperature behavior of fuels. It was found that fuels having similar cloud and pour points may differ in their low temperatures flow characteristics. These laboratory tests were not found to correlate with the field experience.

Dean and Johnson [217] observed that rapid cooling of middle distillates results in the formation of smaller crystals whereas slow cooling gives rise to larger wax crystals that hinders the flow.

Low-temperature properties of distillate fuels depend on the concentration of various n-paraffinic compounds and their distribution in the fuel. These paraffins may be composed of C_{10}–C_{22} hydrocarbons, depending on the crude oil source and refining processes followed. The nature and amounts of n-paraffins in these fractions influence the low temperature behavior. Diesel fuels may have 12–40% paraffinic hydrocarbons (C_{10}–C_{20}), and so have limited solubility in the fuels especially at lower temperatures. The wax can block the fuel system and filters of diesel vehicles (at lower temperature), causing erratic engine operation, loss of power, and even complete engine failure. To avoid these problems, the low-temperature properties of diesel fuel are required to be matched with the climatic conditions under which the engines are operated. This problem can be solved by employing either of the following schemes:

1. In the refinery, the distillates may be undercut and blended with low wax containing streams. This approach can adversely affect refinery economics because lower boiling material is being diverted to a less profitable material. A more economical way to improve the cold flow properties of middle distillates is by catalytic isomerization of n-paraffins to isoparaffins, which have lower freezing points. This process is called iso-dewaxing.

2. Incorporation of flow improvers [218–221] in the diesel fuel. Flow improvers are chemicals that function by several mechanisms that can depress the pour point, inhibit wax crystal growth, and depress the cold filter plugging point (CFPP).

3. Combination of schemes 1 and 2.

The crystallization of paraffin wax under slow cooling conditions affects the viscosity and thus causes a change in the rheological behavior of the fuel. Fuels become non-Newtonian at the initial point of wax crystallization [222,223].

It is clear that in order to have adequate flow characteristics of middle distillates, when the fuel is cooled in the field conditions, the wax crystallization process or the paraffinic molecular structure must be controlled. This can be done by the use of polymeric flow improver and/or by controlling the refinery processes.

5.13.2 Pour Point Depressants

Pour point depressants appreciably reduce the yield stress of frozen fuels and thus improve pumpability at low temperatures. Microscopic observations in polarized light show that at a few degrees above the pour point, wax crystals begin to form in the system. In the absence of the pour point depressant, these crystals continue to grow and form a coherent gel-like structure that restricts the flow of fuel. In the presence of pour point depressants, the wax crystal size is reduced, and the coherent gel-like structure tends to form at significantly lower temperatures.

Presently as pour point depressants, many polymeric materials like polymethylacrylates, ethylene vinyl acetates copolymers, fumarate vinyl acetate copolymers, alkylated polystyrenes, acylated polystyrenes, polyolefins, aliphatic amine oxides, and oxidized waxes are being used. Pour point depressants are essentially wax crystal modifiers. On cooling, the solubility of pour point depressant is such that it comes out of solution before the wax; the depressant may act as a nucleating agent and promote the formation of small wax crystals that are less capable of forming a gel. If the depressant and wax come out of solution together, they may co-crystallize into bulky crystals that have lower tendency to form a rigid gel. If the depressant comes out of solution after the wax, the precipitation on the surface of the normally formed wax may inhibit the adhesion of the crystals necessary to form a gel. The extent of pour point reduction with the use of additives, however, depends on several factors such as type and structure of wax, its concentration in the fuel, nature of the fuel, cooling conditions, and the type of pour point depressant and its concentration.

Alkyl aromatics adsorb on the nascent wax crystal, causing it to grow in new directions and form a compact, multilayer-isotropic wax crystal. Such crystals cannot form the gel-like structure, which causes the oil to congeal at lower temperatures. Polyalkyl methacrylate and polyacrylamide modify wax crystal growth in a similar manner, although the mechanism is through co-crystallization with the wax rather than adsorption on the surface of the wax crystal.

Polyalkyl methacryalates are the most widely used pour point depressants due to the structure and the flexibility of their molecules to design variants of the base oils or middle distillates containing different types of waxes. A typical pour point depressant based on PMA has following structure:

$$
-\left[H_2C - \underset{\underset{\underset{R_1O \quad\; O}{/\quad \backslash\backslash}}{\overset{\overset{CH_3}{|}}{\underset{|}{C}}}{C} - \right]_n - - \left[H_2C - \underset{\underset{\underset{R_2O \quad\; O}{/\quad \backslash\backslash}}{\overset{\overset{CH_3}{|}}{\underset{|}{C}}}{C} - \right]_m -
$$

Here R_1 and R_2 are two alkyl side chains of different carbon lengths; n and m are the degrees of polymerization of two monomers of varying alkyl chain lengths. One of the alkyl side chains R_1 or R_2 is of such length that it is of similar size to that of wax molecules, and thus interacts with the wax crystals. The other alkyl group is of lower carbon length and is neutral to the wax molecules. These shorter side chains act as spacers between the longer chains and wax crystals and try to fit into the wax crystal lattice structure. A polyalkyl methacrylate copolymer having two alkyl groups of C_{11}–C_{15} and C_{16}–C_{30} in a specified ratio provides excellent low-temperature properties in petroleum oils [224].

The efficiency of the pour point depressant (PPD) depends on how the longer chains match the wax crystal structure and interact with the crystallizing molecules of waxes. Wax in oil is a mixture of different carbon chains; therefore, in the pour point depressant or in the flow improvers, mixtures of alkyl chains of matching length are also useful.

5.13.3 Flow Improver Additives

The basic difference between a pour point depressant and the flow improvers lies in their mechanism of action. Pour point depressant additives change the nature of the wax crystals that precipitate from the fuel and thus reduce their tendency to interlock and set into gels. This phenomenon reduces the pourability temperature of the fuel. Pour point depressants will obviously influence the flow characteristics to some extent, but they are not fully satisfactory. Flow improvers, in contrast, interfere with the growing wax crystals, reduce the sizes of the crystals, change the wax crystal morphology, and provide more advantageous wax crystal shapes so that the fuel flow is not impaired [225,226]. Both pour point depressants and flow improvers generally do not change or influence the cloud point of the fuel, although a few reports indicate that a particular ethylene-vinyl acetate copolymer, and a few others, lower the cloud point of diesel fuel slightly [227,228]. Table 5.8 shows the influence's of some additives on the cloud point, the solidification point, and the filterability of a hydrotreated diesel fuel having 1000 ppm of sulfur.

The solidification point is 3 °C lower to the pour point for fuel without additives. These data indicate that the pour point can be reduced to a great extent, but the filterability cannot be reduced to the same level. The flow improvers are therefore to be assessed by their effects on filterability, since this is the most important property

TABLE 5.8 Influence of pour point depressants on diesel fuel properties

Hydrotreated Diesel Fuel (0.1% Sulfur)	Temperature (°C)		
	Cloud Point	Solidification Point	Filterability
Fuel without additive	−1	−5	−2
Fuel with unsaturated naphthalene ester	−2	−36	−6
Fuel with ethylene vinyl acetate copolymer	−2	−38	−7
Fuel with polymethacryalate	−2	−17	−5

for vehicle operability at lower temperatures. Thus the two group of compounds—pour point depressants and flow improvers—belong to two different chemical groups.

The structure of flow improver will greatly vary according to the crude oil source, the nature of the wax present (carbon chain length) in the fuel and also the temperature at which the fuel is used. Distillate fuels are widely treated with polymeric compounds known as flow improvers to improve filterability and pumpability of the fuels at the desired operating temperatures. The requirement will, however, vary from country to country, place to place, depending on the climatic conditions. According to IP 309/EN 116/DIN 51,428, the efficiency of the flow improvers is expressed indirectly by measurement of the cold filter plugging point (CFPP).

5.13.4 Paraffin Dispersants

Although the pour point, pumpability, and filterability of diesel fuels can be improved by the use of suitable flow improvers, it has been observed that the wax of the flow improver treated fuels has the tendency to settle at the bottom during storage at low temperatures. This is because paraffins have higher density than the other liquid components. This phenomenon is a function of both temperature and time. To resolve this problem, another class of additives, called wax dispersants or wax antisettling additives, are required [229]. To evaluate this property, another term, wax dispersion index (WDI), has been introduced. This is defined as the ratio of the volume of the settled wax versus the total volume of the fuel multiplied by 100.

$$\text{WDI} = 100 \times \frac{\text{Apparent volume of the settled or dispersed wax } (A)}{\text{Apparent volume of the fuel } (B)}$$

When A and B are the same, WDI is 100, which means complete dispersion of wax. This is the most desirable phase to achieve. However, the wax dispersant additives must be ensured not to interfere with other additives and properties of the fuel. Both the flow improvers and the wax dispersant act by interacting with the wax crystal coming out of solution at lower temperatures. The only difference is in the crystal sizes controlled by these two additives. The wax dispersant confines the crystal size to a much smaller size so that they remain dispersed

TABLE 5.9 Performances of flow improvers and wax dispersants in diesel fuel (cloud point +2 °C; pour point −3 °C)

Fuel blend	Flow Improver (FI)		FI + Wax Dispersant	
	CFPP °C	WDI	CFPP, °C	WDI
Fuel without additive	+1	10	—	—
Flow improver 1	−1	17	−14	100
Flow improver 2	−6	10	−15	100
Flow improver 3	−8	15	−16	100
Flow improver 4	−4	10	−10	100
Wax dispersant only (without flow improver)	—	—	−8	100

in the oil phase in the colloidal form (less than 1 micron size). Compatible flow improvers and wax dispersant can provide greater synergism and better flow properties and dispersion of wax crystals are obtained.

Several commercial products are available for good results in flow improvement at lower temperatures in different middle distillates [229].

Table 5.9 shows the effects of wax dispersants on CFPP and WDI in the presence of some of the commercially available flow improvers. The data presented in this table show the importance of using compatible flow improvers and wax dispersants in the middle distillates.

Specifications for diesel fuels are therefore, designed according to the seasonal and climatic conditions where the fuel is to be used. Different countries have separate requirements for fuel properties according to the climatic conditions.

Following types of polymers have been reported to be effective in diesel fuel for improving the filterability:

- Ethylene propylene (OCP) copolymer
- Ethylene vinyl acetate copolymer
- Ethylene cyclic amide copolymer
- Alkenyl succinamides
- Alkyl aromatics
- Acylated styrene polymers and copolymers
- Polysaccharides, condensation polymers of polyols and long-chain fatty acids [230].

The common procedure for manufacturing the olefin copolymer is by solution polymerization using a homogeneous Ziegler–Natta catalyst that is soluble in the reaction mixture. In the solution polymerization process, both Ziegler–Natta and Metallocene catalysts are used for synthesizing random polymers. Metallocene catalyzed polymers have lower molecular weight as compared to Ziegler–Natta catalyzed OCP. The process involves mixing ethylene and propylene in an appropriate ratio in a hydrocarbon solvent like hexane along with the soluble catalyst in the reactor, and as the polymerization proceeds, viscosity increases. The polymer concentration is generally kept at 5–6% to maintain good agitation and temperature control. When the desired consistency is achieved, the reaction is terminated and catalyst is washed. The polymer solution is added with an extender oil and antioxidant, when still in solution with the solvent. Solvent and unreacted monomers are distilled off, purified, and recycled for further use. The polymer can be isolated as a solid or mixed with light mineral oil to get the finished product.

OCP can also be produced in a suspension polymerization reaction. Ethylene is contacted with liquid propylene where the later acts as both monomer and the reaction medium. In the presence of a catalyst the polymerization takes place quickly, and polymer is suspended in the liquid propylene. The temperature in this process is controlled automatically by the evaporation of propylene through the generated reaction heat. The suspended polymer is stripped off the propylene, which is recycled. Olefin copolymers of a defined chain length [231] have also been used as diesel cold flow improvers as well as in fuel oils.

Preparations of ethylene and vinyl esters, and acrylates and methacrylates copolymers, have been extensively described in literature. These are prepared by free radical initiation using peroxide in a solvent medium. Acrylic esters are represented by following structure:

$$\overset{\displaystyle R}{\underset{\displaystyle |}{CH_2 = CO \cdot CO \cdot R^1}}$$

When R is H and R^1 is CH_3, it is vinyl acetate. R may be methyl and R^1 may be a straight-chain alkyl group containing 12–18 carbon atoms. Mixtures of alkyl acrylate and methacrylate as well as their partial esters have been found to be effective flow improvers. Lower alkyl acrylate esters can also be used with free radical polymerizable mono-ethylenically unsaturated compounds like mono-vinylidene compounds having one functional group in its structure such as vinyl acetate, styrene, alkyl styrene, and vinyl alkyl ethers. These monomers can also be polymerized with nitrogen containing vinyl compounds such as 2-vinyl pyridine, 4-vinyl pyridine, N-vinyl pyrrolidone, 4-vinyl pyrrolidone. Monomers of unsaturated amides prepared by the reaction of acrylic acid or low molecular weight acrylic ester with butyl amine, hexyl amine, tetra propylene amine, or octyl amine. have also been used. The amides have following structure:

$$\overset{\displaystyle R_1}{\underset{\displaystyle |}{CH_2 = C - CONH\, R_2}}$$

R_1 could be a hydrogen or methyl group. R_2 is hydrogen or an alkyl group containing upto 24 carbon atoms.

Esters of nitrogen containing compounds (US Pat. No. 4,491,455) having polyhydroxyl groups with linear saturated fatty acids have been described as cold flow improvers for fuel oils.

Combinations of additives that function both as wax nucleators and/or wax crystal growth stimulators and as wax growth arrestors have been described in US Pat. No. 3,961,916,—1976 to Ilnyckyj et al. The combination consists of ethylene copolymerized with ethylenically unsaturated mono- or dicarboxylic acid alkyl esters or a vinyl ester of a C_1–C_{17} saturated fatty acid.

A three-component combination flow improver consisting of an ethylene polymer or copolymer, a second polymer of an oil-soluble ester and/or C_3 and higher olefin polymer, and, as a third component, a nitrogen-containing compound has been reported (US Pat. No. 4,211,534,—1980 to Feldman) to have advantages over the two-component system for improving the cold flow properties of distillate fuels.

US Pat. No. 3,982,909,—1976 to Hollyday discloses an additive system consisting of amides, diamides, and ammonium salts alone or in combination with hydrocarbons such as microcrystalline waxes or petrolatums and/or an ethylene backbone polymeric pour depressant.

Nitrogen-containing oil-soluble succinic acid and its derivatives (US Pat. No. 4,147,520,—1975 to Ilynckyj) in combination with ethylene vinyl acetate copolymer wax nucleators significantly improve flow and filterability properties of high-boiling middle distillates.

An oil-soluble amide or amine salt of an aromatic or cycloaliphatic carboxylic acid and an ethylene-vinyl acetate copolymers [232] having a vinyl acetate content of about 10–20% and a number average molecular weight (Mn) in the range of 1500–7000 has also been reported.

The flow improver combination is useful in a broad category of distillate fuels boiling in the range of about 150–400 °C (ASTM D1160). These fuels tend to contain longer chain n-paraffins and generally have higher cloud points. Such wide-cut fuels are more difficult to treat effectively with conventional flow improver additives.

The nitrogen-containing wax crystal growth inhibitors used are generally those having 50–150 carbon atoms and oil-soluble amine salts and amides formed by reacting 1 molar portion of a hydrocarbyl substituted amine with 1 molar portion of the aromatic or cycloaliphatic polycarboxylic acid, preferably dicarboxylic acids, or their anhydrides or partial esters, such as mono-esters, of dicarboxylic acids. The preferred amine is a secondary hydrogenated tallow amine, composed of approximately 4% C_{14}, 31% C_{16}, and 59% C_{18}. Suitable carboxylic acids are isophthalic acid or its anhydride. It is believed that the nitrogen-containing compounds are highly effective in inhibiting the growth of wax crystals.

Generally, as a distillate fuel cools, normal alkanes containing 14 to 22 carbon atoms crystallize out, the longer alkanes crystallizing first. Nitrogen-containing compounds appear to be highly effective in controlling the growth of the bulk of alkane waxes but appear to be slightly less effective in controlling the initial stages of wax precipitation.

Although the optimum polymer properties will vary from one fuel to another, the ethylene-vinyl acetate copolymer containing about 17% vinyl acetate with 8 degree branching that is, 8 methyl terminating alkyl side chains other than those of vinyl acetate per 100 methylene groups is more effective. This has a number of average molecular weight (Mn) of 3400 as measured by vapor phase osmometry. The degree of branching is the number of methyl groups other than those of the vinyl acetate in the polymer molecule per 100 methylene groups as determined by proton nuclear magnetic resonance spectroscopy. It has been found that the relative proportion of the nitrogen-containing compound and the ethylene vinyl acetate copolymer is important in achieving the improvement in flow and filterability.

The nitrogen-containing compound should be used between 60 and 80%, with the balance being the ethylene-vinyl acetate copolymer. The fuel has been evaluated according to the Distillate Operability Test (DOT test), which is a slow cooling test shown to be reasonably accurate compared with actual field conditions.

Mineral oil middle distillate compositions containing a paraffin dispersant that is a reaction product of aminoalkylene carboxylic acids with primary or secondary long-chain amines [233] have been reported. The conventional ethylene copolymers, especially copolymers of ethylene and unsaturated esters, are used as cold flow improvers. DE 1147799 and DE 1914756 describe, copolymers of ethylene with vinyl acetate, containing from 25 to 45% by weight of vinyl acetate or vinyl propionate and having a molecular weight of from 500 to 5000. However, these mixtures are still unsatisfactory with regard to the dispersing properties of the paraffin, which separates out on standing for a long duration. It is therefore necessary that the additives also disperse the separated paraffins. EP 398,101–A describes a reaction product of aminoalkylene polycarboxylic acids with long-chain secondary amines as paraffin dispersants. To achieve good dispersion of the paraffins, however, 0.25 to 40 ppm of a conductivity improver has to be added to the diesel fuel, in addition to the dispersant.

Polysaccharide derivatives have been identified as flow improvers for waxy crude oil and heavy fuels. These additives decrease pour point, reduce yield stress and viscosity, and are therefore suitable for fuels transported through pipe lines. These also reduce wax deposition during transportation and storage. Amylose stearate, dextrin stearate, esters of amylose and dextrin with hydrogenated rape oil fatty acids are reported to be useful flow improvers. Substituted polysaccharides [234] with different molecular weights (12,000–15,500) were synthesized by reacting aliphatic acid chlorides of different carbon numbers (C_{12}–C_{18}) with dextrin in dry pyridine at 100–120 °C for 24 hours. The resulting polymers were purified by dissolving in toluene, filtration, and subsequent precipitation in methanol. These products have been tested as pour point depressants and flow improvers for middle distillates (gasoils), and comparative evaluation with some commercial additives showed their good activity, especially as flow improvers.

Polysaccharides are also synthesized by the etherification of amylose, dextrin, and dipentaerythritol by acids such as stearic acid, hydrogenated rape oil fatty acids and behanic acids. A typical polysaccharide unit is represented by the following structure:

R varies from $C_{11} H_{23}$ to $C_{17} H_{35}$ and n is approximately 16–42.

The polysaccharides derivatives are more active in the lower fraction of distillation and less active in the higher fractions with respect to the flow-improving properties. However, their pour point reduction capabilities remain fairly constant at 15 °C depression.

5.13.5   Distillate Operability Test (DOT Test)

Flow improved Distillate Operability Test is a slow cooling test designed to correlate with the pumping of stored heating oil. The cold flow properties of the described fuels containing the additives were determined by the slow cool flow test as follows: 300 ml of fuel is cooled linearly at 1 °C per hour to the test temperature, then held constant. After 2 hours at the test temperature, approximately 20 ml of the surface layer is removed by suction to prevent the test from being influenced by the abnormally large wax crystals that tend to form on the oil–air interface during cooling. Wax that has settled in the bottle is dispersed by gentle stirring, and then a CFPP filter assembly is inserted. A vacuum of 300 mm of water is applied, and 200 ml of the fuel is passed through the filter into the graduated receiver. A Pass is recorded if the 200 ml are collected within 60 seconds through a given mesh size, or a Fail if the filter becomes blocked and the flow rate is too slow. Filter assemblies with filter screens of 20, 30, 40, 60, 80, 100, 120, 150, 200, 250, and 350 mesh numbers are used to determine the finest mesh number that a wax-containing fuel will pass. The smaller the wax crystals are and the finer the mesh is the greater the effectiveness of flow improver additive. In this test no two fuels will give the same test results at the same treat level of the same flow improver, and therefore actual treat levels will vary somewhat from fuel to fuel.

5.14   DRAG REDUCERS

Liquid hydrocarbon fuels are generally transported through pipelines. Sometimes it becomes necessary to increase the pipeline's capacity without an additional expenditure. Viscous drag experienced in the pipeline is the limiting factor for pipeline capacity. Drag reducers are only used if the products are transported by pipelines whose capacity is less than that required.

The drag reducer molecules have a tendency to reduce the drag created due to the flow of fuel through the pipeline at its surface. Due to this drag reduction, the pipeline capacity can be increased without laying additional pipelines. The high molecular weight polymers, however, have a disadvantage that they readily break down, due to the shear in the pipeline, and therefore the chemical has to be injected at a regular intervals to makeup for the loss. A treat rate of about 2 ppm of polymer is sufficient to obtain a pipeline capacity enhancement of 10–15%.

Drag can be reduced by the use of specific high molecular weight polymers. The high molecular weight polymers confer non-Newtonian behavior on the fuel. US Pat. No. 4,384,089 describes a method for making a very high molecular weight polyalphaolefin polymer for increasing the throughput of hydrocarbon liquids flowing through a pipeline.

Other examples of such a drag reduction polymer are described in US Pat. Nos. 4,508,128; 4,573,488; and 5,080,121. Low concentrations of relatively high molecular weight polymers (4–5 million Daltons) of polyisobutylene (PIB) are known to reduce flow turbulence and have been used as drag-reducing additives. Polyisobutylenes are commercially produced in low molecular weight to very high molecular weight products of 10 million or more. The highest molecular weight products are used as pipeline flow improvers by reducing the onset of turbulence. Because of their high molecular weight, these polymers are able to disrupt eddy currents and allow the laminar flow at flow rates that would be otherwise turbulent.

Polyisobutylenes are produced by the polymerization of C_4-olefins consisting of an isobutene using acidic catalyst (e.g., $AlCl_3$, BF_3). The polymerization of isobutene is sensitive to certain impurities like 1-butene and di-isobutylene. The higher butane content and higher temperature give rise to a low molecular weight product. The polymerization process is exothermic, and low temperatures are required to control the reaction.

$$H_3C-CH_2-CH=CH_2 \qquad CH_3-\underset{\underset{H}{|}}{C}=\underset{\underset{H}{|}}{C}-CH_3 \qquad \underset{\underset{CH_3}{|}}{\overset{\overset{CH_3}{|}}{C}}=CH_2$$

Butene-1   Butene-2   Isobutylene

C4- Olefins hydrocarbons

C_4-Olefins hydrocarbons

Conventional refinery C_4 raffinate contains about 40% isobutylene, 40% N-butene, and about 20% butane. Polymerization of this mixture with $AlCl_3$ yields products where the double bond is located toward the center of the molecule. The conventional polyisobutylene contains about 10% α-olefin, 42% β-olefin and its isomer, and 15% tetra substituted olefin. The product also contains chlorine and is not desirable.

A process using the BF_3 catalyst and concentrated isobutylene has yielded good quality PIB, where the double bond is located toward the end of the chain. The reaction with BF_3 proceeds through protonation of the isobutylene, leading to the formation of a carbonium ion. The chain growth mechanism further leads to the formation of a cationic polymer, which on removal of the proton from the methyl group forms the α-olefin polymer (PIB) in a yield of about 85%.

The molecular weight, the percentage of terminal double bond in isobutylene, and the concentration of the active material determine the molecular structure of the PIB and its reactivity. The terminal double bond can be saturated by hydrogenation.

Polyisobutylene can also be produced by cationic polymerization in solution by using the Ziegler–Natta catalyst.

5.15 ANTI-ICING ADDITIVES

Exuded free water from diesel gasoils (or from heating oils), or water from outside freezes at low temperatures. The formed ice crystals can choke both the fuel pipe and filter, thus inhibit the flow of fuel. For prevention, light molecular weight alcohols or glycols were added to gasoils earlier, but in later years only glycol-ethers are used. The freezing point of the water-additive mixture is much lower than the freezing point of water. For example, these additives are poly(ethylene-glycol)-mono-alkyl-ethers [235]:

$$R - \left(O - CH_2 - CH_2 \right)_n - OH$$

R: 4–6 carbon number of the alkyl-group
$n = 3-8$ whole number

5.16 ANTIFOAM ADDITIVES

Gas oils can foam during pumping to fill the fuel tank. In passenger cars, the relatively small tank and low diametric pipes can cause the foam to spill over. In trucks having multiple fuel tanks, the filling time can be exceedingly long. Antifoam additives are applied to prevent gasoil foaming. These materials decrease the gas bubbles by their low surface absorption, whereby the stability of the liquid phase becomes a separate surface membrane. Antifoam additives are polysiloxanes, like poly(methyl-siloxane) and silica-polyether copolymers, and they are used together with a cosolvent [236,237,238].

The important surface active materials of diesel fuel additive packages have foam stabilizers too. Accordingly, these additive packages containing antifoaming additive(s) assist the formation of separated drops. So there has to be equilibrium (compatibility) between the antifoaming compound and the other compounds of the additive package. The choice compound is an organic compound of modified siloxanes. These are based on a di-methylsiloxane (silica oil) polymer connected with organic chains (e.g., copolymers of ethylene-oxide and propylene-oxide). The silica group assures low surface tension, and the connected organic groups increase the compatibility with the gasoil. The physical-chemical and application properties of the antifoam are determined by the ratio of siloxane to organic groups, and the number and nature of the organic groups.

5.17 BIOCIDES

Fuels get contamined quickly by microorganisms in air or water. These can be bacteria or fungus (mold fungus) [239]. Microorganisms also grow very fast in the presence of water because they accumulate at the interface of water and fuel. Thousands of microorganism do not have a harmful effect, but a colony does, because organic side products are formed during their multiplication. They accelerate the corrosion of the fuel tank, or the "microbiological saliva" formed will choke the filters [239,240]. The most effective biocides inhibit the microbiological degradation by destroying the microorganisms. The biocides can be cyclic imines, other amin derivatives (e.g., succinimide), or imidazoline, like the N,N'-methylene-bis-5-metil-oxaazodiline waiting to be approved by the NATO [241]. Morpholine derivatives are also effective [242].

5.18 COLORING MATTERS AND MARKERS

Colored additives are used to differentiate products with same application, but different quality and properties during their storage [243]. Coloring matters. Yellow, red, blue, green, the colors extended to fuels do not change other properties of fuel. The most used colors are anthraquinone (blue color) and its derivatives and azocompounds (red color) [244]. A new structure coloring matter is ftalociamine and its derivatives [245].

Markers are invisible. They develop color only with another reagent. For example, a marker is added together with an additive into the fuel and its concentration is in proportion with the concentration of the additive package. As an adequate reagent is added to the fuel, it changes its hue. The concentration of the additive can be estimated by the color's intensity [246]. Furthermore it can be identified with HPLC.

5.19 ADDITIVE COMPOSITIONS

Usually automotive fuel additives are not blended individually to the base fuel; they are added, except for some additives, as a prepared concentrated solution of additives with an adequate ratio. This is called the additive composition or additive package [247].

Components of the fuel additive packages are the following [248]:

* Active agent (e.g., detergent/dispersant/antioxidant)
* Solvents (e.g., synthetic oils, polyisobuthilenes, polyethers, polyether amines) and cosolvents (e.g., aromatics, alcohols).

The packaged solution of additives has to be stable and completely soluble in fuel.

During preparation of the additive mixture, interactions between different additive types must be considered:

- Positive interactions (synergy). The intensify effects of each compound, as in the case of using antioxidant and metal deactivator where the desired effect is the summation of individual effects.
- Negative interactions (antagonism). One or more compounds cancel the effects of other compounds, as in the case of same lubricity improver decreasing the effect of an anticorrosion additive or a lubricity improver suppressed by a detergent dispersant from the metal surfaces during adsorption.
- Incompatibility. Acidic and basic additives interact chemical reaction.

In packages, each additives must be carefully chosen, due to the following reasons:

- Surface active additives can disturb adsorption and counter the effects of other additives
- Detergent-dispersant additives can change the emulsion, lessening the efficiencies of the antiwear and antioxidant additives.

Recently additive, packages containing two to three additives with similar effect, but with different chemical structures have come into use. Two examples are detergent-dispersant part packages that contains two to three compounds and flow improver partpackages that have multiple cold flow improvers, such as a cold filter plugging point improver, a paraffin dispersant, and a pour point depressant.

Additive packages are blended to fuels in concentrations of 300–1000 mg/kg depend on the active agent content and composition of the base fuel for performance. The blending of additives or additive compositions in to fuels can occur:

- In the refinery during the blending of fuels
- At the oil terminal
- At the discretion of the user

The first two cases are handled by oil professional. The third option is not advised by the large fuel producers, sellers, and environmental protection offices.

REFERENCES

1. ATC, Technical Committee of Petroleum Additive Manufacturers in Europe. (2004). *Fuel Additives and the Environment*, 2nd ed. ATC, Saint Paul, MN.
2. Hancsók. J. (1999). *Fuels for Engines and JET Engines Part I: Engine Gasolines.* University of Veszprém, Veszprém.
3. Reid, J., Russell, T. (2007). *Fuel additives*. Applications for Future Transport, London, April 16.

4. Russell, T. J., Batt, R. J., Mulqueen, S. M. (2001). The effect of diesel fuel additives on engines. *Proceedings of 3rd International Colloquium, Fuels*, January 17–18, Stuttgart/ Ostfildern.

5. Papachristos, M. (2000). Use of multifunctional additives for achieving fuel for the next century. *Proceedings 2nd International Symposium of Fuels and Lubricants (ISFL-2000)*, (ed. S. P. Srivastava), New Delhi.

6. Keene, P., Browne, B. A. (2011). Preservation strategies for ultra low sulfur diesel, bio-diesel and unleaded gasoline. *Proceedings of 8th International Colloquium, Fuels*, January 19–20, Stuttgart/Ostfildern.

7. Anon. (2011). Feeling the squeeze: flow improvers help refineries to stay competitive, *Infineum Insight*, 2011(9), 9–11.

8. Hancsók. J. (1999). *Fuels for Engines and JET Engines Part II: Diesel Fuels*. Veszprém University Press, Veszprém.

9. R. Merchant (1998). Modern diesel performance additives and their impact on vehicle exhaust emissions. 4th Annual Fuels and Lube Asia conference, Singapore.

10. Adler, K. (2001). Fueling growth. *World Refining*, 12(7), 6–8.

11. Anon. (2005). US speciality fuel additives demand to jump 6%/year. *Hydrocarb. Proc.*, 84(1), 23.

12. Kim, S., Cheng, S. S., Majorski, S. A. (1991). Engine combustion chamber deposits: fuel effects and mechanism of formation. *SAE paper* 912379.

13. Kalghatgi, G. T. (1995). Combustion chamber deposits in spark-ignition engines: a literature review. *SAE paper* 952443.

14. Benson, J. D. (1986). Fuel and additive effect on multi-port fuel injector deposits. *SAE paper* 861533.

15. Nagao, M. (1995). Mechanism of combustion chamber deposit interference and effect of gasoline additives on CCD formation. *SAE paper* 950741.

16. Bitting, W. H. (1996). Combustion chamber deposits. American Chemical Society Symposium on Mechanism of Combustion Chamber Deposits, March 24–29, New Orleans, USA.

17. Aradi, A. A. Colucci, W. J., Scull Jr., H. M., Openshaw, M. J. (2000). A study of fuel additives for direct injection gasoline injector deposit control. *SAE Technical paperw* 2000-01-2020.

18. Arters, D. C., Dimitrakis, W. J., MacDuff, M. J. (2000). Injector fouling in direct injection gasoline vehicles. 6th Annual Fuels and Lube Asia Conf., January 25–28, Singapore.

19. Ohkubo, H., Tomoda, Y., Ahmadi, M. R. (2000). Deposit formation and control in direct injection spark ignition engines. 6th Annual Fuels and Lube Asia Conf., January 25–28, Singapore.

20. Kurdlich, W., Macduff, M., Markó, E. (2002). Utilisation of aftermarket fuel additives to prevent starting problems of gasoline engines. 5th Int. Symp. Motor Fuels, June 17, Vyhne, Slovakia.

21. Haycock, R. F., Thatcher, R. G. F. (2004). *Fuel additives and environment*. Technical Committee of Petroleum Additive Manufacturers in Europe.

22. Beck, Á., Bubálik, M., Hancsók, J., eds. (2011). Development of multifuctional detergent-dispersant additives based on fatty acid methyl ester for diesel and biodiesel fuel. In *Biodiesel Quality, Emission and By-Products* (eds. G. Montero, M. Stoytcheva). Intech, Rijeka, Horvátország.

23. Hancsók, J., Baladincz, J. Magyar, J. (2008). *Mobility and Environment*. Pannon University Press, Veszprém, Hungary (in Hungarian).

24. Reid, J. Russell, T. (2007). Fuel Additives. *Additives 2007: Applications for Future Transport*, April 17, London.

25. Babic, G. T. (1997). Classification system for gasoline detergent performance. *International Symposium of Fuels and Lubricants (ISFL-1997), Proceedings*, (eds. B. Basu, S. P. Srivastava). New Delhi.

26. Gerlach, T., Funke, F., Melder, J.-P. (2006). Preparation of an amine. US Patent 7,034,186, April 25.

27. Funke, F., Wulff-Doring, J., Schulz, G., Siegel, W., Kramer, A., Melder, J.-P., Hohn, A., Buskens, P., Reif, W., Nouwen, J. (2002). Preparing amines. US Patent 6,417,353, July 9.

28. Lin, J.-J., Ho, Y.-S., Ku, W.-S., Shiu, W.-J., Lee, C.-N. (2003). Process for manufacturing gasoline additives of ester-free polyoxyalkylene amide. US Patent 6,627,775, September 30.

29. Melder, J.-P., Blum, G., Gunther, W., Posselt, D., Oppenlander, K. (2005). Preparation of polyalkeneamines. US Patent 6,909,018, June 21.

30. Nelson, A. R., Nelson, M. L., Nelson Jr., O. L. (2002). Motor fuel additive composition and method for preparation thereof. US Patent 6,488,723, December 3.

31. Dever, J. L., Baldwin, L. J., Yaggi, C. J. (2001). Halogen-free, deposit-control fuel additives comprising a hydroxypolyalkene amine, and a process for its production. US Patent 6,262,310, July 17.

32. Rath, H. P., Lange, A., Posselt, D. (2007). Polyisobutenamines. US Patent 7,291,681, November 6.

33. Sabourin, E. T. (2002). Fuel compositions containing hydroxyalkyl-substituted amines. US Patent 6,497,736, December 24.

34. Sabourin, E.T., Buckley III; T. F., Campbell, C. B., Plavac, F. (2002). Fuel compositions containing hydroxyalkyl-substituted amines. US Patent 6,368,370, April 9.

35. Sabourin, E. T., Buckley III, T. F., Campbell, C. B., Plavac, F., Tompkins, M. J. (2002). Fuel compositions containing hydroxyalkyl-substituted polyamines. US Patent 6,346,129, February 12.

36. Bartha, L., Hancsók, J., Bobest, É. (1992). Resistance of ashless dispersant additives to oxidation and thermal decomposition. *Lubr. Sci.*, 4(2), 83–92.

37. Hancsók, J., Bartha, L., Baladincz, J., Auer, J., Kocsis, Z. (1997). Use of succinic anhydride derivatives in engine oils and fuels. *Petrol. Coal*, 39 (1), 21–24.

38. Mach, H., Rath, P. (1997). Highly reactive polyisobutylene as a composition element for a new generation of lubricant and fuel additive. *Proceedings of Symposium, Additives in Petroleum Refinery and Petroleum Product Formulation*. Hungarian Chemical Soc., May 21–23, Sopron, 52–58.

39. Hancsók, J., Bartha, L., Baladincz, J., Kocsis, Z. (1997). Relationships between the properties of PIB-succinic anhydrides and their additive derivatives. *Proceedings, Conference on Additives in Petroleum Refining Practice and Lubricants, Sopron, May* 21–23, 59–66.

40. Hancsók, J., Bubálik, M., Beck, Á., Baladincz, J. (2008). Development of multifunctional additives based on vegetable oils for high quality diesel and biodiesel. *Chem. Eng. Res. Des.*, 86, 793–799.

41. Bubálik, M., Beck, Á., Baladincz, J., Hancsók, J. (2009). Development of deposit control additives for diesel fuel. *Petrol. Coal*, 51(3), 158–166.

42. Stuart, F. A., Anderson, R. G., Drummond, A. Y. (1968). Lubricaing oil composition containing alkeneyl succinimides of tetra ethylene pentamine. US Patent 361673.

43. Meinhardt N. A., Davis K. E. (1980). Noval carboxylic acid acylating agent—Process for their preparation. US Patent 4234435.

44. Weill, J. N., Garapon, J., Sillon, B. (1984). Process for the manufacturing anhydride of alkeneyl dicarboxylic acids. US Patent 4433157.

45. J. K. Pudelski and coworkers. (2002). US Patents, 5885944, 1999, 6077909.

46. Hancsók, J, Bartha, L, Baladincz, J, Kocsis, Z. (1999). Relationships between the properties of PIB-succinic anhydrides and their additive derivatives. *Lubr. Sci.*, 11(3), 297–310.

47. Heddadj, M., Ruhe Jr., W. R., Sinquin, G. P. (2006). Process for preparing polyalkenyl-succinimides. US Patent 7,091,306, August 15.

48. Cherpeck, R. E., Nelson, K. D. (1999). Esters of polyalkyl or polyalkenyl *N*-hydroxyalkyl succinimides and fuel compositions containing the same. US Patent. 5,993,497, November 30.

49. Cherpeck, R. E. (2002). Ethers of polyalkyl or polyalkenyl *N*-hydroxyalkyl succinimides and fuel compositions containing the same. US Patent 6,352,566, March 5.

50. Lange, A., Rath, H. P. (2005). Method for producing polyisobutylphenols. US Patent 6,914,163, July 5.

51. Lange, A., Rath, H. P., Walter, M. (2008). Method of purifying long-chain alkyl phenols and Mannich adducts thereof. US Patent 7,355,082, April 8.

52. Carabell, K. D., Gray, J. A. (2004). Fuel additive compositions containing a mannich condensation product a poly(oxyalkylene) monool, and a carboxylic acid. US Patent, 6,749,651, June 15.

53. Houser, K. R. (2003). Fuel additive compositions containing a mannich condensation product a poly(oxyalkylene) monool a polyolefin, and a carboxylic acid. US Patent 6,511,518, January 28.

54. Ahmadi, M. R., Gray, J. A., Sengers, Henk P. M. (2003). Fuel additive compositions containing a mannich condensation product a poly (oxyalkylene) monool, and a carboxylic acid. US Patent 6,511,519, January 28.

55. Malfer, D. J., Noble, A. T., Colucci, W. J., Sheets, R. M. (2004). Secondary amine Mannich detergents. US Patent 6,800,103, October 5.

56. Colucci, W. J., Zahalka, T.L. (2000). Achieving cost-effective combustion chamber deposit control. *Proceedings 2nd International Symposium of Fuels and Lubricants (ISFL-2000),* (ed. S. P. Srivastava), New Delhi, 77–82.

57. Malfer, D. J., Cunningham, L. J. (1997). Fuel compositions. US Patent 5,697,988, December 16.

58. Malfer, D. J., Colucci, W. J., Franklin, R. M. (1998). Additives for minimizing intake valve deposits, and their use. US Patent 5,725,612, March 10.

59. Hancsók, J., ed. (1997). *Modern Engine and JET Fuels. I. Gasolines.* Veszprém University Press, Veszprém.

60. Baladincz, J., Szirmai, L., Bubálik, M., Hancsók, J. (2005). Modern additives of gasolines. *J Hun. Chem.*, 2005, 60(11), 396–403.

61. Su, W.-Y., McKinney, M. W., Lambert, T. L., Marquis, E. T., (2000). Process for the production of etheramine alkoxylates. US Patent 6,060,625, May 9.

62. Su, W.-Y., Nelli, C. H. (2006). Preparation of secondary amines. US Patent 7,074,963, July 11.

63. Daly, T. J., Clumpner, M., O'Lenick Jr., A. J. (2002). Ether diamines amine oxides. US Patent 6,417,401, July 9.

64. Oppenlander, K., Gunther, W., Posselt, D., Massonne, K., Rath, H. P. (2003). Polyalkene alcohol polyetheramines and their use in fuels and lubricants. US Patent 6,548,461, April 15.

65. Bennett, J. (2010). Application of fuel additives to direct injection spark ignition engines. *International Symposium or Fuels and Lubricants-ISFL-2010*, New Delhi, Paper P067.

66. Matsushita, S. (1997). Development of direct injection S.I. engine (D-4). *Proceedings of JSAE (Japanese Society of Automotive Engineers)*, No. 9733440, March.

67. Ahmadi, M. R. (2002). Method for controlling engine deposits in a direct injection spark ignition gasoline engine. US Patent 6,475,251, November 5.

68. Campbell, C. B. (1991). Synergistic fuel composition. EU Patent EP 0452328; US Patent 4877416.

69. Lange, A., Rath, H. P. (2005). Method for producing polyisobutylphenols. US Patent 6,914,163, July 5.

70. Macduff, M. G. J., Jackson, M. M., Arters, D.C., McAtee, R. J. (2002). Fuel additive compositions and fuel compositions containing detergents and fluidizers. US Patent 6,458,172, October 1.

71 Arters, D., Daly, D. T., Jackson, M. M. (2001). Fuel additives and fuel compositions comprising said fuel additives. US Patent 6,193,767, February 27.

72. Daly, T. J., Clumpner, M., O'Lenick Jr., A. J. (2001). Ether amines and derivatives. US Patent 6,331,648, December 18.

73. Jackson, M. M., Corkwell, K. C. (2007). Gasoline additive concentrate composition and fuel composition and method thereof. US Patent 7,195,654, March 27.

74. Lubrizol Bulletin 2010, 11.009.

75. Bennett, J. (2012). Impact of fuel additives on durability and emissions. *UNITI Mineral Oil Technology Forum*, March 20–21, Stuttgart.

76. Ofner, H. (2001). Gas based fuels—an alternative approach to clean propulsion technology. *3rd International Colloquium* (ed. W. J. Wartz). January 17–18, Esslingen, 293–302.

77. Barbour, R., Panesar, A., Arters, D., Quigley, R. (2010). Controlling deposits in modern diesel engines. International Symposium or Fuels and Lubricants—ISFL, New Delhi, p 078.

78. Kocsis, Z., Holló, A., Hancsók, J., Szirmai, L., Resofszki, G. (2003). Detergents for diesel fuels to improve air quality and fuel economy at lower operating costs. 4th International Colloquium on Fuels 2003, Technische Akademie, January 15–16, Esslingen, Ostfildern, Germany.

79. Hancsók, J., Bubálik M., Törő, M., Baladincz, J. (2006). Synthesis of fuel additives on vegetable oil basis at laboratory scale. *Eur. J. Lipid Sci. Technol.*, 108(8), 644–651.

80. Beck, Á., Bubálik, M., Hancsók, J. (2009). Development of a novel multifunctional succinic-type detergent-dispersant additive for diesel fuel. *Chem. Eng. Trans.*, 17, 1747–1752.

81. The Lubrizol Corporation, 110003 Bulletin.

82. The Lubrizol Corporation, 101094 Bulletin.

83. Evaluation of the effect of fuel compostion and gasoline additives on combustion chamber deposits. (1996). *SAE Technical Paper* 962012.

84. Additive evaluation in Venezuelan diesel formulations. (1999). *SAE Technical Paper* 1999-01-1480.

85. *Fuel Additives, Deposits Formation and Emissions.* (1999). European Commission/ ACEA/ EUROPIA—European Commission Industry—Technical Group 1.

86. *Fuel Quality, Vehicle Technology and Their Interactions.* (1999). CONCAWE Report 99/55.

87. Macduff, M. G. J., Corkwell, K. C., Arers, D. C., Jackson, M. M. (2001). The use of gasoline additives to enhance lubricant performance. 3rd *International Colloquium, Fuels* (ed. W. J. Bartz), January 17–18, 343–350.

88. Phillips 66 Reference Data for Hydrocarbons and Petro-Sulfur Compounds. (1962). Phillips Petroleum Company, Bulletin 51.

89. Cummings W. M. (1977). Fuels and lubricant additives-1. *Lubrication*, 32(1), 6.

90. Hancock, E. G., ed. (1989). *Technology of Gasoline.* Blackwell Scientific, Oxford.

91. Owen K. (1989). *Gasoline and Diesel Fuel Additives.* Wiley, New York.

92. Campbell, K., Russel, T. J. (1982). The Effect on Gasoline Quality of Adding Oxygenates. *Associated Octel Publication, OP* 82(1).

93. Palmer, F. H. (1987). The vehicle performance of gasolines containing oxygenated supplements. European Fuel Oxygenates Association Conference, October 22–23, Rome.

94. College of Petroleum and Energy Studies. (1993). *Distillate Fuels: Quality, Specifications and Economics.* Oxford.

95. Owen, K., Coley, T., eds. (1990). *Automotive Fuels Reference Book.* Society of Automotive Engineers, Troy, MI.

96. Zeiner, W., Böhme, W. (1997). Valve seat protection additive as alternative to leaded gasoline. 1st International Colloquium Fuels, January 16–17, Esslingen, Germany.

97. Hutchenson, R. C. (2000). Valve seat recession—an independent review of existing data. *Diesel and Gasoline Performance and Additives*, 2000-01-2015.

98. Whitcomb, R. M. (1975). *Non-lead Antiknock Agents for Motor Fuels.* Noyes Data Corporation, Park Ridge, NJ.

99. Apostolov, I. G. (1995). Production of unleaded gasolines with manganese antiknock additives. *Petrol. Coal*, 37(3), 21–24.

100. Dagani R. (2001). Fifty years of ferrocene chemistry. *Chem. Eng. News* 79 (49): 37–38.

101. Samson, S., Stephenson, G. R. (2004). Pentacarbonyliron. *Encyclopedia of Reagents for Organic Synthesis* (ed. L. Paquette). Wiley, New York.

102. Knifton, J. F., et al. (1991). Ethylene glycol-dimethyl carbonate cogeneration. *J. Molecul. Chem.* 67, 389–399.

103. Pacheo, M. A., Marshall, C. L. (1997). Review of dimethyl carbonate (DMC) manufacture and its characteristics as a fuel additive. *Energy Fuel*, 11, 2–29.

104. Hobson, G. D., ed. (1984). *Modern Petroleum Technology.* Wiley, Chichester.

105. Al-Shahrani, F., Dabbousi, B. O., Martinie, G. D. (2003). Study of oxygenate and metallocene octane enhancers in Saudi Arabian gasolines. *Petrol. Chem. Div., Preprints*, 48(2), 87–89.

106. Royse, S. (1992). Eternal rise for gasoline additives. *Eur. Chem. News*, (October), 5–7.

107. Brown, J. F. (1993). Additive usage in reformulated fuels. *Fuel Reform.*, 3(2), 59–62.

108. Mukhopadhyay, R. (2011). Examing biofuels policy, *CEN*, 89(33), 10–15.

109. Owen, K., Coley, T., eds. (1995). *Automotive Fuel Reference Book*, 2nd ed., Automotive Engineers Inc., Troy, MI.

110. Haycock, R. F. (2004). *Fuel Additives and the Environment*. ATC, Saint Paul, MN.

111. Russell, T. J., Batt, R. J., Mulqueen, S. M. (2001). The effect of diesel fuel additives on engine performance. *3rd International Colloquium Fuels 2001*, TAE, 333–340.

112. King, S. C., Russell, T. J. (1989). Premium diesel fuels in Europe. *Co-ordinating European Council. 3rd International Symposium on Performance Evaluation for Automotive Fuels and Lubricants*, May 19–21, Paris.

113. Ickes, A. M., Bohac, S. V., Assanis, D. N. (2009). Effect of 2-ethylhexylnitrate cetane improver on NO_x emissions from premixed low-temperature diesel combustion. *Energy Fuels*, 23, 4943–4948.

114. Bornemann, H., Scheidt, F., Sander, W. (2002). Thermal decomposition of 2-ethylhexyl nitrate (2-EHN). *Int. J. Chem. Kinet.*, 34(1), 34–38.

115. Stein, Y., Yetter, R., Dryer, F., Aradi, A. (1999). The autoignition behavior of surrogate diesel fuel mixtures and the chemical effects of 2-ethylhexyl nitrate (2-EHN) cetane improver. Society of Automotive Engineers Paper, 1999-01-1504.

116. Castedo, L., Marcos, C. F., Monteagudo, M., Tojo, G. (1992). New one-pot synthesis of alkyl nitrates from alcohols. *Synthetic Commun: An Int. J. Rapid Commun. Synthetic Org. Chem.*, 22(5).

117. Wiles, Q. T., Bishop, E. T., Devlin, P. A., Hopper, F. C., Schroeder, C. W., Vaughan, W. E (1949). Di-terc-butyl-peroxid, and 2,2-bis(terc-butylperoxid)butane. *Ind. Eng. Chem.*, 41(8), 1679–1682.

118. Nandi, M., Jacobs, D. C., Kesling, H. S., Liotta, F. J. (1994). The performance of a peroxide based cetane improvement additive in different diesels fuels. *SAE Paper 94201 9*, October.

119. Kesling, H. S., Liotta, F. J., Nandi, M. K. (1994). The thermal stability of a peroxide-based cetane improvement additive *SAE Paper 941017*, March.

120. Liotta. F. J. (1993). A peroxide based cetane improvement additive with favorable fuel blending properties. *SAE Paper 932767*, October.

121. Green, G. J., Henly, T. J., Stocky, T. P. (1997). The economy and power benefits of cetane improved fuels in heavy duty diesel engines. *SAE Paper 972900*.

122. Kulinowski, A., Henly, T. J., Stocky, T. P. (1998). The effect of 2-ethyl hexyl nitrate cetane improver on engine durability. *SAE Paper 981364*.

123. Schwab, S. D., Gunther, G. H., Henly T. J., Miller, K. T., (1999). The effect of 2-ethyl hexyl nitrate and ditertiary butyl peroxide on exhaust emission from heavy duty diesel engine. *SAE Paper 1999-01-1478*.

124. Owen, K., Coley, T., eds. (1990). *Automotive Fuels Reference Book*. Society of Automotive Engineers, Danvers, USA.

125. Resofszki, G., Szirmai, L., Holló, A. (2002). Gas oils in long term storage—do their properties change? *Proceedings of Interfaces 2002*, September 19–20, Budapest.

126. Pedley, J. F., Hiley, R. W., Hancock, R. A. (1989). Storage stability of petroleum-derived diesel fuel. *Fuel*, 68, 27–31.

127. Bacha, J. D., Tiedermann, A. N. (2000). Diesel fuel stability and instability a simple conceptual model. *Proceedings of 7th International Conference on Stability and Handling of Liquid Fuels*, September 24–29, Graz, Austria.

128. Schwartz, F. G., Whisman, M. L., Albright, C. S., Ward, C. C. (1972). Storage stability of gasoline—development of a stability prediction method and studies of gasoline composition and component reactivity. US Department of Interior, *Bureau of Mines, Bulletin*, 660.

129. D'Ornellas, C. V., de Oliveira, E. J. (2003). Gasoline oxidation stability—influence of composition and antioxidant response. *Proceedings of 4th International Colloquium Fuels*, January 15–16, Esslingen, Germany.

130. Schober, S., Mittebach, M. (2004). The Impact of antioxidants on biodiesel oxidation stability. *Eur. J. Lipid Sci. Technol.*, 106(6), 382–289.

131. Dunn, R. O. (2005). Effect of antioxidants on the oxidative stability of methyl soyate (biodiesel). *Fuel Proc. Technol.*, 86(10), 1071–1085.

132. Hobson, G. D., ed. (1984). *Modern Petroleum Technology*. Wiley, Chichester.

133. Backstrom, H. (2001). Oxidation stability from the right mix of inhibitors. Naphtenics R&D 1/01.

134. Vadekar, M. (2002). Oxygen contamination of hydrocarbon feedstocks. *Petrol. Technol. Quart.*, 2002/3, 87–93.

135. Scott, G., ed. (1993). *Atmospheric Oxidation and Antioxidants*. Elsevier, Amsterdam.

136. Yu Zhang, H. (1998). Selection of theoretical parameter characterizing scavenging activity of antioxidants of free radicals. *J. Amer. Oil Chem. Soc.*, 75, 1705–1709.

137. Potterat, O. (1997). Antioxidants and free radical scavengers of natural origin. *Curr. Org. Chem.*, 1, 415–440.

138. Kancheva, V. D. (2009). Phenolic antioxidants—radical-scavenging and chain-breaking activity: a comparative study. *Eur. J. Lipid Sci. Technol.*, 111, 1072–1089.

139. Denisov, E. T., Denisova, T. G., eds. (2001). *Handbook of Antioxidants, Bond Dissociation Energies, Rate Constants, Activation Energies and Enthalpies of Reactions*. CRS Press, Boca Raton, FL.

140. Mushrush, G. W., Speight, J. G., eds. (1995). *Petroleum Products: Instability and Incompatibility*. Taylor & Francis, London.

141. Hamlon, J. V., Culley, S. A. (1991). Alkyl phenol stabilizer compositions for fuels and lubricants. *US Patent* 5,024,775.

142. Rudnick, L. R., ed. (2009). Lubricant Additives–Chemistry and Applications. Taylor & Francis, New York.

143. Nagata, T., Kusuda, C., Wada, M., (1996). Process for the preparation of diphenylamine or nucleus substituted derivatives thereof. US Patent 5545752.

144. Synthesis of alkylated aromatic amines. (1998). US Patent 5734084.

145. Mortier R. M., Orszulik S. T., eds. (1997). *Chemistry and Technology of Lubricants*. Blackie Academic, London.

146. Jensen R. K., Korcek, S., Zinbo, M., Gerlock, J. L. (1995). Regeneration of Amine in Catalytic Inhibition of Oxidation. *J. Org. Chem.* 60, 5396–54000.

147. Hemighaus, G., Boval, T., Bacha, J., Barnes, F., Franklin M., Gibbs, L., Houge, N., Jones, J., Lesnini D., Lind, J., Morris, J. (2006). *Aviation Fuels Technical Review*. Chevron Corporation, Houston,

148. Pande, S. G., Hardy, D. R. (1998). Effects of extended duration testing and time of addition of *N,N'*-disalicylidene-1,2-propannediamine on jet fuel thermal stability as determined the gravimetric JFTOT. *Energy Fuels*, 12, 129–138.

149. Gernigon, S., Sicard, M., Ser, F., Verduraz, F. B. (2007). Hydrocarbon liquid fuels thermal stability. Antioxidants Influence and Behaviour. *Proceedings of the International Conference on Stability, Handling and Use of Liquid Fuels*. Tucson, Arizona, USA.

150. B. Beaver, (1992). *Fuel Sci. Technol. Int.*, 10, (1), 1.

151. Banavali, R. Karki, S. B. (1997). Examination of tertiary alkyl primary amines as lubricant additives. *Prep. Pap—Amer. Chem. Soc. Div. Petrol. Chem.*, 42(1), 232.

152. Banavali, R., Chheda, B. (1999). Diesel fuel stabilization by tertiary alkyl primary amines: performance and mechanism. *Proceedings of 2nd International Colloquium Fuels*, January 20–21, Esslingen, Germany.

153. Bart, R. J., Henry, C. P., Whitesmith, P. R., (1995). *Proceeding of 5th International Conference on Long Term Storage Stability of Liquid Fuels* (ed. H. N. Giles), vol. 2, 761.

154. Henry, C. (1986). *Proceeding of 2nd International Conference on Long Term Storage Stability of Liquid Fuels*. SWRI, San Antonio, TX, 807.

155. Sharma, Y. K., Agrawal, K. M., Khanna, R., Singh, I. D. (1997). Studies on the stability of middle distillate diesel fuel: effect of composition and additives. *International Symposium of Fuels and Lubricants (ISFL-1997), Proceedings* (ed. B. Basu, S. P. Srivastava), New Delhi, 91–96.

156. Hudec, P., Smieskova A., Zidek, Z., Daucik, P., Jakubik, T., Ambro, J., Sabo, L. (2001). Influence of nitrogen compounds on color degradation of motor fuels. *Petrol. Coal*, 43(3–4), 173–176.

157. Bergeron, I., Charland, J. P., Ternan, M. (1999). Color degradation of hydrocracked diesel fuel. *Energy Fuel*, 13, 689–699.

158. Waynick, J. A. (2001). The development and use of meal deactivator in the petroleum industry: a review. *Energy Fuels*, 15(6), 1325–1340.

159. Banavali, R., Chheeda, B. (1997). Tertiary alkyl primary amin sas multifunctional additives for middle distillate fuels. *International Symposium of Fuels and Lubricants* (ISFL-1997), Proceedings (ed. B. Basu, S. P. Srivastava), New Delhi.

160. Haycock, R. F., Thatcher, R. G. F. (2004). *Fuel Additives and the Environment*. ATC (Technical Committee of Petroleum Additive Manufacturers in Europe).

161. Stansbury, E. E., Buchanan, R. A. (2000). *Fundamentals of Electrochemical Corrosion*. ASM International,

162. Dörr, N., Karner, D., Litzow, U. (2009). Effect of water in diesel fuels on corrosion of 100Cr6. *Proceedings of 7th International Colloquium Fuels*, January 14–15, Ostfildern, Germany.

163. NACE. (1965). Glossary of terms. *Mat. Proc.*, 4(1), 79.

164. Miksic, B. A. (1993). Reviews on corrosion inhibitor science and technology. NACE Conf., Houston TX, II-16-1.

165. French, E. C., Dougherty, J. A., Martin, R. L. (1993). *Reviews on Corrosion Inhibitor Science and Technology*. NACE. Houston, USA.

166. Russel, T. (1990). Diesel Fuel Additives, Diesel Fuel Quality Trends—The Growing Role of Additives. College of Petroleum Studies, Course RF. Oxford, England.

167. Germanaud, L. R., Guy, E. D. (2000). Detergent and anti-corrosive additive for fuels and fuel composition. US Patent 6,083,287.

168. Wei, D., Spikes, H. A. (1986). The lubricity of diesel fuels. *Wear*, 111, 217–235.

169. Song, C. (2003). An overview of new approaches to deep desulfurization for ultra-clean gasoline, diesel fuel and jet fuel. *Catal. Today*, 86, 211–263.

170. Steynberg, A. P., Dry, M. E., eds. (2004). *Fischer–Tropsch Technology*. Elsevier, Amsterdam.

171. Owen, K. (1989). *Gasoline and Diesel Fuel Additives*. Wiley, New York.

172. Wilkes, M. F., Duncan, D. A., Carney, S. P. (2004). Anti-static lubricity additive ultra-low sulfur diesel fuels. US Patent 6,793,695, September 21.

173. Matzke, M., Litzow, U., Jess, A., Caprotti R., Balfour, G. (2009). Diesel lubricity requirements of future fuel injection equipment. *SAE Paper 2009-01-0848*.

174. Chhibber, V. K., Nagpal, J. M., Anand, O. N., Aloopwan, M. K. S., Gupta, A. K. (2004). The compositional aspects of diesel fuel lubricity. 4th International symposium on fuels and lubricants (ISFL- 2004), October 27–29, New Delhi.

175. Pappy, S., Bhatnagar, P., Kumar, A., Marin, V., Sharma, G. K., Raje, N. R., Srivastava, S. P., Bhatnagar, A. K. (2000). Lubricity of low sulphur Indian diesel fuels. *Proceedings International Symposium on Fuels and Lubricants* (ISFL-2000), March 10–12, New Delhi.

176. Padmaja, K. V., Chhibber, V. K., Bhatnagar, A. K., Nagpal J. M., Gupta, A. K. (2004). Enhancement of lubricity of HSD and LCO with Salvadora methyl esters. 4th International symposium on fuels and lubricants (ISFL-2004), October 27–29, New Delhi.

177. Connor, D. S., Burckett-St., L., Charles, J., Roger, T., Cripe, T. A. (2008). Synthetic jet fuel and diesel fuel compositions and processes. US Patent, 7,338,541, March 4.

178. Quigley, R. (1999). Fuel additive. US Patent 6,001,141, December 14.

179. Quigley, R., Jeffrey, G. C. (2000). Fuel additives and compositions. US Patent 6,086,645, July 11.

180. Lin, J. (2002). Fuel additive and fuel composition containing the same. US Patent 6,458,173, October 1.

181. Vrahopoulou, E. P., Schlosberg, R. H., Turner, D. W. (1999). Polyol ester distillate fuels additive. US Patent 5,993,498, November 30.

182. Ball, K. F., Bostick, J., G., Brennan, T. J. (1999). Fuel lubricity from blends of a diethanolamine derivative and biodiesel. US Patent 5,891,203, April 6.

183. Cross, C. W., Chang, Z.-Y., Pruett, S. B, Goliaszewski, A. E. (2005). Lubricity additives for low sulfur hydrocarbon fuels. US Patent 6,872,230, March 29.

184. Ribeaud, M., Dubs, P., Rasberger, M., Evans, S. (2001). Liquid polyfunctional additives for improved fuel lubricity. US Patent 6,296,677, October 2.

185. Barbour, R. H., Rickeard, D. J., Schilowitz, A. M. (2003). Fuel composition with improved lubricity performance. US Patent 6,656,237, December 2.

186. Caprotti, R., Le Deore, C. (2001). Lubricity additives for fuel oil compositions. US Patent 6,293,976, September 25.

187. Caprotti, R., Le Deore, C. (2001). Lubricity additives for fuel oil compositions. US Patent 6,277,159, August 21.

188. Dilworth, B. (1999). Additives and fuel oil compositions. US Patent 5,882,364, March 16.

189. Dillworth, B., Caprotti, R, (1999). Additives and fuel oil compositions. US Patent 5,958,089, September 28.

190. Dillworth, B., Caprotti, R. (2001). Additives and fuel oil compositions. US Patent 6,280,488, August 28.

191. Aradi, A. A. (2005). Friction modifier additives for fuel compositions and methods of use thereof. US Patent 6,866,690, March 15.

192. Sanduja, M. L., Horowitz, C., Mukherjee, S., Thottahil, P., Olliges, W., Foscue, C. T. (2002). Liquid hydrocarbon fuel compositions containing a stable boric acid suspension. US Patent 6,368,369, April 9.

193. Sanduja, M. L., Horowitz, C., Mukherjee, S., Thottahil, P., Olliges, W., Foscue, C. T. (2003). Liquid hydrocarbon fuel compositions containing a stable boric acid suspension. US Patent 6,645,262, November 11.

194. Norman, F. L. (2005). Fuel additive containing lithium alkylaromatic sulfonate and peroxides. US Patent 6,858,047, February 22.

195. Fuentes-Afflick, P. A., Gething, J. A. (2001). Fuel composition containing an amine compound and an ester. US Patent 6,203,584, March 20.

196. Daly, D. T., Adams, P. E., Jackson, M. M. (2001). Additive composition. US Patent 6,224,642, May 1.

197. McLean, G. (2001). Additive concentrate for fuel compositions. US Patent 6,277,158, August 21.

198. Tadros, T. H., Vincent, B. (1983). *Encyclopedia of Emulsion Technology*, Vol. 1 (ed P. Becher). Dekker, New York.

199. Kikabhai, T., Gope, S. (1997). Dehazer and demulsifier technology for fuels and lubricant applications. *International Symposium on Fuels and Lubricants (ISFL1997)*, December 8–10, New Delhi, 103–110.

200. Wilkinson et al., eds. (1982). Pergamon 1982, *Comprehensive Organic Chemistry*, Vol. 4:475–494 and Vol. 8:1014–1043. Pergamon, Oxford.

201. Vincent, M., Richards, W., Cook, P. J., Leonard, S. (2002). Fuel additives. US Patent 6,488,725, December 3.

202. Hebekeuser, H.-P., MacKowiak, H.-P., Gottlieb, K., Jungbluth, H., Neitsch, H, (2001). Ferrocene dicarboxylic acid diesters and solid composite propellants containing the same. US Patent 6,313,334, November 6.

203. Elliott, A. F. (2008). Method for producing ferrous picrate. US Patent 7,335,238, February 26.

204. Elliott, A. F., Stewart, D. M. (2007). US Patent 7,157,593, January 2.

205. Stewart, D. M. (2005). Fuel additive containing ferrous picrate produced by a process utilizing wire. US Patent 6,969,773, November 29.

206. Stewart, D. M. (2003). Process for producing ferrous picrate and a fuel additive containing ferrous picrate from wire. US Patent 6,670,495, December 30.

207. Hazarika, R., Morgan, B. L. (2007). Fuel additive. US Patent 7,195,653, March 27.

208. Wallace, G. M. (1999). Hydrocarbonaceous fuel compositions and additives thereof. US Patent 5,944,858, August 31.

209. Howard, J. B., Kausch Jr, W. J. (1979). Soot control by fuel additives—a review. Report ADA074870. Dayton University, Ohio.

210. Podlipskiy, V. Y., (2002). Fuel reformulator. US Patent 6,482,243, November 19.

211. Foote, A. R., Lakin, M. (2001). Fuel additive formulation and method of using same. US Patent 6,319,294, November 20.

212. Willis-New, J. D. (2000) Combustion catalyst. US Patent, 6,156,081, December 5.

213. Brown, E., Resch, A. (1999). Magnetic fuel enhancer. US Patent, 5,943,998, August 31.

214. Colucci, W. J. (2006). In-tank time release ion exchange resin containing a fuel additive. US Patent, 7,097,771, August 29.

215. Jakob, C. P., Bahr, C., Schwahn, H., Posselt, D. (2001). Solid fuel additive. US Patent, 6,312,480, November 6.

216. Srivastava, S. P. (2007). *Modern Lubricant Technology.* Technology publication, Dehradun, India, 39–45.

217. Dean, H. E., Johnson, E. H. (1963). Pour depressed fuels are a field success. NPRA meeting, September 18–19, Cleveland.

218. Sutton, D. L., (1986). Investigation into diesel operation with changing fuel property. International Congress and Exposition, February 24–28, Detroit.

219. Steere, D. E., Marino, J. P., (1981). Low temperature field performance of flow improved diesel fuels. *SAE Paper 810024.*

220. Muzatko, J. W., (1980). Reducing low temperature wax plugging in fuel system of diesel passenger cars. *SAE Paper 800222.*

221. Reddy, S. R., McMillan, M. L. (1981). Understanding the effectiveness of diesel fuel flow improver. *SAE Paper 811181.*

222. Chernov, A. A. (1962). About kinetics of crystal formations. *Kristallografia*, 7, 895.

223. Anderson, J. S. (1963). *Trans. AIME*, 227, 248.

224. Liesen, G. P., Jao, T. C. Li, S. (2001). Methacrylate copolymer pour point depressants. US Patent 6,255,261.

225. Holder, G. A., Winkler, J. (1965). Wax crystallization from distillate fuels. *J. Inst. Petrol.*, 51 (498), 228.

226. Coley, T., Rutishauser, L. F., Ashton, H. M. (1966). New laboratory test for predicting low- temperature operability of diesel fuels. *J. Inst. Petrol.*, 52 (510), 173.

227. Veretennikova, T. N., Énglin, B. A., Nikolaeva, V. G., Mitusova, T. N. (1980). Mechanism of action of pour point depressant in diesel fuels. *Chem. Tech. Fuels Oils*, 16 (6), 392–395.

228. Tolstova, G. V., Shor, G. I., Énglin, B. A., Lapin, V. P. Mitusova, T. N. (1980). Mechanism of action of pour point depressant additive in diesel fuels. *Chem. Tech. Fuels Oils*, 16 (2), 122–126.

229. Chiang, W. (1983). Wax dispersant for distillate fuels. March 7–10, Cairo.

230. Abou El Naga, H. H., et al. (1983). Polysaccharides as low temperature middle distillate additives. 4th International Seminar on Development of Fuels, Lubricants, Additives and Energy Conservation, March 7–10, Cairo.

231. Tack, R. D., Dilworth, B., Peiffer, D. G. (2006). Fuel oil additives and compositions. US Patent 7,067,599, June 27.

232. Tack, R. D., Lewtas, K., Davies, B. W. (1982). Two component flow improver additive for middle distillate fuel oils. US Patent 4481013.

233. Dralle-Voss, G., Oppenlaender, K, Barthold, K., Wenderoth, B, Kasel, W. (1994). Mineral oil middle distillate compositions. US Patent 5376155.

234. El AbouNaga, H. H., El AbdAzim, W. M., Mahmoud, M. (1994). Some substituted polysaccharides as low temperature middle distillate additives. US Patent 5376155.

235. Hancsók J. (1999). *Fuels for Engines and JET Engines Part II: Diesel Fuels.* Veszprém University Press, Veszprém.

236. Kugel, K. (1999). Antifoams in diesel fuel. 2nd International Colloquium on Fuels, January 20–21, Esslingen, Germany.

237. Venzmer, J., Hänsel, R. (2005). Next antifoam generation for multi-functional diesel fuel additive packages. 5th International Colloquium Fuels, January 12–13, Esslingen, Germany.

238. Venzmer, J., Hänsel, R., Stadtmüller, S. (2007). *Antifoams for diesel fuel: What is so special about organomodified siloxanes—or why are there no organic antifoams for diesel fuel?* 6th International Colloquium on Fuels, January, Esslingen, Germany.

239. Parker, S. M., White, G. F., Hill, E. C., Hill, G. C., Moupe, A. P., Becker, R. F. (1999). *Origins of the polymers wick foul middle distillate fuels.* 2nd International Colloquium Fuels, January 20–21. Esslingen, Germany.

240. Duda, A., Skret, I., Hill, E. C. (1999). Industrial experience in control and elimination of microbial contamination of petroleum fuels. 2nd International Colloquium Fuels, January 20–21, Esslingen, Germany.

241. Siegert, W. (1999). Biocidal treatment and presentation of liquid fuels. 2nd International Colloquium Fuels, January 20–21, Esslingen, Germany.

242. Edchus, A. C., Passman, F. J. (2005). Applications of a morpholine-derivative product in fuel preservation. 5th International Colloquium Fuels, January 12–13, Esslingen, Germany.

243. Slotman, W., Vamvakaris, C. (2002). Sudan dyes and markers for the mineral oil industry. Interfaces'02, Budapest, Hungary, September.

244. Dyes and markers for the petroleum industry, Morton International SA, 1997.

245. Albert, B., Beck, K. H., Meyer, F., Vamvakaris, C., Wagenblast, G. (1998). Phthalocyanines and their use as marker. US Patent 5804447.

246. Sweeney, E. G., Schmidt, C. H., Zirzin, A., Caputo, P. A., Anderson, P. M. (1999). Determination of dyes in diesel fuels. *SAE, SP-1056*, 1–9.

247. Hancsók J. (1997). *Fuels for Engines and JET Engines. Part I: Gasolines.* Veszprém University Press, Veszprém.

248. Hancsók J. (1999). *Fuels for Engines and JET Engines. Part II: Diesel Fuels.* Veszprém University Press Veszprém.

CHAPTER 6

Blending of Fuels

The blending of fuels is performed with the use of blending components, as was already discussed in previous chapters. The blending is followed by admixtures of additive packages or single additives.

6.1 BLENDING OF GASOLINES

Table 6.1 shows the average share of the main blending components in gasolines used in the United States and in the European Union. Some blending components (Table 6.2) are produced by the oil refineries but most are provided by other hydrocarbon processing plants. For example, hydrogenated pyrolysis naphtha comes from petrochemical plants that combine bioethanol from ethanol producers.

During the blending of gasolines not only the physical and chemical properties of each blending component has to be considered but those contributions that may be harmful material emissions. The data in Table 6.2 show the octane numbers, at given shares of the blending components, along with their compositions and quality measures. These concerns are substantiated by the use of reformates, FCC-naphtha, alkylate, ethanol, and bio-ETBE. The vapor pressure value is determined mainly by n-butane fraction, light FCC-naphtha, isomerates, and ethanol. The main source (ca. 80%) of the benzene content is the reformate, but the benzene content of the C_5–C_6 fraction of the cocker, as well as of light FCC-naphtha, light straight-run, and hydrocracking gasolines, is also significant. The quantity of reformate and light FCC-naphtha determines definitely the other aromatic content (ca. 65%). The olefin content depends definitely on the used quantity of light FCC-naphtha (ca. 90%). In many refineries, the polymer naphthas and naphthas from variants of thermal cracking processes have different effects on the olefin content. Basically, the sulfur content is determined by the fraction of heavy FCC-naphtha (ca. 97%; see Figure 6.1).

The main sources of the volatile organic compounds (VOC) in gasolines are n-butane (ca. 25%), ethanol (ca. 12%), alkylate (ca. 8%), reformate (ca. 15%), heavy

Fuels and Fuel-Additives, First Edition. S. P. Srivastava and Jenő Hancsók.
© 2014 John Wiley & Sons, Inc. Published 2014 by John Wiley & Sons, Inc.

TABLE 6.1   Composition of gasoline pool in the United States and European Union

Component	Share (v/v %)	
Gasoline blending component	United States	European Union
FCC-naphtha	33.5	34.0
Reformate	33.0	35.0
Straight-run naphtha	4.0	7.5
Alkylate	12.0	4.4
Isomerizate	4.5	5.5
Butanes	5.5	5.7
Ethanol/MTBE/bio-ETBE	6.0	4.0
Polymer naphtha	0.5	1.5
Others	1.0	2.4
Total (%)	100.0	100.0

FCC-naphtha (ca. 5%), light FCC-naphtha (ca. 23%), and coking C_5–C_6 fraction (ca. 1%). The reformate and FCC-naphthas favor the formation of nitrogen oxides (reformate ca. 21%; heavy FCC-naphtha: ca. 40%; light FCC-naphtha: ca. 30%; n-butane: ca. 5%; isomerate: ca. 4%; coking C_5–C_6 fraction: ca. 2%). The formation of toxic materials and their emission quantities depend on mainly the proportions used of reformate and the FCC-naphthas (reformate ca. 60%; heavy FCC-naphtha: ca. 14%; light FCC-naphtha: ca. 16%; n-butane: ca. 5%; isomerate: ca. 2%; coking C_5–C_6 fraction: ca. 1%; alkylatum: ca. 2%).

Beyond the emissions from the blending components, any harmful material emissions from the additives must be considered during the preparation of the blending "recipe" of gasolines, as was discussed in detail in the preceding chapter.

6.2   BLENDING OF DIESEL GASOILS

Diesel gasoils are blended from desulphurized gasoil streams, from desulphurized and "dewaxed" gasoil streams, or from desulphurized, dearomatized, and "dewaxed" gasoil streams. The gasoil boiling points range is based on the ratios of three components:

- Biodiesels
- Biogas oils
- Fischer–Tropsch gasoils

The blending process is followed by the admixture of additives or additive packages. It is possible that before this final blending some additives have already been introduced into the blending component streams (e.g., flow improvers, stabilizers) for reasons of easier storage and handling.

TABLE 6.2 Main properties of some gasoline blending components

Blending Component	Benzene Content (v/v%)	Sulfur Content (mg/kg)	Olefin Content (v/v%)	Aromatic Content (v/v%)	RON	MON	Vapor Pressure (kPa)	Oxygen Content (%)
Reformate	1.0–10	≤1	<0.1	60–75	99	88	25–50	—
FCC–naphthas	0.7–1.0	100–2000	30–40	5–45	91–96	78–84	48–52	—
Light fraction	0.9–1.5	15–300	20–55	1–2	93–96	80–82	58–63	—
Heavy fraction	0.1–1.1	350–3500	2–14	40–60	91–96	78–84	1–4	—
Isomerates	<0.1	<1	<0.0	<1	87–92	84–90	100–140	—
Alkylates	<0.1	<1	<1	<0.01	94–96	92–93	40–42	—
Polymer naphthas	<0.01	<1	>98	<0.1	94–96	80–82	—	—
Ethanol	0	0	0	0	130	96	16.2	34.8
MTBE	<0.10	0	<1	<0.01	118	100	52–54	18.2
Bio-ETBE	<0.10	0	<1	<0.01	115–120	98–120	29.8	15.7
Steam-reforming naphthas	0–6	0–600	25–35	75–88	98	84	—	—
Light fraction	0–6	0–50	55	5	96	80	15–20	—
Heavy fraction	<1	0–600	<0.1	86	99	84	<5	—
Straight-run naphthas	1–4	10–300	<0.01	3	65	64	78–82	—
C₉-aromatics	<0.1	<1	<0.1	100	105	93	<20	—
n-Butane fraction	0	<1	0	0	96	95	453	—

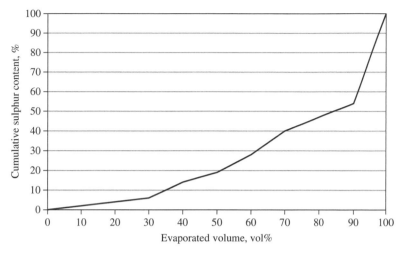

FIGURE 6.1   The typical distribution of sulphur content in FCC naphtha

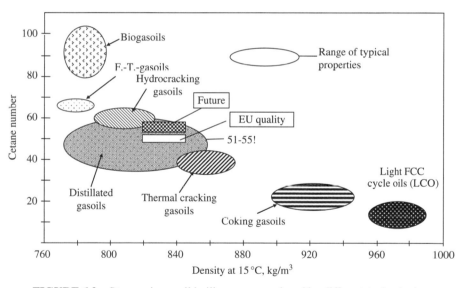

FIGURE 6.2   Streams in gasoil boiling range produced by different technologies

The blending components can be produced at the refineries, and large companies produce their own additive packages. The smaller ones buy partial or complete packages and make their own blends by using one or more additives in the diesel gasoils. The ratios of the most important properties—cetane number, sulfur content, and density—of gasoil streams produced with the different technologies are presented in Figures 6.2 and 6.3.

During the blending of diesel gasoils, not only the physical and chemical properties of each blending component has to be considered, but its contribution to harmful

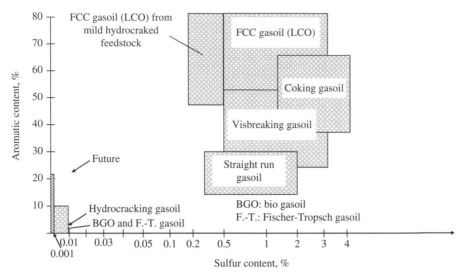

FIGURE 6.3   The sulphur and aromatic content range of different gasoil streams

material emissions. Of course, the additives have already been considered during the preparation of the blending "recipe" of diesel gasoils, as was discussed in the preceding chapter. The positive effects of the additives must override the negative effects in every case.

The blending of fuels is carried out in a computer-controlled system. The algorithms for the applied program values are reviewed continuously or periodically, and also evaluated by the computer. The computer can approve or even modify the blending process by changing the ratios of stream quantities.

The blending optimalization is carried out at minimal production costs in order to maximize profits. As with other high-quantity crude oil industrial products, the blending of fuels is carried out according to two basic principles. First, the stream ratio of blending components is set in the blending system so that the quality properties of mixture are specified at the entrance to the storage tank. Second, the quantity ratios of the blending components are chosen at the mixing point so that the given quality properties of the fuel mixture are reached at the point of filling the storage tank.

The design and connection system of a fuel-blending system is shown in Figure 6.4. At the juncture of the blending management and control technology come short deadline timings of more products; the multivariate control and verification of diesel gasoils, gasolines, and other fuels; and the versatility of NIR (near infra-red technology) spectroscopy and/or other modern analytical processes to maximize profits.

The main components of the blending technology package are the following:

- Interface for monthly linear programmed refinery models for middle-period recipes
- Timing system for optimalizing future products and blending orders

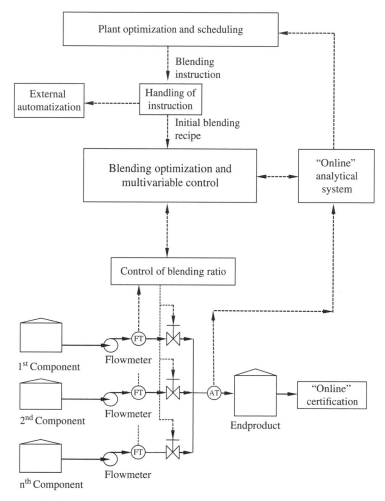

FIGURE 6.4   Management and control of motor fuel blending

- Online multivariate control and optimalization system for feedback from control equipment to enable inline certification and transport of products.

The blend optimization and supervisory system (BOSS) shown in Figure 6.5 determines high the optimal blends of the refinery products (e.g., diesel gasoils and gasolines) and in the case of inline blends supervises the blending applications.

The controlled blending of fuels assures consistent profits for the refineries, and the application of suitably admixtured products having favorable hydrocarbon compositions means numerous advantages for the users as well:

- Smooth performance of vehicles
  - Easy cold start
  - Smooth idle

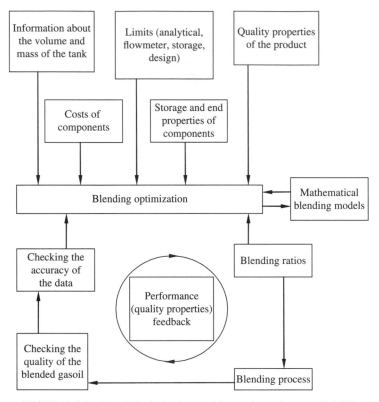

FIGURE 6.5   Blend Optimization and Supervisory System (BOSS)

- ○ Good combustion
- ○ Optimal track behavior (no vibration, engine stop, etc.)
- ○ Excellent acceleration
- ○ Low noise pollution
- • More efficient fuel use:
  - ○ Reduction of fuel consumption
  - ○ Reduction of exhaust gas
  - ○ Emission exhaust gas with more preferable composition
- • Lower maintenance needs, longer engine life, lower maintenance cost

CHAPTER 7

Properties of Motor Fuels and Their Effects on Engines and the Environment

In this chapter the main physical, chemical, and application properties of fuels and their environmental impacts are discussed, first for the gasolines and then for the diesel gasoils.

7.1 EFFECTS OF GASOLINE PROPERTIES ON ENGINES AND THE ENVIRONMENT

Gasoline is a mixture of hydrocarbons—alkanes (paraffins: saturated hydrocarbons), alkenes (olefins: unsaturated hydrocarbons), and aromatics (arenes)—along with sulfur- and nitrogen-containing organic compounds. A typical gasoline may contain more than 500 organic compounds (C_4–C_{14}). Most of these compounds can be identified by gas chromatography (GC) and more recently by a combined GC-MS method. The hydrocarbons are volatile, and their physicochemical properties vary widely. Because gasoline is exposed to different physical, chemical, and mechanical influences, the properties of gasoline must be carefully balanced to obtain satisfactory engine performance over a range of operating parameters. Engine design, driving style, environment, and engine maintenance, and properties like octane number and volatility of the fuel, affect the efficiency of gasoline. The composition of gasoline has changed considerably over a period of years, primarily due to the replacement of straight-run naphtha with cracked stocks obtained from secondary refinery processes like FCC and coker units, and from catalytic reforming, alkylation, isomerization, ether synthesis, and the like. The straight-run naphtha is now being mostly applied as a petrochemical feedstock in light olefin production.

The quality of the gasoline and the technical features of the Otto engine determine what fuel(s) can be introduced into the combustion chamber. The properties of

Fuels and Fuel-Additives, First Edition. S. P. Srivastava and Jenő Hancsók.
© 2014 John Wiley & Sons, Inc. Published 2014 by John Wiley & Sons, Inc.

gasolines affect the carburation, the ignition, combustion processes, fuel efficiency, and the emission characteristics. [1–13].

Among the many properties of the gasolines that must meet standards or certain regulations, the most important are the following:

- Octane number (combustion process)
- Volatility (e.g., vapor pressure, boiling point curve)
- Stability
- Corrosive effect
- Chemical composition
  - Hydrocarbons (benzene, total aromatic and olefin content)
  - Oxygen-containing compounds (bioethanol, methanol, MTBE, ETBE, TAME, etc.)
  - Sulfur content
  - Lead content
  - Halogen content
- Density
- Other properties
  - Energy content
  - Lubricity
  - Flashpoint, ignition point
  - Electrical conductivity
  - Water content

7.1.1 Combustion Process (Octane Number)

One way to increase the efficiency of an engine is to raise the compression ratio. However this has undesired consequences. Above certain limits using the same fuel quality, the normal combustion process changes; the engine starts to knock ("tinkle"). During the normal combustion process, the combustion of the fuel is initiated by a spark, and the resulting flame front spreads throughout the combustion chamber at a steady speed. The unburned mixture ahead of the flame front may explode, which is incited by the high pressure and temperature of the burning gas. The combustion speed greatly increases and the engine makes an unpleasant noise, called knocking [1–5,7,10,12,13]. The high pressure wave induced by the high-temperature explosion decreases the performance and efficiency of the engine, and damages the engine.

During the heating of the fuel–air mixture before ignition, there is a pre-period, when intermediate products of the oxidation reaction (e.g., aldehydes, peroxides) begin to form. Aldehydes tend to dissociate, and any formed molecular fragments facilitate a flame reaction, which accelerates the pre-flame reaction that causes the explosion. The knock of the gasoline is fundamentally determined by the type of hydrocarbons in the gasoline and the engine design. Among the hydrocarbons

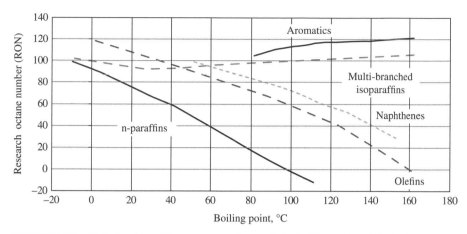

FIGURE 7.1   Relationship of the octane number and the boiling point of the hydrocarbons

that have knock resistance [14], the most resistant ones are the aromatics (e.g., benzene, toluene, xylenes) and the multibranched paraffins (e.g., 2,2,4-trimethyl-pentane), followed by the olefins, the naphthenes, the mono-branched paraffins, and finally the n-paraffins (Figure 7.1) [15]. As for the operating conditions, the temperature increase enhances the knocking tendency, but the compression increase has the highest effect.

The knock resistance and the knocking (autoignition) tendencies of the gasolines are characterized by the octane number. The *octane number* is the percentage, by volume, of isooctane in a mixture of isooctane (2,2,4-trimethyl-pentane) and n-heptane that gives the fuel the same antiknock qualities—under standard conditions—as the given gasoline. The research octane number of the 2,2,4-trimethyl-pentane is 100 while that of the n-heptane is 0 [16,17]. The physical and chemical properties of some of these compounds are shown in Table 7.1.

The knock resistance of fuels used in engines of on-road vehicles is characterized by the *research octane number* (RON, ASTM D-2699, EN ISO 5164) and the *motor octane number* (MON, ASTM D-2700, EN ISO 5163). Online octane analyzers are also available in refineries for continuous determination of these values (ASTM D-2885). The research octane number is determined at low rpm without loading, and the motor octane number at high temperature with partial loading. The measurements are carried out with a one-cylinder CFR test engine. The research octane number is determined at constant speed (600 rpm) and the ignition angle with air pre-heated to 50 °C. The motor octane number is determined at 900 rpm using an adjustable ignition setting, with a mixture pre-heated to 149 °C (Table 7.2). Due to the stringent conditions (particularly the higher temperature of the mixture), the motor octane number of the commercially available gasolines is always lower than the research octane number.

The difference between the research and the motor octane number is called *sensibility* (knocking sensibility), which gives information about how sensible the fuel is to the operational changes of the engine. The smaller the difference, the better

TABLE 7.1 Physical properties of some hydrocarbons

Compound Group	Formula	Structural Formula	Molecular Weight	Boiling Point, °C (101.3 kPa)	Density (d_4^{15}, g/cm³)	Octane Number RON	MON
n-Paraffins and isoparaffins							
n-Butane	C_4H_{10}	$H_3C-CH_2-CH_2-CH_3$	58.1	−0.5	0.585	95	92
i-Butane	C_4H_{10}	$H_3C-CH(CH_3)-CH_3$	58.1	−11.7	0.563	>100	99
n-Pentane	C_5H_{12}	C-C-C-C-C	72.1	36.1	0.631	61.7	61.9
2-Methyl-butane (isopentane)	C_5H_{12}	C-C(C)-C-C	72.1	27.8	0.625	92.3	90.3
n-Hexane	C_6H_{14}	C-C-C-C-C-C	86.2	68.7	0.664	24.8	26.0
3-Methyl-pentane	C_6H_{14}	C-C-C(C)-C-C	86.2	63.6	0.669	74.5	74.3
2,2-Di-methyl-butane	C_6H_{14}	C-C(C)(C)-C-C	86.2	49.7	0.654	91.8	93.4

n-Heptane	C_7H_{16}	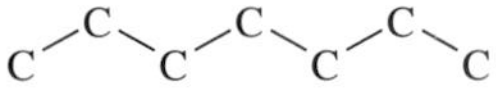 100.2	98.4	0.688	0.0	0.0
2,2,4-Tri-methyl-pentane	C_8H_{18}	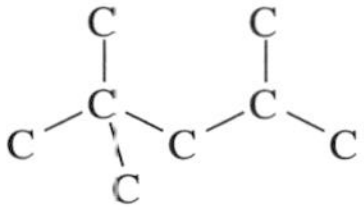 114.2	99.2	0.969	100.0	100.0

Naphthenes

Cyclopentane	C_5H_{10}	70.1	49.3	0.750	101.3	—
Cyclohexane	C_6H_{12}	84.2	80.7	0.783	83.0	77.2
Methyl-cyclo-hexane	C_7H_{14}	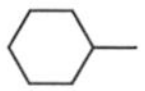 98.2	100.9	0.774	74.8	71.1

Aromatics

Benzene	C_6H_6	 78.1	80.1	0.871	117.0	114.8
Toluene	C_7H_8	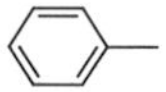 92.1	110.6	0.871	120.0	103.5
Ethyl-benzene	C_8H_{10}	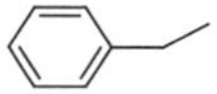 106.2	13.62	0.871	107.4	97.9
Meta-xylene	C_8H_{10}	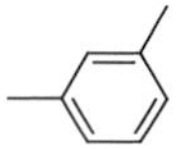 106.2	139.1	0.868	117.5	115.0
Para-xylene	C_8H_{10}	 106.2	138.4	0.865	116.4	109.6

(continued)

TABLE 7.1 (Continued)

Compound Group	Formula	Structural Formula	Molecular Weight	Boiling Point, °C (101.3 kPa)	Density (d_4^{15}, g/cm³)	Octane Number RON	MON
Olefins							
1-Pentene	C_5H_{10}	C=C—C—C—C	70.1	−30.0	0.646	90.9	77.1
1-Hexene	C_6H_{12}	C=C—C—C—C—C	84.2	63.5	0.678	76.4	63.4

Note: Heating values of paraffinic and olefinic hydrocarbons: 44.3–46.7 MJ/kg, aromatics: 41.0–41.8.

TABLE 7.2 Main conditions for determining the octane number

Properties	Test Method	
	Research (F-1)	Motor (F-2)
Fuel	Gasoline	
Motor	CFR with one cylinder	
Bore diameter, mm	114.3	
Displacement, cm³	610	
RPM, 1/min	600	900
Pre-ignition angle, deg	13	changing
Preheating of air, °C	52	38 ± 14
Temperature of mixture, °C	—	149
Temperature of cooling, °C	98–100	

is the quality of the gasoline or blending component. Generally, the isoparaffins have the best sensibility (between −2 and 3). The sensibility of engine gasolines does not have to be higher than 10. Some countries specify an average of RON and MON to indicate octane rating, denoted as AKI (antiknocking index).

Gasoline engines are constructed and calibrated to operate with a gasoline of appropriate octane number. If lower octane fuel is used, knocking will set in, which can lead to engine damage. Modern engines are fitted with knock sensors, and spark timing can be adjusted to lower octane fuel by retarding, while performance decreases and fuel consumption increases. Using higher octane fuel than recommended does not provide any additional benefit to the user, but no harm to the engine is expected. For older engines, the octane number requirement changes with the sea-level height because of the temperature and pressure changes. Lower octane at high altitudes and higher octane at sea level are preferred. In modern engines, an electronic control system adjusts to the temperature and pressure changes, so octane number requirement remains the same at all altitudes.

Besides the high research and motor octane numbers, gasolines must have a good (steady) octane number distribution over the whole boiling range. Wrong and relatively steady octane number distributions are illustrated in Figure 7.2 *a* and *b* [14,18]. (In order to determine the octane number distribution, the gasoline is distilled into 5 to 10 narrow boiling range fractions. Then the two octane number is separately measured, and plotted in a column-diagram as a function of volume percent.)

The octane number distributions of gasolines are often characterized by giving the octane number of the lighter fraction (distilled until 100 °C or 75 vol%) too. This is called the front octane number: $R_{100\,°C}$ or $R_{75\%}$, or $FEON_{100\,°C}$ (front end octane number) or $FEON_{75\%}$. These data give information about cold start possibilities and the accelerations from low speed (e.g., to estimate the time of acceleration). The main reason is that the act of "stepping on the gas pedal" causes the lighter hydrocarbons to evaporate faster and constitute the predominant mixture in engine [19].

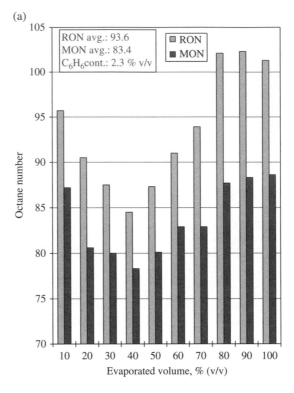

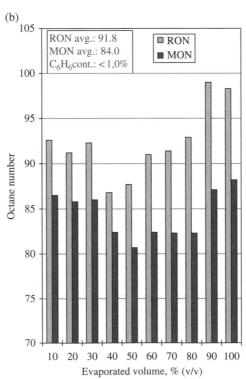

FIGURE 7.2 Gasoline with wrong and flat distributions of octane number.

The octane number distribution of gasoline is simply estimated as the difference between the octane number of the total fraction and the front octane number (ΔR) [20,21].

There is another octane number that is used to evaluate the on-road driving behavior of fuels, called the *road octane number*. The testing is carried out in stock cars. The value of the road octane number is usually between that of the research and motor octane numbers. The road octane number is often calculated by including other properties of the fuel. An average of the RON and MON numbers is frequently used to characterize gasoline quality, especially in the United States [2,13].

For the estimation of the road octane number, we suggested some equations. In the case of leaded gasoline [2,13],

$$RdON = 16.34 - 0.873 \times RON - 0.387 \times \Delta R + 2.051 \times C_{Pb} - 0.08 \times d_{API}$$

where,

RON = research octane number

ΔR = difference of the full octane number and the front octane number

C_{Pb} = lead content of gasoline, g/dm^3

d_{API} = API gravity

To evaluate the acceleration and steady speed, a so-called antiknock quality achievement index is recommended [2]:

$$RPI_{AC} = 0.51 \times RON + 0.4 \, MON + 0.09 \times FEON_{100°C} \quad \text{or}$$
$$RPI_{SS} = 0.25 \times RON + 0.75 \times MON$$

where,

RPI_{AC} = road achievement index at acceleration

RPI_{SS} = road achievement index at steady speed

RON = research octane number

MON = motor octane number

$FEON_{100°C}$ = front octane number

Engine gasolines are manufactured to handle different refinery gasoline streams and additives. But, after the blending, the blended octane number cannot be calculated from the octane numbers of the individual components and their percentage shares. Instead of these values, a *blending octane number* (RONB – RON blending, MONB – MON blending) is to be used, and it shows that the investigated component still behaves as a specific component in the blend. The octane number of the blended gasoline is determined as follows: a defined volume (20–25 vol%) of a known octane number gasoline (basic gasoline) is mixed with the sample, and the octane number of the mixture is measured. The blending octane number can be calculated with the equation [2]

$$RONB = RON_{Ab} + \sqrt[4]{\frac{(RON_E - RON_{Ab})100}{V}}$$

where,

RONB = blending octane number of the component calculated from experimental octane number data

RON_{Ab} = blending octane number of the base gasoline calculated from experimental octane number data

RON_E = blending octane number of the mixture calculated from experimental octane number data

V = volume share of the component, %

The blending octane number significantly depends on the octane number of the base gasoline, the mixing ratio, and the chemical composition of the sample [22–27]. The distinction and marking of engine gasolines in commercial practice is based on the research octane number. To fulfill the demands of drivers throughout the world, gasolines with different octane numbers (e.g., 91, 95, 98) are sold at filling stations.

7.1.2   Volatility of Engine Gasolines

The volatility characteristic of engine gasolines has a fundamental influence on the performance of Otto engines. Volatility is characterized generally by the gasoline's Reid vapor pressure and distillation curve [1–4,13]. The vapor–liquid ratio is often considered as well [1–4].

Good volatility has important consequences also for emissions and cold start and warm-up performances. Control of vapor pressure at higher temperatures reduces evaporation losses. Highly volatile gasoline can cause vapor lock in the gasoline pipeline, tank overload, and higher emission. Vapor lock occurs when a large amount of fuel vapor is formed that inhibits or even makes it impossible for the fuel to pump to the engine. This can lead to power loss and impede engine function.

The *vapor pressure* indicates the amount of low boiling components in the engine gasoline. It gives information about the formation of vapor in the fuel line in use, as well as about the amount of that can easily evaporate during travel and storage. The Reid vapor pressure (RVP) is determined at 38 °C when the vapor–liquid volume ratio is 4:1, in a pressure bomb (Reid equipment) that is immersed into a waterbath at standardized conditions. Its value must not be high; otherwise, hydrocarbon gases and vapors form in the gasoline pipelines and in the fuel pump, so the gasoline supply becomes uneven. Accordingly, the injection system or the carburetor does not get enough liquid gasoline because of the break in gasoline flow. All these phenomena

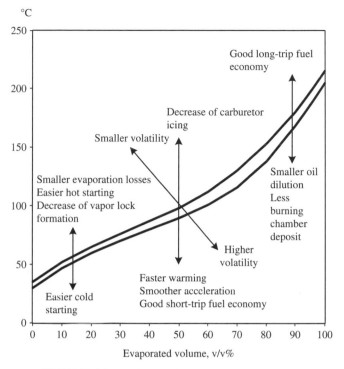

FIGURE 7.3   Typical distillation curve of gasolines

affect negatively engine performance. In addition, the higher vapor pressure increases travel and storage losses.

The maximum vapor pressure is based on the seasons, the climate of the region, and atmospheric pressure. In winter the vapor pressure limit is higher than in the summer. For these reasons distributed engine gasolines are graded according to volatility (e.g., 1, 2, EN 228:2009 + A1:2010 standard).

The other essential measured used to control the volatility of engine gasolines is the *distillation curve*, also known as *boiling point curve*. Remember, an engine's gasoline is a mixture of many hydrocarbons, so it cannot be characterized by a single, well-defined boiling point, but only by a boiling point range attached to the distilled volume ratio. This range is determined using "Engler" distillation, also known as ASTM distillation, at atmospheric pressure and under standardized conditions.

A typical distillation curve (the boiling temperature of the distilled fraction in vol% of the engine's gasoline) is shown in Figure 7.3 [2]. Notice how the range and the course of the distillation curve (characteristic curve) highly affect the behavior of the motor fuel in the engine.

The fuel–air combustion mixture produced is appropriate for both cold and hot engines. (That is why the engine gasoline consists of lower and higher boiling point compounds.) The performance of the engine—both in cold and hot running—is

TABLE 7.3 Gasoline fractions and effects of different boiling points on an engine's operation

Operation Parameters	Front Fraction, 10–30% Point	Middle Fraction, 50–70% Point	End Fraction, 90% Point, End Boiling Point
Cold startup	Determinative	Insignificant	Insignificant
Evaporation loss	Determinative	Insignificant	Insignificant
Gas and vapor formation	Determinative	Considerable	Insignificant
Hot starting	Determinative	Considerable	Insignificant
Idle with hot engine	Determinative	Considerable	Insignificant
Icing	Considerable	Determinative	Insignificant
Hot running	Insignificant	Determinative	Considerable
Acceleration	Insignificant	Determinative	Considerable
Energy content	Insignificant	Determinative	Considerable
Oil dilution	Insignificant	Insignificant	Determinative
Combustion residue	Insignificant	Insignificant	Determinative

generally characterized by startup readiness, the steadiness of the idle, and the throttle response. However, besides the distillation curve and the aforesaid vapor pressure, engine performance significantly depends on the design of the engine, the carburetion, the ignition, and last but not least the ambient temperature. (The distillation curve—like the vapor pressure—is standardized depending on the season. In winter, the temperature of some points of the distillation curve is lower by 5–15 °C than in warm periods, namely the distillation curve is below the regular curve of the summer gasoline. The easy start-up and the fast reach of the running temperature is ensured only this way.) As the abovementioned facts show, the smooth running of a vehicle depends closely on some particularly important measures and ranges of the distillation curve (see Figure 7.3). These relationships are listed in Table 7.3 [2].

At low temperatures icing appears in the carburetor (vaporizer) as gasoline evaporates. Then the water vapor in the fed air condensates and freezes next to the inlet. Consequently the air stream is blocked, and the mixture enriches with fuel, which leads to incomplete combustion and power loss.

Icing generally occurs only between 0 and 10 °C when the relative humidity is at least 85%. At temperatures higher than 10 °C, the carburetor does not cool below 0 °C, and in ambient temperature below 0 °C, the humidity of the air is lower, namely it is not high enough for icing. It has generally been found that icing increases with increased gasoline volatility.

For a homogeneous fuel–air mixture formation, the composition of all the gasoline fractions should be appropriate. When mixtures of different compositions are in the cylinders, the combustion processes will be irregular as well.

For the characterization of gasoline volatility the *vapor–liquid ratio* is also used sometimes. (The method is described in standard ASTM D 2533.) The ratio is determined at a given temperature and pressure; otherwise, the temperature belonging to the given vapor–liquid volume ratio is determined.

A volatility index (VLI) is also used to analyze volatility. This index is calculated from the vapor pressure and volume fraction of the Engler distillation range up to 70 °C (e.g., 1, 2, EN 228:2009 + A1:2010 standard):

$$VLI = 10 \times RVP + 7 \times E_{70},$$

where,

RVP = Reid vapor pressure, kPa

E_{70} = amount of fraction of Engler distillation up to 70 °C, v/v%

Eight different classes can be distinguished in this way. The volatility index of Western European gasolines is between 850 and 1200.

The correlation between the volatility of the fuel and engine performance is best described by the driveability index (DI) (e.g., EN 228:2009 + A1:2010 standard):

$$DI = 1.5 \times t_{10} + 3.0 \times t_{50} + t_{90}$$

where,

$t_{10,50,90}$ = temperatures at 10, 50, 90 volume fractions of Engler distillation in degrees Fahrenheit

This expression is extended by a term that takes into account the oxygen content called the *corrected driveability index* (CDI) (e.g., EN 228:2009 + A1:2010 standard):

$$CDI = 1.5 \times t_{10} + 3.0 \times t_{50} + t_{90} + 20 \times C_{O2}$$

where,

$t_{10,50,90}$ = temperatures at 10, 50, 90 volume fractions of Engler distillation in degrees Fahrenheit

C_{O2} = oxygen content of gasoline, %

The highest permitted CDI value is usually 1200; good performance of the vehicle and low hydrocarbon emissions are reached at this value. Difficulties can arise in spring and autumn.

The required volatility of gasoline causes vaporization losses at storage, fuel uptake, and during driving—in summer weather the warming of the gasoline tank and injector, contributes further to hydrocarbon emissions: solvents: ca. 40%; exhaust gas of vehicles: ca. 25%; evaporation losses of vehicles: ca. 10%; filling of vehicles: ca. 2%; other: ca. 25% [2].

Recent changes to the engine distribution system have introduced a special float cap fuel tank, with closed filling and drawer systems and hydrocarbon recovery units

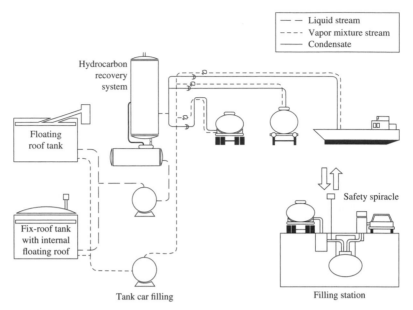

FIGURE 7.4   Scheme of closed distribution system of hydrocarbons

(see Figure 7.4). The hydrocarbon vapor and gas emissions of rail, road, and ship-filling stations, and gasoline stations will be decreased significantly (90–95%) [2], since the new setup enables recirculation and condensation of hydrocarbon vapors.

7.1.3   Stability of Gasolines

During storage, gasoline can undergo oxidation, forming "rubber-like" materials (nonvolatile, insoluble gums) and their precursors that are deposited in the engine, adversely affecting the operation, driveability, and exhaust emissions. The modern, high–performance and low-emission engines—mainly due to new injection systems—require the proper mix of fuel and air for optimum operation, and they are very sensitive to deposits in the fuel system. Sometimes the insoluble gums settle as sludge in the lower section of the fuel tank. Because oxidized gasolines contain hydroperoxides (gasolines with acidic pH), the different elastomers in the fuel systems get damaged. The fuel sludge can even permeate the elastomer, causing the fuel tank to leak and pollute the environment [1,2,4,7,13].

Oxidation stability depends on the composition of the gasoline, the manufacturing of the blending components (cracking, hydrogenation), and the storage conditions (e.g., tank, presence of metals and metal oxides, ambient and storage temperature, storage time, mixing). Storage stability is especially important in case of military and strategic reserves that are not used for a long period of time.

The types of hydrocarbons used also affect the stability of gasolines. Molecules with double bonds are much more sensitive to oxidation than aromatics or paraffins. Olefinic blending components are formed in refineries by different cracking processes in the absence of hydrogen (thermal and catalytic cracking, visbreaking, coking, etc.) and steam cracking of liquid hydrocarbons (gasoline pyrolysis). Furthermore diolefins and acetylene derivates can be formed as well. These are more sensitive to oxidation and easily form high molecular weight gums during storage. In certain regions of the world with high temperatures, the increased use of cracking products has brought about stability and marketing problems.

"Autooxidation" of gasolines can be promoted significantly by some sulfur, nitrogen, and oxygen compounds. The resulting peroxides reduce the octane number.

Therefore refineries have to avoid blending components that enhance undesired oxidation, and the oxidation stability of cracked gasolines has to be improved with hydrogenation and/or with the antioxidants.

Some metals (e.g., copper) can catalyze oxidation reactions; they have to be avoided in fuel systems. If these metals are present, their effects should be suppressed by metal deactivators.

The gasoline quality is significantly affected by storage conditions, so high temperatures, too much aeration and contacting with air, and small stored amounts reduce the product quality.

The oxidation of hydrocarbons takes place through peroxide and hydroperoxide radicals forming acidic compounds and gum-like materials. The formation of the latter proceeds through complex reactions. The first step is the formation of peroxides. Before the gasoline quality decreases their presence can be detected only in a small amounts. The first step of the forming of hydroperoxides and peroxides is the reaction of olefins with oxygen [1,2,28–34]:

$$C_nH_{2n} + O_2 \rightarrow C_nH_{2n-1} - O - OH \qquad \text{(hydroperoxide)}$$

In the next step the gum-like polymer forms from the hydroperoxides. The equation of gum formation is

$$n\left(C_nH_{2n}\right) + nO_2 \rightarrow \left(-C_nH_{2n} - O - O -\right)_n$$

The reactions take place by a chain mechanism:

1. At the chain-initiation step, free radicals (electron acceptor intermediates) form:

$$R - H \rightarrow R^{\cdot} + H^{\cdot}$$

2. At the chain-propagation step, the hydrocarbon free radical (R⁺) reacts with oxygen and peroxide radical as chain carrier is formed:

$$R^{\cdot} + O_2 \rightarrow R - O - O^{\cdot}$$

This product can react with another hydrocarbon molecule forming a new hydrocarbon free radical and hydroperoxide:

$$R - O - O^{\cdot} + R'H \rightarrow R - O - OH + R^{\cdot}$$

This chain reaction proceeds until the chain termination step or an antioxidant is introduced. The free radicals—besides the oxidation—promote the polymerization, resulting in the formation of high molecular weight substances.

3. At the chain-termination step, the chain reaction, in the absence of antioxidants, recombines the free radicals:

$$R^{\cdot} + R^{\cdot} \rightarrow R - R$$
$$R - O - O^{\cdot} + R^{\cdot} \rightarrow R - O - O - R$$

The oxidation or the formation of gum-like compounds takes place at a high rate when initiator-forming compounds are present. These are, for example, the very reactive olefins and diolefins. $R - O - O^{\cdot}$ free radicals that have long life and their concentration is high.

Antioxidants (e.g., gasoline additives) break the chain reaction or block the chain. The antioxidant (inhibitor) donates a hydrogen atom to the peroxide radical that is the chain carrier:

$$R - O - O^{\cdot} \ + \ AH \ \ \rightarrow \ R - O - OH \ + \ A^{\cdot}$$
$$\quad\quad\quad\quad\ \text{Inhibitor} \quad\quad \text{Hydroperoxid} \quad \text{Inhibitor radical}$$

The efficiency of the chain terminating antioxidants is affected by the stability of radical $A^{\cdot}$ and by the rate of its recombination or by its reaction with another free radical to form a stable compound.

The application of alcohols as oxygen-containing compounds in gasolines does not deteriorate the stability if the rate of free radical formation in the gasoline is under the detection limit. In FCC gasolines or other gasolines that are susceptible to oxidation, where the formation of free radicals and peroxide radicals is significant, the presence of alcohol enhances their formation. The radicals formed initiate the oxidation of alcohols to aldehydes and acids. Generally, these effects can be prevented by applying the proper inhibitors in the adequate amount.

To evaluate the storage stability of the engine gasolines, the most frequently used methods are the following:

- *Induction period* (EN ISO 7536/ASTM D 525). The gasoline sample is oxidized in a bomb at 100 psi (689 kPa) and 120 °C in oxygen. The pressure is recorded until the pressure drop is more than 2 psi (13.8 kPa) in 15-minute; this is the breakpoint. The induction period is the time required for the sample to reach the breakpoint. The characteristic curves of the induction period are shown in Figure 7.5 [2].

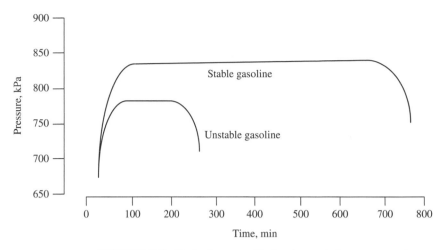

FIGURE 7.5 Typical curves of induction periods

- *Gum content determination* (EN 6246/ASTM D 381). The gasoline sample is evaporated in air stream at a specified temperature and the residue ("unwashed" gum content) is extracted with n-heptane. The insoluble residue is the gum content. The maximum 5 mg/100 cm^3 value is satisfactory according to most of the standards.
- *Potential residue method* (ASTM D 873). This is usually a 16-hour measurement under conditions of induction period measurement. The sample is filtered, and after the evaporation, the residual gum (soluble and insoluble) is the potential gum content.
- *Evaporation residue determined in copper beaker glass.* This gives information about the effects of the metals catalyzing the oxidation.

7.1.4   Corrosive Properties

Gasolines are in contact with the metal construction materials of the vehicles does not have any harmful effect. Pure hydrocarbons are not corrosive. Active sulfur-containing and acidic compounds in gasolines are very aggressive toward copper and its alloys, and they can cause significant wear. Solid and/or gel materials that are formed in chemical reactions can deposit in different parts of the engine and cause disturbances [1,2,13].

Gasoline corrosion is determined by a copper strip test (EN ISO 2160). A 12.5-mm wide, 75-mm long, and 1.5–3-mm thick, previously buffed and polished, copper strip is immersed into a water-free gasoline at 50 °C for 3 hours. The degree of corrosion is determined by a comparison with standard strips.

7.1.5 Chemical Composition

The chemical composition of fuel determines numerous other properties, for example, the octane number, stability, corrosivity, and lubricity. These have been discussed in detail previously. However, there remains to discuss how their individual components and mixtures can damage the environment and human health. For this we focus on the hydrocarbon composition of the exhaust gases [27,35,36].

Tables 7.4 and 7.5 show the properties of harmful material emissions from vehicles before and after treatment with catalysts [1]. Effect of gasoline quality to harmful material emission can be summarize for non-catalyst vehicles in the following. Decreasing of vapor pressure from 70 kPa to 60 kPa only the volatile hydrocarbon quantity decreases with more than 20%, CO, NOx, benzene, butadiene and aldehydes and particles do not change. This tendency is similar to vehicles with catalyst. Increasing the value of E100 from 50% to 60% quantity of hydrocarbons in the exhaust gas decreases a low. Increasing the value of E150 from 85% to 90% low quantity aldehyde and butadiene and NOx emission increasing can be observed. In case of vehicles without catalyst, benzene content of exhaust gas decreased with 10–20% at changing benzene content from 5 v/v% to 3 v/v% and after 1 v/v%. Decreasing of total aromatic content from 40% to 25% reduced the benzene, particle and CO content of exhaust gas, changing the olefin content from 10% to 5%, only the butadiene and particle emission decreased unequivocally in case of non-catalyst vehicles; in case of vehicles with catalyst the value of emissed NOx decreased with 4–7%. Blending oxygen content compounds decreased unequivocally the carbon monoxide emission (10–30%) and particle emission in case of both vehicle types, but at the same time increased the concentration of aldehydes in the exhaust gas (10–20%). Decreasing the sulphur content from 300 mg/kg to 10 mg/kg reduced the particle emission (ca. 10%) in case of vehicles without catalyst, while in case of vehicles with catalyst all type of harmful material emission are reduced between 2% and 20%, except the volatile hydrocarbons, what does not change.

Smog forms by a photochemical transformation of emissions of harmful materials. The potential forming index of smog is determined by the amount of different hydrocarbon groups present, using the following expression [2,35,36]:

$$PSPFI = 1.6 \times C_A + 1.27 \times C_{OL.} + 0.82 \times C_P$$

Where

PSPFI = photochemical smog potential forming index

C_A = concentration of aromatic hydrocarbons, %

C_{OL} = concentration of olefinic hydrocarbons, %

C_P = concentration of paraffinic hydrocarbons, %

As the expression shows, the main source of photochemical smog is a group of aromatics, followed by the olefins and paraffins. Aromatics are 25% more

TABLE 7.4 Harmful emissions of gasoline in cars without catalysts

Property	Change	CO	Hydrocarbon in Exhaust Gases	Evaporation of Hydrocarbons	NO$_x$	Benzene	Butadiene	Aldehyde	Particles[b]
Decrease of vapor pressure	70→60 kPa	0	±0	↓↓↓	0	0?	0?	0	0
Increase of E$_{100°C}$	50→60%	+0?	↓?	±0	0	0	0	0	0
Increase of E$_{150°C}$	85→90%	0	↓↓?	0	↑?	0	↓?	↓?	0
Decrease of benzene content	3→1%(v/v)	0?	0?	–0	0	↓↓	0	0	0
Decrease of aromatic content	40→25%(v/v)	↓	↓	0	→	↓↓	0	0	↓↓
Decrease of olefin content	10→5%(v/v)	±0	↑	–0[a]	→	0	↓↓	0	0
Use of oxygenates	0→2.7% O$_2$	↓↓↓	→	0 – ↑	±0	0	0	↑↑	→
Decrease of sulfur content	300→100(→10) mg/kg	0	0	0	0	0	0	0	↓(↓↓)

Note: *Impact*: 0: no; ± 0: from −2 to +2%; ↑ or ↓ : 2–10%; ↑↑ or ↓↓ : 10–20%; ↑↑↑ or ↓↓↓ : >20%; ? : no evidence of impact.

[a] Small decrease of reactivity.

[b] Direct injection.

TABLE 7.5 Changes in harmful emissions of gasoline in cars with catalysts

Property	Change	CO	Hydrocarbon in Exhaust Gases	Evaporation of Hydrocarbons	NO_x	Benzene	Butadiene	Aldehyde	Particles[b]
Decrease of vapor pressure	70→60 kPa	0	–0	↓↓[c]	0	0	0	0	0
Increase of $E_{100°C}$	50→60%	+ 0?	↓?	0	0	0	0	0	0
Increase of $E_{150°C}$	85→90%	0	↓↓	0	↑?	0	↓?	↓?	0
Decrease of benzene content	3→2%(v/v)	0	0	0	0	↓↓	0	0	0
Decrease of aromatic content	40→25%(v/v)	↓	↓?	0	↑?	↓↓	+0	↑?	→
Decrease of olefin content	10→5%(v/v)	0	+0	–0[a]	–0	0	↓↓	0	→
Use of oxygenates	0→2.7% O_2	↓↓	→	0 – ↑	+0	0	0	↑↑	↓?
Decrease of sulfur content	300→100 mg/kg	↓?	↓?	0	↓?	↓?	↓?	→	→
	100→10 mg/kg	→	→	0	→	→	→	↓↓	→

Note: The symbols are the same as described in Table 7.4.
[a] Small decrease of reactivity.
[b] Direct injection.
[c] Decrease of emission from very small value.

reactive than olefins and 200% more reactive than paraffins. The compensation for the octane number deficiency that resulted from the significant decrease of lead-containing compounds by increasing the aromatic concentration has not been an environment friendly solution. This has contributed to the relatively high ozone destructive effect and the atmospheric reactivity of these compounds in general [37–39]. This latter gives information about the rate of the gas phase reaction with the hydroxyl radical of the discussed components. Human health studies have shown that this index is at a level capable of damaging the mucosal membrane.

The benzene concentration has been rigorously limited by national standards as a result of a worldwide effort to decrease the aromatic and especially the benzene content of gasolines (≤0.4–1.0 v/v%). Benzene gets out to the atmosphere by evaporation, and through exhaust gas emissions is highly carcinogenic. It damages blood-forming organs (causing leukemia); toluene and xylenes significantly damage the nervous system.

Benzene is a health hazard, and the higher the aromatic content of gasoline, the higher are the carbon particulates and carcinogenic polyaromatic emissions.

The influence of aromatic compounds (compounds containing at least one benzene ring) has also been studied exhaustively with regard to the performance of gasoline. Aromatics are carbon-rich, high energy density, and high octane number fuel components but increase engine deposits and tailpipe emissions that include carbon dioxide. The boiling point of benzene (80 °C) is in the distillation range of gasoline and thus cannot be removed, so gasoline will contain the carcinogenic benzene. The importance to control benzene in gasoline has been recognized in most countries. The combustion of higher aromatics in engines also forms benzene. The US AQIRP (Air Quality Improvement Research Program) program determined that in reducing the aromatics from 45 to 20%, the total amount of toxins in exhaust can be reduced by 28% (74% of the toxic emission is benzene). A European study showed that the aromatic content has a linear relationship with CO_2 emission. By reducing the aromatics from 50 to 20%, CO_2 emissions were reduced by 5%. Most common the C_9 and C_{10} alkylbenzenes have a boiling point in the range of 306 to 424 °F (152–218 °C). The CARB gasoline specifications restrict the aromatic content, and T90 ASTM distillation that automatically limits the presence of C_{10} aromatics. The most desirable gasoline aromatic is toluene (boiling point 110 °C) from the point of view of combustion and emissions, but this component is carcinogenic as well.

7.1.6 Other Properties

In addition to the already discussed, important properties, some other physical, chemical, and performance properties of gasolines can be important commercially, or during storage and use.

As discussed in Chapter 5, the lubricating properties have become important due to the reduced sulfur content of gasolines to 50 mg/kg and later to 10 mg/kg. Because of the deep desulfurization, concentration of components of the natural lubricating properties decreased to the extent that significant wear has been observed in dosage pumps. The practically sulfur-free, modern gasolines have to contain lubricity improver additives to control pump wear.

The *sulfur content* of gasolines came from hydrogen sulfide, organic sulfur compounds (e.g., mercaptanes), and from free sulfur. Sulfur causes corrosion problems, when it is in higher concentration than 10–20 mg/kg. Sulfur dioxide formed in combustion produces an acid when combined with water below the dew point. This acid increases cylinder wear at cold start, and in the case of a not sufficiently warm engine, it corrodes the exhaust pipe. Moreover the sulfur dioxide and sulfur trioxide in the environment, significantly contribute to acid rain. Some sulfur compounds change the color and stability of gasolines [1–13].

Modern vehicles are even more sensitive to sulfur levels in fuel than older vehicles. The newer emission regulations prescribe long life compliance with highly efficient after treatment catalysts that requires ultra-low sulfur content in the fuel and lubricant. Vehicle manufacturers are also developing engines with improved fuel economy and reduced CO_2 emission by operating at a lean air–fuel ratio. These technologies can reduce fuel consumption by 15–20% but are very sensitive to sulfur content in the fuel. Therefore the important target in all fuel quality improvements is the reduction of the sulfur content. For fuel-efficient lean mixture operation and ultra-low emission constructions, sulfur-free gasoline is essential.

With the reduction of sulfur in gasoline other harmful emissions will also be lowered considerably. For example, the emissions computed with different EUDC methods on decreasing the sulfur content from 382 to 18 mg/kg are as follows: CO: −42%; hydrocarbons: −52%; NO_x: −21%. The results of the EUDC studies were published by EPEFE (European Programme on Emission, Fuels and Engine Technology), a subcommitte of the European Automobile/Oil Program. The same organization found that, hydrocarbon emissions do not decrease linearly with the sulfur content. In addition, the sulfur content of gasolines (sulfur dioxides that form during combustion) decreases very significantly the activity of the after-treatment catalysts [14,20,40–45]. As a result in developed industrial countries the permitted sulfur content is at most 10 mg/kg (European Union, EN 228:2009 + A1:2010)— 15 mg/kg (United States, California).

Liquid fuels have the advantage of *high energy content*. Engine performance is determined not exclusively by the heating value of the fuel, but by the internal energy of a theoretically correct fuel–air mixture. This oscillates within narrow margins, and for gasolines, it is about 3.8 MJ/dm^3. The heating value is computed by the fuel's elemental composition.

Knowing the exact *density* of a fuel is useful primarily at the point of gasoline receipt. The flow of fuel into the engine influences the operation of the carburetor. The volumetric mass of the flow to the carburetor is set generally at 0.700 g/cm^3. If the density of the gasoline is lower, such as when the ambient temperature is high, then the carburetor malfunctions.

The oxygen content of gasolines, raised by adding alcohols and/or ethers, has to be in line with strict standards for the last 10 to 15 years. This is because these compounds have a high octane number (ca. 110–118), and by boosting the internal oxygen content, they assure improved combustion.

The water content of gasolines influences unfavorably both the engine's operation and storage. Water dissolves poorly in hydrocarbons. Solubility depends on the type of hydrocarbon and increases with rising temperatures. Aromatics have the highest water uptake capacity, followed by naphthenes and paraffins. The saturation limit at 20 °C is about 0.005–0.05%. Alcohol solubility in water is practically unlimited. Therefore they are used in limited concentrations in order to avoid phase separation [46–48].

The flammability of fuels is determined by their *flashpoint*. Gasolines have low flashpoint and are the most dangerous class. The *self ignition temperature* relates to some extent to compression tolerance, and it increases in the order of paraffin, iso-paraffin, and aromatic. In gasolines the probability of self-ignition is actually low. It can happen when the gasoline vapors being heavier than the air are concentrated (near to the soil, and take fire from a far ignition source).

Electric conductivity. Static electricity, can be quite dangerous. Refining is by a chemical method that generates friction electricity, which is accumulated in gasoline. The chemical is positively changed; the gasoline is negatively charged. Contact with a conductor can cause a spark and so a fire. Streaming fuels can be charged as well. To avoid a spark generation, tanks and pipes are grounded. The charging is also influenced by the humidity of air.

The heat of evaporation influences the carbureting. Very high evaporation heat of low carbon number alcohols is utilized in racing cars where the cylinder is cooled with the compressed mixture containing alcohols.

7.2   EFFECTS OF PROPERTIES OF DIESEL GASOILS ON ENGINES AND THE ENVIRONMENT

Early diesel engines were large, slow in speed, and ran on low-quality fuels. As the engine has been improved, and became lighter and capable of higher speed, the viscous and heavy fuel had to be changed to lighter fuel of a well-defined quality. Diesel engine performance is a function of injection timing, compression ratio, certain mix of atomized air and fuel, and ignition delay. The cetane number and other properties of diesel fuel become important in the high-speed engines. High-speed diesel fuel is the main transport fuel for highway, off-highway engines, tractors, and railroad engines.

The diesel gasoil quality and diesel engine parameters determine together the introduction of fuel(s) into the combustion chamber, the mixture forming, the ignition, and the combustion characteristics, and so the emission properties. The properties of diesel gasoils affect [1,3,15,50]:

- Mixture forming
- Ignition and combustion processes

- Performance
- Lubricating oil
- Environment and emission

Beside these listed effects the diesel gasoil's quality can influence more or less other properties, pertaining to storability, flammability, and so forth. The critical properties are:

- Density and energy content
- Distillation
- Chemical composition (hydrocarbon group, heteroatom content):
  - Stability
  - Corrosion effect
  - Lubricity
  - Cold flow

Less critical properties are:

- Viscosity
- Foaming
- Flashpoint
- coking ability
- electrical conductivity

7.2.1   Ignition and Combustion Properties of Diesel Gasoils

Ignition delay is determined by an engine's design, operating conditions, and the cetane number of diesel gasoil. The Diesel gasoil with a lower cetane number causes higher ignition delay. This means delayed cold starting, consequently higher emissions, often as "white smoke," and engine noise caused by the higher pressure peaks. With higher cetane number diesel engines run in hot operation mode, so the combustion is better and smoother, which decreases the emissions of all harmful materials, the noise level, and fuel consumption and increases the engine's life.

The cetane number of diesel gasoils depends on the hydrocarbon composition: the more paraffinic and lower aromatic compounds there are in the fuel, the higher is the cetane number (see Chapter 3). The value of the cetane number can, however, be improved with additives.

7.2.2   Density and Energy Content of Diesel Gasoils

The density of diesel gasoils influences directly the volumetric energy content (lower heating value), which in turn influences the driving moment as a function of the revolution. With higher density and the same injection volume, engines gives higher performance as a result of the higher energy content, so the acceleration of the vehicle will be higher [1,13,51].

With increase in the density, particle emission increases, but at the same time the volumetric fuel consumption decreases if the vehicle manufacturer has set a maximal injection volume [1,2,50–57]. A decrease of density does the opposite.

7.2.3 Distillation Properties of Diesel Fuels

A fundamental parameter of diesel fuels is the distillation curve, which is a curve of boiling points. Because diesel gasoil is a mixture of several hydrocarbons, it cannot be characterized by a single, well-defined boiling point, but only with a boiling range, and by some temperature data related to the amount of distilled fraction. The determination is carried out with Engler distillation, also known as ASTM-distillation, at standard conditions and ambient pressure (ASTM D86, EN ISO 3405). The boiling point curve determined by this method is physically not exact, but reflects boiling point distribution which is close to the practical conditions of rapid evaporation. With the increase of the temperature, more volatile components may be reduced, while the heavier compounds may be caught by lighter ones. Therefore the real initial boiling point is kept lower, and the final boiling point higher than the measured values. The results of a standard test are used to compare these parameters of diesel gasoil.

A typical distillation curve (the boiling temperature as a function of the distillated fraction, v/v%) of a diesel gasoil is shown in Figure 7.6. In principle, hydrocarbons out of the presented boiling range are suitable to combust in diesel engines as well. This relates to both the lighter (ca. 118–190 °C boiling range) and the heavier (ca. 370–500 °C boiling range) fractions. For the previous case, the examples are the so-called city diesel gasoils, and for the latter, the ship's diesel-engine fuel. It should be noted that the boiling range of the city diesel gasoils is lower than that of the conventional diesel gasoils. Accordingly, they are of favorable hydrocarbon composition (see later), volatility, and of low sulfur content, so the emissions are much lower (e.g., 50, 58). However, because of their lower density, their energy content is lower, so the fuel consumption (usually in volume) can be higher by 3–6% than that of conventional diesel gasoils. (The production and sale of city gasoils have lower importance because of the very strict specifications used for standard diesel fuels, e.g., ≤ 10 mgS/kg; ≤ 8% PAH.)

Diesel fuel has a wide boiling point range, which can be divided into three sections: light fraction, medium fraction (*50% distilled T50*) and heavy end (*T90, T95, and final boiling point*). The lighter fraction affects the flashpoint of the fuel and also the starting ability of the vehicle. The medium fraction influences density and viscosity, which in turn affect engine performance. The heavy ends have profound influence on the emission and are subjected to thorough investigations. However, clear directions have not emerged from the European EPEFE studies. In heavy-duty diesel engines, decreasing T95 from 375 °C to 320 °C were observed to cause lower NO_x and higher HC emissions. In light-duty engines, the same change brought on a

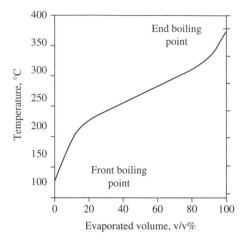

FIGURE 7.6   Typical distillation curve of gasoils.

reduction in particulate matter and an increase in NO$_x$ emissions. No clearcut conclusions emerged out of these studies, except that the heavy ends are responsible for the formation of soot, smoke, and particulate matter. There is a trend to lower the heavy ends and thus improve fuel quality.

In motor vehicles the boiling range of the fuel basically affects other properties such as viscosity, density, ignition, flashpoint, and flow at low temperatures. The range and course of the boiling point curve determine the applicability of the diesel gasoils and their behavior in the engine. One of the preferred points on the curve— but not prescribed because of the foregoing reason—is the initial boiling point, which is closely related to the flashpoint in being a safety requirement. Therefore the initial boiling point of the diesel gasoils should be at least as high as that of the closed-cup flashpoint, 55–60 °C. So the amount of lighter components in diesel gasoil is regulated by the initial boiling point and the flashpoint. The value of the closed-cup flashpoint is limited by standards, whereas the flashpoint is indirectly regulated by the initial boiling point.

The fraction distilled below 250 °C is responsible for the low density and energy content; the boiling fraction up to 350 °C, and higher up to 370 (360) °C (at least 85% and 95%), sets the limit for the heavier components. Fractions in this temperature range contain polynuclear aromatics in high concentration and are mainly responsible for particulate emissions. They significantly contribute to the formation of carbonaceous deposits on the nozzles as well. Of course, at the top of the boiling range, the concentration of the higher molecular weight n-paraffins can be so high that they significantly degrade the cold flow properties of diesel gasoils (see Chapter 3).

It was experimentally proved that a distillation point of 90v/v% correlates well with the freezing point as an independent variable, pour point with a concentration of

C_{22}–C_{23} n-paraffins, and the cold filter plugging point with the two previous independent variables [50].

The amount, distribution, and dominant effects of the unfavorable properties in hydrocarbons are essentially determined by the source of the crude oil, from which gasoil is produced.

7.2.4 Chemical Composition

The chemical composition of diesel gasoils determines several important properties, such as the already discussed ignition and combustion properties, the density, the energy content, and even the shape of the distillation curve. The chemical composition of diesel gasoils also affects the stability, the corrosive effects, the low temperature flow properties, and the composition of the gas mixture formed during the combustion, and so the emissions.

7.2.5 Stability of Diesel Gasoils

Oxidation stability of diesel gasoil is a problem for consumers and engine manufacturers alike. Diesel gasoils can only be partially oxidized and polymerized, so rubber-like materials (insoluble resins) and their precursors are formed. They are responsible for forming deposits in the fuel system (or be corrosive), which causes the fouling of filters, and if they are transmitted through the filter, they can form deposits on the upper cylinder wall of the engine and on the nozzles. These resin compounds greatly decrease the performance of the engine, impair driving, and increase the emission as well.

The stability of diesel gasoils depends on:

- Fuel composition. The production technology of the blending component is the main factor. For example, FCC gasoil or cocker gasoil not sufficiently hydrogenated contains a lot of unsaturated, aromatic, and heteroatomic compounds, unlike hydrocracked gasoil.
- Storage conditions. The tank's construction (fix or float caped), the presence of metals and metal-oxides, the ambient and storage temperatures, and the duration of storage are all important factors, and, of course, these can all combine.

Double bond containing molecules of diesel gasoils are more sensitive to oxidation than paraffins and aromatics. Sulfur-, nitrogen-, and oxygen-containing compounds significantly promote "autooxidation." The oxidation of hydrocarbons takes place through peroxide and hydroperoxide radicals with the forming of acidic and rubber-like products. Reactions start with the forming of free radicals and proceed by a chain mechanism (as has been discussed for gasoline).

It must be emphasized that all the nitrogen-containing compounds significantly decrease the stability of diesel gasoils [1,4,50,59]. For example, during the

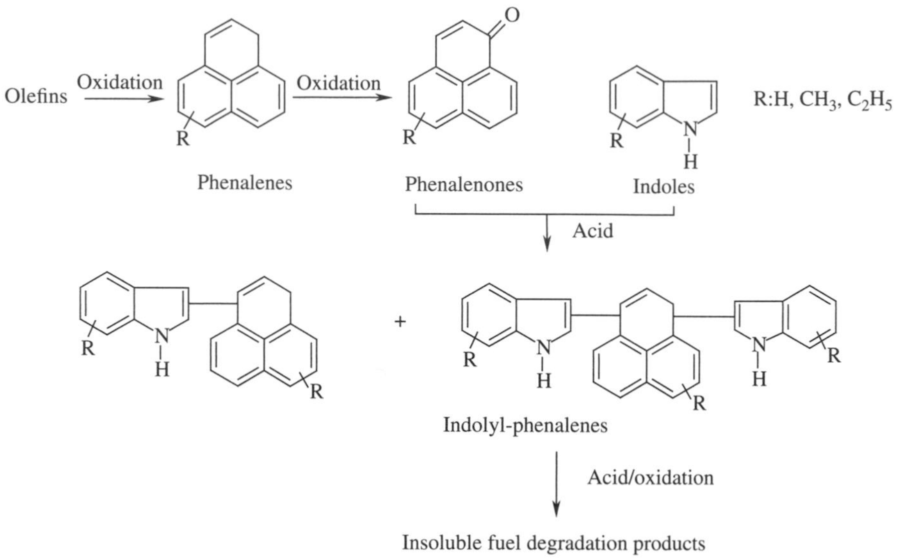

FIGURE 7.7   Decreasing storage stability due to nitrogen compounds

polymerization of pyrrole derivatives or the reactions of indoles with some heterocyclic compounds (Figure 7.7), high molecular weight compounds are formed, and these are hardly soluble in further chemical reactions (e.g., oxidation) [50]. Other nitrogen- and/or sulfur-containing compounds, organic acids, and the above-mentioned reactive olefins behave as indoles as well. Their oxidation reactions are catalyzed by some metals (e.g., by copper).

Phenalenes formed by the oxidation of reactive olefins as shown in Figure 7.7, and they make complex indophenalene salts with indols, like diesel gasoil blending component constituting compounds, in the presence of an acid catalyst. An organic acid is initially in the gasoil, or it is formed during the oxidation of merchaptanes to sulfonic acid. However, the presence of acids should be restricted in any case, for example, by the removal of the acidic components or their precursors from the fuel, by transformation of the precursors (sulfur- and nitrogen-containing compounds, acids, olefins, etc.) in the process of heteroatom removal, or by the application of antioxidant or basic stabilizer additives. The solution to the presented problems is not so easy. The stability-decreasing compounds of each blending component are to be considered, since in unfavorable circumstances, despite the blending of two or more components that are stable, while in storage reactions may take place to make them unstable, and heavy molecular weight rubber-like compounds may form.

It should be noted also that the thermal stability of gasoils is adversely influenced by some additives. For example, the cetane number improver (ignition improver) 2-ethyl-hexyl-nitrate assists the formation of heavy molecular weight intermediate products [50].

The level of oxidation stability of diesel gasoils as desired by vehicle manufacturers and customers, can be attained with appropriate hydrogenation of the blending components using thermal technologies, and with the application of special additives in a suitable concentration (see Chapter 5).

7.2.6 Corrosion Properties

Diesel gasoils come in contacts with the metallic structural materials of vehicles, but they must not exert any corrosive effect on them. Pure hydrocarbons are noncorrosive materials. The acidic and active sulfur-containing compounds present in gasoil (e.g., naphthenic acids) are very aggressive toward copper and its alloys and can cause significant "wear," especially at long shutdowns. Solids and/or gels formed in the chemical reactions can be deposited and cause trouble in various parts of the engine [1,3,13,50].

With long-term storage in changing temperatures, water condensation can corrode tank walls. The corrosion can be prevented by applying anticorrosion additives.

However, in engines and exhaust systems, corrosion and other undesired effects cannot be prevented with addititives, since they are brought by the acidic compounds (sulfur dioxide, sulfur trioxide, nitrogen oxides, acids) formed during the combustion of sulfur- and nitrogen-containing compounds. These can reduce the reserves of the engine oils under certain conditions, and in the environment engine oils contribute to the formation of the acid rains and the ozone depletion. Furthermore these sulfur-containing combustion products significantly decrease the activity of after-treatment catalysts. Consequently, as we discussed in the preceding chapter, in the last couple of years the sulfur content of diesel gasoils was drastically reduced in several steps. In the developed, industrial west, the maximum sulfur content is 10 mg/kg (European Union) and 15 mg/kg (United States, California). The nitrogen content has not yet been limited. The reasoning may be that deep hydrodesulfurization takes place with the reduction of the nitrogen content. Some countries are, however, setting up standards to limit the nitrogen content.

7.2.7 Lubricating Properties

Recently the significant loss of lubricity in diesel gasoils, due to their decreased sulfur and nitrogen contents, and reduced content of polar compounds (e.g., aromatics), has become apparent in the wear to the cam disc of the injection pump and the injector. The wear on the cam disc has a direct effect on the vehicles performance causing higher fuel consumption, higher emissions, and a shorter pump lifetime (see Chapter 3).

Several laboratory methods have been developed to test injection pump wear (HFRR: high frequency reciprocating rig; BOCLE: ball on cylinder lubricity evaluator; BOTS: boll on three corner seats). While investigations using different vehicle pumps give more reliable results, especially using vehicle fleet experiments, these are very expensive and have a high energy demand. Nevertheless, the detrimental effects of diesel gasoils with low lubrication capacity is supported by standardized laboratory lubrication investigations, and the maximal wear values are prescribed accordingly in product standards. In Europe, the high frequency reciprocating rig (HFRR) testing is prevalent (EN ISO 12156–1). Current standards limit the scar values obtained by this test.

Essentially, in an HFRR test, sliding friction in the injection pump is simulated by pressing a 6-mm diameter ball with constant force on a polished steel sheet, which is under diesel gasoil. The degree of wear is measured by the average wear scar (μm) on the ball after 75 minutes and at the temperature used during the test.

In last 5 to 8 years some lubricity improver additives have been developed for preventing harmful wear, and so far they have proved satisfactory.

7.2.8 Low-Temperature Flow Properties

The high cetane number of the diesel gasoil is mainly due to the straight-chain paraffin hydrocarbons. But these compounds adversely affect cold flow properties of diesel gasoil, because they have a high freezing point and settle out by forming crystals (see Chapter 3).

The crystals are large enough to hinder pumping, to block up the fuel filters, and to settle in the tanks. These phenomena, their mechanisms, and the possibilities of prevention with additives (flow improvers, paraffin dispergators, etc.) were presented and discussed earlier. Additionally, the cold flow properties can be improved in other ways, for example, by electric heating of the fuel filter.

In the test to determine the CFPP (cold filter plugging point) (EN 116), which is the lowest temperature for continuous fuel feed, a filter with 45-μm pores is used, while the test standard for diesel engine vehicles use filters with 5-μm pores. So it is important to apply additives that enable continuous fuel feed through such small pore filters.

7.2.9 Effects of Chemical Composition on Emissions

The relations between the chemical composition and the properties of diesel gasoils have been discussed earlier. In this chapter, the emissions harmful exhaust gases due to the chemical composition of fuel will briefly be reviewed [50–57,60].

The compositions and changes in some properties of diesel gasoils as measured in their emissions are shown in Table 7.6 for light diesel vehicles, and in Table 7.7 for heavy diesel vehicles [50]. The errors in these estimations are significant, but because they are based on the results of a fleet investigation, they may differ depending on their sources.

Clearly, the sulfur content decrease is not reflected directly in the decrease of carcinogenic components in the exhaust particles, which contain a share of carbon-rich particles and polycyclic aromatics. So, besides developing a more perfect engine, work still need to be done on reducing the polycyclic aromatics content of diesel gasoils, as these emissions are carcinogens. Ideally, the reduction of aromatic content (including polycyclic aromatics) will proceed simultaneously with the

TABLE 7.6 Effects of diesel fuels on the emissions (light vehicles)

Property Changes	Change	CO	Hydrocarbon	NO$_x$	Particle
Decreasing of sulfur content	2000 → 500 mg/kg	0	0		↓
Decreasing of density[a]	850 → 820 kg/m³	↓	↓	0	↓↓↓
Decreasing of polyaromatics	8 → 1 v/v%				
Increasing of cetane number	50 → 55	↓↓	↓↓	±0	↓
Decreasing of T$_{95}$	370 → 330 °C	−0	−0	−0	−0

Note: 0 = no impact and ±0 = from −2 to +2 %; ↓ or ↑ = 2–10 % impact; ↓↓ or ↑↑ = 10–20 % impact; ↓↓↓ or ↑↑↑ = > 20 % impact.
[a]The difference was significant only when studying the combined effects; further studies are needed.

TABLE 7.7 Effects of diesel fuels on emissions (heavy vehicles)

Property Changes	Change	CO	Hydrocarbon	NO$_x$	Particle
Decrease of sulfur content	2000 → 500 mg/kg	0	0		↓↓
Decrease of density[a]	850 → 820 kg/m³	0	0	↓	↓
Decrease of polyaromatics	8 → 1 v/v%				
Increase of cetane number	50 → 55	↓	↓	↓	↓
Decrease of T$_{95}$	370 → 330 °C	−0	−0	−0	−0

Note: 0 = no impact and ±0 = from −2 to +2 %; ↓ or ↑ = 2–10 % impact; ↓↓ or ↑↑ = 10–20 % impact; ↓↓↓ or ↑↑↑ = > 20 % impact.
[a]The difference was significant only when studying the combined effects; further studies are needed.

reduction of the sulfur content of diesel gasoils, so that the total particle emissions are reduced significantly.

Emissions from Aromatics Aromatics do have a big role in diesel engine emissions. During combustion, the higher aromatic content increases the flame temperature and consequently the NO$_x$ emission. Since diesel fuel contains a higher boiling fraction, certain polyaromatics are present in it, and these influence the formation of particulate and PAH emissions. The ACEA follow-up program to EPEFE has shown that by reducing total aromatics 30–10%, NO$_x$ is reduced by about 4–5% in both heavy-duty and light-duty diesel engines. Similarly, when polyaromatics were reduced 9–1%, particulate emissions were reduced by about 4–6 %. The presence of polyaromatics (tri$^+$) in diesel fuel has a direct correlation with PAH emissions in diesel engine exhaust [61]. Diesel engine manufacturers have achieved great success by incorporating new technologies to control engine emissions like NO$_x$ and particulates.

The influence of the aromatic content of diesel fuel on soot formation has been investigated by varying aromatics in the fuel at 0–27 % (total of mono-, di-, and tri-nuclear aromatics to max 25%). Aromatics in diesel fuel are also correlated to unburned hydrocarbons, particulate formation, and NO$_x$ in the emissions. Thus the high boiling polynuclear aromatics in diesel fuel were limited. In another study, the sulfur content and density have been found to be the most important parameters affecting the particulate emissions in both DI and IDI engines [62]. A soluble organic fraction (SOF) of PM was found to be 17–22% due to higher soot loading in the engines. Higher cetane numbers in diesel fuels were found to be beneficial in reducing gaseous emissions of NO_x, CO, and hydrocarbons (HC).

Vehicles with oxidation catalysts convert a large amount of SO_2 into SO_3; thus increasing the particulate matter substantially. US studies on heavy-duty diesel engines showed particulate reduction in the range 12–27% for sulfur reduction from 0.3–0.05%. For similar sulfur reduction, European studies showed benefits in the range 7–12% [63]. Some studies showed no change. As the emission regulations become tighter with respect to NO_x, PM, and CO_2 emissions and fuel consumption, ultra-low sulfur fuels will need to be used to meet these requirements. There is compromise between PM fuel consumption and NO_x. The deNOx catalyst seems to remove this emission from oxygen-rich exhaust. The oxides of sulfur compete with the oxides of nitrogen for adsorption on the deNOx catalyst, thus decreasing its catalytic efficiency. In NO_x storage catalysts, a platinum catalyst converts NO to NO_2, which is stored as barium nitrate. Rich fuel operation provides hydrocarbons to reduce NO_x over a standard three-way catalyst. The presence of sulfur in the fuel results in storage of sulfur as barium sulfate, which exhausts the storage system. The catalyst thus requires sulfur-free fuel for good performance. Oxides of sulfur poison the NO_x adsorption sites of the catalyst. Fuels containing 30 ppm sulfur show relatively poor performance in NO_x conversion efficiency as compared to diesel fuel containing 3 ppm sulfur. Thus sulfur reduction is absolutely necessary in employing these catalyst systems.

For reducing emissions, combustion improvement technologies, such as high-pressure injection, exhaust gas recirculation (EGR) systems, and exhaust gas treatment technologies, like oxidation catalysts, diesel particulate filters (DPF), and NO_x storage reduction catalysts, have been developed. DPFs are used for capturing and removing particulate matter in exhaust gas with filters of various structures. For example, a wall-flow type DPF has a structure where the two ends of a honeycomb monolith are alternately plugged. However, in this type of particulate filter, as the captured PM accumulates, pressure loss increases and engine power decreases, as does the fuel efficiency. Thus, DPFs have to be replaced after certain mileages, or have the particulate matter removed by oxidation and combustion. Recently, continuous regenerative DPFs have been developed. Common rail engines are being developed with electronic injections that increase the exhaust gas temperature to burn off trapped particulates. Through the combination of post–combustion fuel injection in the cylinder, pre-filter hydrocarbon combustion, and a catalytic additive mixed on board with the fuel, the modern diesel engine could be enabled to continuously regenerate diesel particulates filters. Sulfur in diesel fuel is converted to SO_2 and then to a sulfate by the oxidation catalyst in the DPF. Sulfate particulates are emitted into the atmosphere. It is therefore necessary to use diesel fuels with ultra-low sulfur content (3 ppm) to obtain the maximum advantage of advanced low emissions engines. Ash from the fuel and lubricants also has a profound effect on the life of diesel particulate filters, and this has to be kept to a low level as well.

The particle filter contains catalytically active coating on the filter body to accelerate burn-off during a regeneration phase of the soot particles accumulated on the filter. The particle filter's catalytic coating contains compounds of barium, magnesium, and an element of the platinum group metals. The soot particles accumulated on the filter are burned by the exhaust gas of the diesel engine driven using a lean mixture. Soot particles have an ignition temperature and the particle filter is effectively regenerated from time to time by raising the temperature of the particle filter above the soot ignition temperature.

The effect of fuel additives containing sodium and lithium of an aliphatic alcohol on the regeneration behavior of diesel particulate filters was studied [64]. The efficiency of these additives was compared with ferrocene, which was found to be effective in particulate filter regeneration. The additives were tested under steady-state conditions, on a transient dynamometer, and finally in a real driving test. As a result of the addition of alkali metal salts, the ignition temperature of the soot in the particulate filter proved to be considerably lower than the ignition temperature without the additive. The results with the alkali metal additives are similar to those of ferrocene and other additives with transition metals like manganese and copper. However, the regeneration intervals of the particulate filter are shorter under real operating conditions. This way the critical filter loading with soot can be avoided, since it can lead to filter damage during regeneration. The ash accumulation in the particulate filter caused the filter pressure to continuously drop during the driving test. Besides this positive outcome, the ash formed is water soluble and can easily be removed by rinsing the filter with water. Engine bench tests and vehicle programs on light-duty and heavy-duty diesels have confirmed that a bimetallic platinum and cerium diesel fuel borne catalyst (FBC) can reduce engine emissions and improve the efficiency of diesel oxidation catalysts and diesel particulate filters [65]. Particulate emission reductions were up to 25% for FBC alone, up to 50% for the FBC and oxidation catalyst, and up to 95% for FBC and a diesel particulate filter. Vehicle and engine bench tests confirm that the regeneration of the filters at exhaust gas temperatures can be as low as 280–320 °C. Field trials with commercial fleets have confirmed the engine bench test, which showed the fuel saving to be around 6%. Measurements have also confirmed that ultra-low additive dose rates of 4 to 8 ppm do not result in an increase of ultra-fine particulate emissions and significantly reduce the ash loading of diesel particulate filters. Combined with exhaust gas recirculation, the fuel-borne catalyst and a diesel particulate filter have demonstrated NO_x + HC and particulate emissions to be below 2.5 g/bhp-h (3.3 g/kWh) and 0.01 g/bhp-h (0.013 g/kWh).

Ultra-Fine Particulate Emissions   Many studies have suggested that the emission of ultra fine nanoparticles (smaller than 50 nm in size) have the ability to penetrate into human lung tissue and cause serious respiratory disorders and acute pulmonary disease [66,67]. A study found that 6% of all deaths in Austria, France, and Switzerland are due to the air particulate pollution [68]. Other studies have shown that particle emissions of gasoline injection engines are much lower than those of diesel engines [69–71]. These emissions increase during cold starting and at high loading. Fuel additives also influence particulate emissions [72]. Different additives have shown different results; for example, polyether amines have been shown to reduce particle number emissions (by 35% for PFI and by 28% for GDI engines at 100 km/h).

7.2.10 Other Properties

Besides the effects of diesel gasoil properties, discussed so far in this chapter, there are yet some important physical properties that we summarize below.

Viscosity   The work on perfecting the fuel-feeding system and the mixture-forming system of diesel engines, and investigations into lubricity improver additives, has overshadowed the significance of viscosity in diesel gasoils. Viscosity also has a role in good injection and assists in the lubrication of the fuel supply system [1,3,13,50]. Too high viscosity makes the pumping difficult at a low temperature, resulting in cold start problems, and too low viscosity can lead to failure at hot starting and at a high temperature it can cause performance to deteriorate and eventually to wear out the pump. During injection, too high viscosity causes larger diesel gasoil drops to form ("poor atomization"), so these are fed to the burning zone before having been finely mixed with air. Lower performance results with higher harmful material emissions, and, under some conditions, the dilution of the engine oil. The industry standards for the kinematic viscosity range of diesel gasoils have to be considered. For example, with the decrease of temperature of ordinary fuel 40–20 °C or with the increase of the pressure to ca. 600 bar, the viscosity doubles [50].

Compatibility with Seals   It has been found that in diesel gasoils of low aromatic content especially during long term storage the seals shrink, and consequently dripping fuel can be observed. Improved seals with appropriate resistance (e.g., fluoroelastomers) were developed. They have a tight sealing in the range of −40 °C to +130 °C [50,68,73].

Foaming   Foaming of diesel gasoils occurs especially in cool or cold weather. Because at this time the viscosity of the diesel gasoil is higher, during the filling of the vehicle fuel tank the compact foam can reach the loading point, which will turns off the fuel flow in the nozzle. This increases the filling time, and can cause the overflow of fuel, thus polluting environment and can cause inconvenience to the consumer. However, foaming rarely occurs because all modern diesel gasoils contain antifoaming additives.

Coke Residue   The coke residue indicated in fuel by the Conradson number contains both organic and inorganic (e.g., residue from metal-containing additives) components. The index helps give information about the fuel coke forming liability on jets and in combustion chambers. Since metal-containing combustion improvers can significantly affect the Conradson number, and lead to inaccurate conclusions, therefore, coke residue should be determined from the base (additive-free) diesel gasoil.

Water Content The water gets into gasoil with blending components, such as when the initial boiling point is established by stripping or when ammonia and hydrogen sulfide (produced by heteroatom removal) are removed by blowing with steam. After the subsequent drying processes the diesel gasoil still contains a small amount of water. The maximum amount of dissolved water in the diesel gasoils depends on the hydrocarbon composition and the temperature. At a low temperature and low aromatic content, the diesel gasoil dissolves less water. Therefore, when the temperature decreases, water precipitates from the diesel gasoil. In winter it may form ice crystals that plug the fuel filter or hinder filling by settled paraffin crystals. To prevent or avoid this process, deicing additives are used.

Electric conductivity The presence and accumulation of the static electricity can be dangerous. Flowing engine fuels can be charged up. To avoid spark formation, tanks and pipelines must be grounded. Charging is also affected by humidity. The use of the antistatic additives helps solve this problem because they improve the electric conductivity.

Flashpoint During the storage and transport of diesel gasoils, the flashpoint is an important safety parameter. The flashpoint can easily be found in crude oil refineries by establishing the initial boiling point of blending components or measuring by a flash point tester.

REFERENCES

1. Guibet, J. C. (1999). *Fuels and Engines Technology Energy Environment*. Edition Technip, Paris.
2. Hancsók, J. (1997). *Modern Engine and Jet Fuels. I. Engine Gasolines*. Veszprém University Press, Veszprém.
3. Basshuysen, R., Schäfer F., eds. (2007). Handbuch Verbernnungsmotor, Vieweg Verlag, Wiesbaden.
4. Lucas, A. G. (2000). *Modern Petroleum Technology*. Wiley, Chichester.
5. Maples R. E. (2000). *Petroleum Refinery Process Economics*. PennWell Corporation, Tulsa.
6. Gary, J. H. Handwerk, G. E. (1984). *Petroleum Refining—Technology and Economics*. Dekker, New York.
7. Speight, J. G. (1990). *Fuel Science and Technology Handbook*. Dekker, New York.
8. Valer, M., Song, M. M. (2002). *Enviromental Challenges and Greenhouse Gas Control for Fossil Fuel Utilization in the 21st Century*. Kluwer Academic, New York.
9. Elvers, B. (2008). *Handbook of Fuels*. Wiley, Weinheim.
10. Kuo, K. K. (2005). Principles of Combustion, Wiley, Hoboken, NJ.

11. Westbrook, S. R., Shah, R. J. (2003). *Fuels and Lubricant Handbook: Technology, Properties, Performance, and Testing*, 1–5. ASTM International, West Conshohocken, PA.

12. Sobiesiak, A., Uykur C., Ting D.S.-K., Henshaw P. F. (2002). Hydrogen/oxygen additives influence on premixed iso-octane/air flame. *Res. Altern. Fuel Dev.*, SP-1716, 87–94.

13. Hobson, G. D., ed. (1984). *Modern Petroleum Technology*. Wiley, Chichester.

14. Hancsók, J., Holló A., Forstner J., Gergely J., Perger J. (2000). Production of environmentally friendly engine gasolines with increased isoparaffin content. *Petrol. Coal* (3–4), 166–170.

15. Hancsók, J., Kasza, T. (2011). The importance of isoparaffins at the modern engine fuel production. *Proceedings 8th International Colloquium Fuels 2011*, January 19–20, Stuttgart/Ostfildern, 361–373.

16. Phillips Chemical Company. (1974). Reference data for hydrocarbons and petro-sulfur compounds. *Bulletin* 521.

17. ASTM (1971). Physical constants of hydrocarbons C_1 to C_{10}. *ASTM Data Series DS 4A*, Baltimore.

18. Hancsók, J., Holló, A., Perger, J., Gergely, J., eds. (1998). Environmentally friendly possibilities to compensate octane deficiency resulting from benzene content reduction of motor gasolines. *Petrol. Coal*, 40 (1), 33–38.

19. Hancsók, J., Magyar, S., Holló, A. (2007). Importance of isoparaffins in the crude oil refining industry. *Chem. Eng. Trans.* 11, 41–47.

20. Magyar, Sz., Hancsók, J., Holló, A. (2007). Key factors in the production of modern engine gasolines. *Proceedings 6th International Colloquim, Fuels 2007*, January 10–11, Németország, Esslingen, 273–284.

21. Marshall, E. L., Owen, K. (1995). *Motor Gasoline*. SCI, Oxford.

22. Stewart W. E. (1959). Predict octanes for gasoline blends. *Petrol. Refiner*, 38, 135–139.

23. Rusin, M. H., Chung, H. S., Marsall, J. F. (1981). A "transformation" method for calculating the research and motor octane numbers of gasoline blends. *Ind. Eng. Chem. Fund.*, 20, 195–204.

24. Zahed, A. H., Mullah, S. A., Bashir, M. D. (1993). Predict octane number for gasoline blends. *Hydrocarb. Proc.*, 5, 85–87.

25. Murty, B. S. N., Rao, R. N. (2004). Global optimization for prediction of blend composition of gasoline of desired octane number and properties. *Fuel Proc. Technol.*, 85, 1595–1602.

26. Morris, W. (2008). Aromatic increase rvp blending value. *OGJ*, 106, 56–58.

27. Holló, A., Csernik, K., Hancsók, J. (2008). Car industry developments—challenges for oil industry. *Interfaces'08, Sopron*, 9, 24–26.

28. Nagpal, J. M., Joshi, G. C., Singh, J. (1994). Gum forming olefinic precursors in motor gasoline: model compound study. *Fuel Sci. Technol. Int.*, 12(6), 873–894.

29. Kinoshita, M. Saito, A., Matsushita, S., Shibata, H., Niwa, Y. (1998). Study of deposit formation mechanism on gasoline injection nozzle. *JSAE*, 19, 355–357.

30. Nagpal, J. M., Joshi, G. C., Singh, J., Kumar, K. (1998). Studies on the nature of gum formed in cracked naphtas. *Oxid. Commun.*, 21(4), 468–477.

31. Zainer, A. (1997). Thermal-oxidative stability of motor gasoline by pressure d.s.c. *Fuel*, 77(8), 865–870.

32. Nagpal, J. M., Joshi, R. S. N. (1995). Stability of cracked naphtas from thermal and catalytic process and their additive response. Part I. Evolution of stability and additive response. *Fuel*, 74(5), 714–719.

33. Nagpal, J. M., Joshi, R. S. N. (1995). Stability of cracked naphthas from thermal and catalytic processes and their additive response. Part II. Composition and effect of olefinic structures. *Fuel*, 74(5), 720–724.

34. Zádor, J., Taatjes, C. A., Fernandes, R. X. (2011). Kinetics of elementary reactions in low-temperature autoignition chemistry. *Prog. Energy Combust. Sci.*, 37, 371–421.

35. MacKinven, R., McArragher, J. S., Fredrikson, M. (1997). Review of the European Auto/Oil Programme and EPEFE. 1st International Colloquium on Fuels, Ostfildern, Germany, 15–34.

36. McArragher, J. S., Becker, R. F., Goodfellow, C. L. (1996). The influence of gasoline benzene and aromatics contetnt on benzene exhaust emissions from non-catalyst and catalyst equipped cars a study of European data. CONCAWE Report No. 96/51, Brussels, 1–55.

37. Trithart, P. (1991). Requirements for petrol and diesel fuels. *Proceedings of the Conference on Engine and Enviroment'91*, July 23–24, Graz, Austria, 125–138.

38. Davis, B. C. (1991). Oxigenates in reformulated gasoline. NPRA, AM-91-47.

39. Unzelmann, G. H. (1992). Refining options and gasoline composition. NPRA, AM-92-05.

40. Farrauto, R. J., Heck, R. M. (1999). Catalytic converters: state of the art and perspectives. *Catal. Today*, 51, 351–360.

41. Heck, R. M., Farrauto, R. J. (2001). Automobile exhaust catalysts. *Appl. Catal. A: General*, 221(1–2), 443–457.

42. Kaspar, J., Fornasiero, P., Hickey, N. (2003). Automotive catalytic converters: current status and some perspectives. *Catal. Today*, 77(4), 419–449.

43. Meeyoo, V., Trimm, D. L., Cant, N. W. (1998). The effect of sulphur containing pollutants on the oxidation activity of precious metals used in vehicle exhaust catalysts. *Appl. Catal. B: Environmental*, 16(2), L101–L104.

44. Matsumoto, S. (2004). Recent advances in automobile exhaust catalysts. *Catal. Today*, 90(3–4), 183–190.

45. ACEA. (2002). Data of the sulphur effect on advanced emission control technologies. Executive Summary, Brussels.

46. Hofmann, P., Geringer, B., Holub, F., List, R., Winter, S., Urbanek, M. (2007). Potenzial von Ethanol Blends in modernen Ottomotoren, 11. Tagung der Arbeitsprozess des Verbrennungsmotors, September 20–21.

47. Dedl, P., Hofmann, P., Geringer, B., Karner, D., Lohrmann, M. (2010). Biogenous gasoline—suitability and potential of alcohols, ether and BTL-gasoline for engine operation and performance. F2010-a-058. FISTA World Automotive Congress, May 31, Budapest.

48. Dedl, P., Hofmann, P., Bernhard, G., Dieter, K., Lohrmann, M. (2011). Suitability and potential of alternative fuels for the use in spark ignition engines. 8th International Colloquium Fuels Conventional and Future Energy for Automobiles, January 19–20.

49. Nagy, G., Hancsók, J. (2009). Key factors of the production of modern diesel fuels. *Proceedings 7th International Colloquium Fuels, Mineral Oil Based and Alternative Fuels*, January 14–15, Stuttgart/Ostfildern, 483–500.

50. Hancsók, J. (1999). *Modern Engine and Jet Fuels. II. Diesel Fuels*, Veszprém University Press, Veszprém.

51. Juva, A., Rautiola, A., Saikkonen,P., Brenton, D. (1991). Influnce of diesel fuel composition on performance and exhaust emissions of diesel engines. 13th WPC, October 20–25, Buenos Aires.

52. Unzelman, G. H. (1993). Review and outlook for additives. *Fuel Reform.* (January–February), 16–18.

53. Virk, K. S., Lachowicz D. R., Mitchell, E. (1992). Diesel fuel cetane number, aromatic content and exhaust emmissions. Symposium on Octane and Cetane Enhancement Processes for Reduced-Emissions Motor Fuels, April 5–10.

54. Signer, M. (1998). Fuel quality influence on engine performance and emissions. 3rd Annual World Fuels Conference, June 3–5.

55. Lange, W. W., Reglitzky A. A., Krumm, H., Cowley, L. T. (1993). Einfluß des Dieselkraftstoffes auf die Abgasemissionen von Nutzfahrzeugen. Vortrag, gehalten auf der Technischen Arbeitstagung Hohenheim, March 23.

56. Sieverding, R. (1999). Möglichkeiten der Reduzierung von Dieselmotoremissionen durch reformulierte Kraftstoffe. 2nd International Colloquium on Fuels, January 20–21.

57. Sarvi, A., Fogelholm, C-J., Zevenhoven, R. (2008). Emissions from large-scale medium-speed diesel engines: 2. Influence of fuel type and operating mode. *Fuel Proc. Technol.*, 89, 520–527.

58. Röj, A. (1997). Enviromental class fuel in Scandinavia in the light of the European Auto- Oil Program. 1st International Colloquium Fuels, January 16–17.

59. Zeuthen, P., Knudsen, K. G., Whitehurst, D. D. (2001). Organic nitrogen compounds in gas oil blends, their hydrotreated products and the importance to hydrotreatment. *Catal. Today*, 65, 307–314.

60. Thielemans, G. L. B. (1993). Manufacturing of low aromatic diesel fuel. 1993 European Oil Refining Conference, June 21–22.

61. McCarthy, C. I., Slodowske, W. J., Sienicki, E. J., Jass, R. E. (1994). Diesel fuel property effects on exhaust emissions from heavy duty diesel engine that meet 1994 emissions requirements. *SAE Paper, 922267.*

62. Singhal, S. K., Singh I. P. (2000). The study of fuel quality on emissions in DI and IDI engines. International Syposium on Fuels and Lubricants, March 10–12.

63. Concawe. (1990). The sulphur content of diesel fuel and its relationship with praticulate emissions from diesel engines, Report No. 90/54. Brussels.

64. Krutzsch, B., Wenninger, G. (1992). Effect of sodium- and lithium-based fuel additives on the regeneration efficiancy of diesel particulate filters. Daimler-Benz AG Document Number: 922188, October.

65. Valentine, J. M., Peter-Hoblyn, J. D., Ares, G. K. (2000). Emissions reduction and improved fuel economy. Performance from a Bimetalic Platinum/Cerium Diesel Fuel Additive at Ultra Low Dose Rates, *SEA Paper 2000-01-1934.*

66. Season, A. MacNee, W., Donaldson, K., Godden, D. (1995). Particulete air pollution and acute health effect. *Lancet*, 345, 176–178.

67. Ferin, J., Oberdorster, G., Penny, D. P. (1992). Pulmonary retention of ultra fine and fine particles in rats. *Amer. J. Resp. Call molecul. Bio.* 4. 535–542.

68. Kunzli, N. (2000). public impact of outdoor and trafic related air pollution: a European assessment. *Lancet*, 356, 795–801.

69. Kayes, D., Hochgreb, S. (1998). Investigation of the dilution process for the measurment of particulate matter from spark ignition engines. *SAE paper, 982601.*

70. Graskow, B. R., Kittelson, D. B., Abdul-Khalek, I. S., Ahmadi, M. R., Morris, J. E. (1998). Characterization of exhaust particulate emissions from a spark ignition engine. *SAE Paper 980528.*

71. Graskow, B. R., Kittelson, D. B., Ahmadi, M. R., Morris, J. E. (1999). Exhaust particulate emissions from two port fuel injected spark ignition engine, *SAE Paper 99-01-1144.*

72. Graskow, B. R., Ahmadi, M. R., Sengers, H. P. M. (2001). The infuence of fuel additives on ultrafiner particulae emissions from spark ignition engines. *3rd International Colloquium, Proceedings*, January 17–18, Esslingen, 211–224.

73. Streit, G., Achenbach M. (1999). Safe sealing at low temperature—an innovative sealing system. *2nd International Colloquium Fuels, Proceedings*, 561–570.

CHAPTER 8

Aviation Fuels

8.1 AVIATION GASOLINES

The aviation industry uses two types of fuels: gasoline for piston or Wankel engines and jet fuel (kerosene type) for the jet engines (turbine engines) aircrafts. The former is known as aviation gasoline (Avgas) and the later is as aviation turbine fuel (ATF). Early aircrafts used low octane straight-run gasoline, which resulted in overheating and engine failure. With the discovery of tetraethyl lead as an octane improver, it was realized that higher octane gasoline increases engine power output. Soon after, aviation gasoline of 92 octane containing maximum 6 ml of tetraethyl lead (TEL)/gal (2.6 gal/liter) was introduced. This was further improved to 100 octane to reduce engine size and improve engine power during World War II. Several gasoline grades of varying octane numbers were in use. These ranged as 68, 80, 82, 87, 92, 100, and 100 low lead (LL). Various test methods were also formulated (ASTM D-614, ASTM D-2700, ASTM D-909) to determine the octane number.

Aviation gasolines are characterized by energy content (heat of combustion), knock resistance, volatility (fuel tendency to vaporization, distillation characteristics), low freezing point, high stability, cleanliness, and high electrical conductivity. They usually contain an antioxidant, anti-icing, corrosion inhibitor and an antistatic agent. A small dosage of metal deactivator is also incorporated to meet the corrosion characteristics and to provide synergisms with the antioxidant. The most important properties of aviation gasoline are octane rating (especially rich rating) and volatility. These properties along with the electrical conductivity degrade during transportation and storage, therefore special care has to be taken in working out the specifications of products [1].

The demand for these fuels is, however, now limited due to the widespread use of jet engines in both military and civil aircrafts. The annual US usage of Avgas was about 700,000 m³ in 2008, which was approximately 0.14% of the motor gasoline consumption.

Aviation gasolines are the mixtures of C_4–C_9 hydrocarbons (mainly isoparaffins). They contain C_8 isoparaffins as a major constituent. The reason for the high isoparaffin

Fuels and Fuel-Additives, First Edition. S. P. Srivastava and Jenő Hancsók.
© 2014 John Wiley & Sons, Inc. Published 2014 by John Wiley & Sons, Inc.

content is that these hydrocarbons are the most favorable regarding the three main characteristics of aviation fuels: lean-mixture antiknock, rich-mixture antiknock, low-temperature fluidity. The main petroleum derived component used in the blending of Avgas is alkylate, which is essentially a mixture of various isooctanes and is produced by the catalytic reaction of C_4 olefins and isobutane (alkylation and the properties of alkylates have been discussed in Chapter 3). Some refineries also use reformates.

Avgas has lower and more uniform vapor pressure than automotive gasoline to keep it in the liquid state at high-altitude and to prevent vapor lock formation. Currently Avgas is available in several grades with variation in maximum lead concentrations.

8.1.1 Aviation Gasoline Grades

In the past the aviation gasoline was generally identified by two numbers (e.g., 100/130) associated with its motor octane number (MON). The first number indicates the octane rating of the fuel tested to "aviation-lean" standards, which is similar to the antiknock index given to automotive gasoline in the United States. The second number indicates the octane rating of the fuel tested to the "aviation-rich" standard, which tries to simulate a supercharged condition with a rich mixture, at elevated temperatures, and a high manifold pressure. Nowadays, only the lean-mixture octane rating is used. Avgas is generally dyed to identify different grades. Table 8.1 gives the key specifications for four grades of Avgas according to the ASTM D-910 specifications. These quality specifications are applied in Europe as well.

Most piston aircraft engines require 100LL Avgas, and an equivalent unleaded grade of fuel has not yet been developed for these engines. Aircraft is often purchased with engines that use 100LL because many airports only serve 100LL. Efforts are being made to develop a lead-free product by using a combination of following components [2]: alkylate: 12–18 vol%; super alkylate: 28–42%; toluene: 20–30%; toluidines: 3–5%; ethanol: 0–5%; C_5 stream 10–20%.

ASTM D-910-1998 specification lists chemical additives that can be used in appropriate dosages. Russian Avgases are specified in GOST 1012–72 and TU N 4-60-67 standards.

8.1.2 Aviation Gasoline Additives

Antiknock compounds Tetraethyl lead plus ethylene bromide provide two atoms of bromine per atom of lead. Ethylene bromide acts as scavenging agent for lead. When Avgas is burned in an engine, the lead in the tetraethyl lead is converted to lead oxide. These deposits quickly collect on the valves and spark plugs, and can damage the engine. Ethylene dibromide reacts with the lead oxide as it forms, and converts it to a mixture of lead bromide and lead oxybromides. Because these compounds are volatile, they are exhausted from the engine along with the rest of the combustion product. The dosage of TEL depends on the grade.

TABLE 8.1 Key specifications of aviation gasolines

Properties	Method	Grade 80	Grade 100	Grade 100LL
Motor octane no., lean mix., min.	ASTM D2700	80	100	100
Motor octane no., rich mix., min.	ASTM D2700	87	—	—
Performance no., (supercharged) min.	ASTM D-909	—	130	130
Color	ASTM D-2392	Red	Green	Blue
Blue dye, mg/L max.		0.2	2.7	2.7
Yellow dye, mg/L max.		Nil	2.8	Nil
Red dye, mg/L max.		2.3	Nil	Nil
TEL/L max., gPb/L, max.	ASTM D 5159	0.14	1.12	0.56
Distillation				
10% vol, °C max.	ASTM D-86	75		
40% vol, °C min.		75		
50% vol, °C max.		105		
90% vol, °C max.		135		
FBP, °C max		170		
Sum of 10 vol% + 50 vol%, evaporated temperatures, °C min.		135		
Recovery, % min.		97		
Residue, % max.		1.5		
Loss, % max.		1.5		
Vapor pressure at 37.8 °C, kPa				
Min.	ASTM D-5191	38.0		
Max.		49.0		
Freezing point, °C max.	ASTM D-2386	−58		
Corrosion, copper strip 2 hours at 100 °C, max.	ASTM D 130	1		
Potential gum mg/100 ml max.	ASTM D-873	6		
Sulfur, % max.	ASTM D-2622	0.05		
Lead precipitate, mg/100 cm³	ASTM D-873	3		
Electric conductivity, pS/m	ASTM D-2624	50–450		

Note: For complete specification of Avgas 80, 100, and 100LL grades, refer to ASTM D-910 or Def Stan 91–90 (DERD 2485). MIL-G-5572P provides corresponding specifications for Avgas.

Antioxidants, 12 mg/L max

2,6-ditertiary butyl phenol (DBP)

2,6-ditertiary butyl 4-methyl phenol (DBPC or BHC)

2,4-dimethyl-6-tertiary butyl phenol

75% min. 2,6 ditertiary butyl phenol, plus 25% mixed tertiary and tritertiary butyl phenols

75% min. di- and tri-isopropyl phenyl, plus 25% max. di- and tritertiary butyl phenols

72% min. 2,4-dimethyl-6-tertiary butyl phenols, 28% max. monomethyl and dimethyl tertiary butyl phenols

N, N'-di-isopropyl *para*-phenylene diamine

N, N'-di-secondary butyl *para*-phenylene diamine

Anti-icing Agents

Isopropyl alcohol (IPA) at dosage recommended by aircraft manufacturers

Diethylene glycol monomethyl ether (Di-GME) at 0.10–0.15% vol

These are often mixed at the point-of-sale so that users do not have to bear the extra cost.

Antistatic Agent/Electrical Conductivity Improver Additives

Stadis™ 450 1–5 ppm max.

Stadis 450, with dinonylnaphthyl sulfonic acid (DINNSA) as the active ingredient, is an example. These additives are soluble in fuel and ionize to provide electrical conductivity in the system.

Dyes

Blue: 1,4-Dialkylamino anthraquinone

Yellow: *p*-Diethylamino-azobenzene or 1,3-benzenediol 2,4-bis[(alkylphenyl) azo-]

Red: Alkyl derivative of azobenzene-4-azo-naphthol

The combinations of these dyes provide different colors to the fuel; for example, a blue and red dye combination gives purple color. A blue and yellow combination gives green color.

The following additives are also incorporated for additional advantages:

Corrosion inhibitors. For example, Dow Corning's DCI-4A is used for civilian and DCI-6A used for military fuels.

Biocide. This can be added if evidence exists of bacterial colonies inside the fuel system.

8.1.3   Automotive Gasoline for Aircraft

Automotive gasoline (known as Mogas or Autogas in the aviation industry) that does not contain oxygenates may be used in certified aircraft that have a Supplemental type certification for automotive gasoline and in experimental aircraft. However, for most aircraft, automotive gasoline is not a replacement for Avgas because automotive gasoline quality control standards do not meet Avgas standards. Recently, in small air planes the possible use of gasoline having an ethanol content up to 15% was investigated. It was concluded that the use of ethanol-containing fuel would be safe provided that there would be frequent and careful checkups of the fuel supply system [3].

The EPA has previously named Avgas as one of the most "significant sources of lead." In view of these restrictions, the 100LL Avgas phaseout has been planned and work is being carried out to develop a suitable substitute without tetraethyl lead.

8.2   JET FUELS

Jet aircrafts are powered by gas-turbine engines. In jet engines, air flows through the compressor (driven by the turbine) to the combustion zone, where fuel is injected and burned. The exhaust gases are expanded in the turbine and exhausted into the atmosphere. The aircraft is thus pushed by jet propulsion (gas jet propulsion). These jet engines have significantly higher performance than piston engines. Consequently, after the invention of their first type, they spread rapidly. They are continuously under development, which has allowed the growth of worldwide air transportation. Air travel is an indispensable part of mobility in modern societies. The quantity of fuel used in the world was 205×10^6 t (70×10^6 t in the European Union), which was ca. 8% of the transportation fuels used in the world in 2010. Predictions show the consumption of Jet fuels would be increasing [4,5].

8.2.1   Main Quality Requirements and Properties of Jet Fuels

The most important properties of jet fuels are the energy content and combustion quality. Next in importance to performance are stability (oxidation and heat), fluidity (cold flow properties), volatility, purity, and noncorrosivity. Jet fuel is also used as a coolant for fuel systems. The important properties of jet fuels from the aspect of safety are flashpoint and conductivity. In general, the fuel's volumetric energy content (energy per unit volume of fuel; MJ/L) is more important than the mass energy content (energy per unit weight of fuel; MJ/kg). The relationship between these variables is very complex, and other influences must be considered.

In a jet engine, combustion is continuous, whereas in a piston engine, combustion is intermittent. In the jet engines particles formed at the beginning of combustion must be totally burned away during the pass through the flame. Otherwise, they form deposits and cause erosions. Hence particle-forming components (e.g., aromatics, especially naphthalene) of jet fuels must be restricted.

Storage and heat stability basically depend on the composition of fuel. But storage and operational conditions (coolant for engine, etc.) have effects too. For the proper working, particle and gum formation must be avoided. These can cause fuel supply problems, filter fouling, burning product deposits, and the like. Storage stability can be improved by appropriate hydrocarbon compositions (of highly saturated content) and by using an antioxidant additive.

Lubricity is an important property of jet fuel because the fuel lubricates some moving parts in the fuel pumps. This property is assured by the heteroatom content (oxygen, nitrogen, and sulfur content) of the fuel and/or the lubricity improver additives.

Among the flow properties, suitable values of viscosity and freezing point—at very low temperatures to about −60 °C—are especially important for good pumpability.

Jet fuel volatility is determined by distillation and vapor pressure values, which are critical to achieving proper vaporization before combustion. Vapor pressure that is too high can cause vapor lock in the fuel system, and transport and storage losses.

The special flammability or flashpoint at which jet fuel ignites provides good conductivity that serves to prevent the static charges and explosions.

8.2.2   Aviation Turbine Fuel Specifications

Civilian jet fuels and military jet fuels specification are different. The American ASTM and the British MOD (United Kingdom Ministry of Defence) have taken the lead in setting and maintaining specifications. The specifications of many other countries are similar or identical to the ASTM or MOD specifications. In some Eastern European countries, the jet specifications are those issued by the Russian GOST.

The four commercial US (ASTM D 1655) and Russian (GOST 10227) jet fuels are Jet-A, Jet-A1, and GOST TS-1 (kerosene-type fuels). There is further a grade Jet-B (wide-cut fuel), as listed in Table 8.2.

Jet A and Jet A-1, the most common jet fuels, must meet an internationally standardized set of specifications. Jet A and Jet A-1 are kerosene-type fuels. The primary difference between them is the freezing point (the temperature at which wax crystals appear in a laboratory test). Like Jet A-1, Jet A has a fairly high flashpoint of 38 °C (100 °F), with an autoignition temperature of 210 °C (410 °F). Jet A, is mainly used in the United States and has a freezing point of −40 °C or below, while Jet A-1 has a freezing point of −47 °C or below. Jet A does not normally contain an antistatic additive, while Jet A-1 often requires this additive. Jet A and Jet A-1 are produced according to the requirements of ASTM D1655 and Def Stan 91-91 standards, respectively.

The only other jet fuel that is commonly used in civilian turbine engine powered aircrafts is called Jet B and is used for its enhanced cold weather performance. Jet B is mixture of naphtha-kerosene and therefore contains lighter compounds, which makes its handling more difficult due to its high flammability. Jet B is a civil version of JP 4. (as explained later)

Jet fuels are mixture of a large number of different hydrocarbons. The range of carbon numbers is restricted by the requirements for the product, for example, freezing point or smoke point. Kerosene-type jet fuels (including Jet A and Jet A-1) have a carbon number distribution between about 8 and 16. Wide-cut or naphtha-type jet fuels (including Jet B) contain hydrocarbons of about 5 and 16 carbon numbers.

For the military jet fuels in the United States and in other countries, special specifications are required because military aircraft must meet higher performance standards than civilian aircraft.

TABLE 8.2 Actual Jet specifications

Fuel Properties	Jet-A	Jet-A1	Jet-B	GOST TS-1	ASTM Test Method
Appearance		Clear, transparent and sediment free			Report
Acid number, max., mgKOH/g	0.10				D 3242 / ISO 6618
Aromatics, max., vol%	22(25)	—		22 m/m%	D1319 / EN 15553
Sulfur, mercaptan, % (m/m)	0.003				D3227 / ISO 3012
Sulfur, total, % mass, max.	0.30				D 1266, D1552, D 2622, D 4294 or D 5453 / EN ISO 8754 / EN ISO 20846 / EN ISO 20884
Distillation, °C, vol% recovered:					D 86 / EN ISO 3405
Initial boiling point				<150	
10	205			165	
20			145	195	
50			190	230	
90			245	250	
Final boiling point, max.	300				
Residue, max., vol%	1.5				
Loss, max., vol%	1.5				
Flashpoint (TAG), °C, min., °C	38			28	D 56 D 3828 / EN ISO 3679 / EN ISO 2719

	0.775–0.840	0.751–0.802	0.775 (min)	Test method
Density; 15 °C, kg/dm³	0.775–0.840	0.751–0.802	0.775 (min)	D 1298 or D 4052; EN ISO 3675; EN ISO 12185
Vapor pressure, kPa, max.	—	21		D 323 or D 5191
Crystallization point (chilling point), °C, max.	−40 / −47	−50	−50	D 2386, D 4305, D 5501 or D 5972
Viscosity on −20 °C, mm²/s, max.	8.0		1.25–8.0	D 445; EN ISO 3104
Net heat of combustion, MJ/kg, min.	42.8			D 4529, D 3338 or D 4809
Luminometer number,* min.	45			D 1740
Smoke point,* min., mm	25			D 1322
Smoke point,* min., mm	18			D 1840
Naphtalenes,* max., vol%	3.0			D 130
Copper strip, 2 h at 100 °C, max.	No. 1			EN ISO 2160
Thermal stability				D 3241
Filter pressure drop, kPa, max.	3.3			
Tube deposit, less than	Code 3			
Existent gum, mg/100 cm³, max.	7.0			D 381; EN ISO 6246
Water reaction, interface rating, max.	1b			D 1094
Electrical conductivity, pS/m	50–450			D 2624; ISO 6297

Note: *Quality properties not yet fulfilled.

United States jet fuels have been coded since 1944 as JP-1, JP-2, JP-3, JP-4, JP-5, JP-6, JP-7, JP-8, and JP-8+100 (JP=Jet propulsion). In kerosene-type fuels (JP-1, JP-5-JP8, and JP-8+100) the differences are in the freezing points and/or thermal stability, and also in the additives. JP-8 and JET A-1 are quiet comparable in performance.

8.2.3   Production of Aviation Turbine Fuels

In the production of jet fuels, the type of hydrocarbon compound groups and the available raw materials and their costs, for example, are the main considerations. Fuels of jet engines can be naphtha-kerosene type, or kerosene type only. Some such compounds that are in the boiling point range of jet fuels are summarized in Table 8.3 [6,7].

As data in Table 8.3 indicate, the preferable compounds are the isoparaffins and naphtenes, due to their oxidation stability and combustion properties. Some properties of jet fuels are determined by compounds that are present in very low concentrations (e.g., lubricity, corrosion, and conductivity). In jet fuels water can only be present in solution, as free or emulsified water is not permitted.

The classification of jet fuel production from conventional and alternative raw materials is shown on Figure 8.1 [6]. For the crude oil kerosene fraction, a suitable boiling point range (initial boiling point ca. 140–160 °C and final boiling point ca. 220–260 °C) is obtained by distillation. If the quality properties satisfy the specifications of the country's standard, the kerosene fraction can be used as jet fuel (straight-run jet) after suitable additivation. However, in most cases quality improvement is necessary in order to partially or totally reduce the heteroatom- and/ or aromatic contents [6–9].

The heteroatom compounds of kerosene fractions may contain sulfur, nitrogen, or oxygen atoms. For example:

- Sulfur compounds:
  - Mercaptans:

$$R-S-H, \quad C_{10} \leq R \leq C_{14}$$

  - Sulfide chain:

$$R_1 - S - R_2$$

Sulfides

$$R_1 - S - S - R_2$$

Disulfides

TABLE 8.3 Compound types in jet fuels

Compound name	Formula	Hydrocarbon Group	Density at 20 °C (g/cm³)	Boiling Point, °C	Freezing Point, °C	Net Energy Content at 25 °C (MJ/kg)
n-Octane	C_8H_{18}	n-Paraffin	0.7027	125.7	−56.8	44.42
2-Methylheptane	C_8H_{18}	Isoparaffin	0.6979	116	−109.0	44.38
1-Methyl-1-ethylcyclopentane	C_8H_{16}	Naphthene	0.7809	72	−143.8	43.57
Ethylcyclohexane	C_8H_{16}	Naphthene	0.7879	103	−111.3	43.40
o-Xylene	C_8H_{10}	Aromatic	0.8801	144	−25.2	40.81
p-Xylene	C_8H_{10}	Aromatic	0.8610	138	+13.3	40.81
cis-Decalin	$C_{10}H_{18}$	Naphthene	0.8967	169	−43.0	42.62
Tetralin	$C_{10}H_{12}$	Aromatic	0.9695	207	−35.8	40.52
Naphthalene	$C_{10}H_8$	Aromatic	1.1750	217.7	+80.3	40.12
n-Dodecane	$C_{12}H_{26}$	n-Paraffin	0.7488	216.2	−9.6	44.11
2-Methylundecane	$C_{12}H_{26}$	Isoparaffin	0.7458	210	−46.8	44.08
n-Hexylbenzene	$C_{12}H_{18}$	Aromatic	0.8602	226.1	−61.0	41.80
1-Ethyl-naphthalene	$C_{12}H_{12}$	Aromatic	1.008	259	−15	n.d.
2-Methyl-pentadecane	$C_{16}H_{34}$	Isoparaffin	n.d.	282	−7	n.d.
n-Hexadecane	$C_{12}H_{34}$	n-Paraffin	0.7735	287	18	43.95

Note: "n.d." means no data.

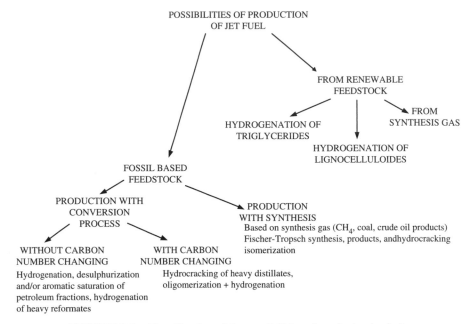

FIGURE 8.1 Classification of the possibilities of producing jet fuels

○ Sulfide cyclic:

Thiophenes

Benzothiophenes
R: alkyl group

• Nitrogen compounds:

Aniline

Quinolin

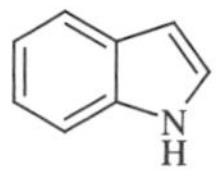

Indole

- Oxygen compounds:

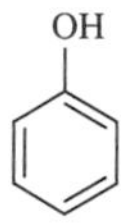

Phenol

$$R - COOH \quad C_5 \leq R \leq C_7$$

Carboxyl acids
R: alkyl-group

The mercaptan content of jet fuel can be at most 0.003% (in several countries it is lower than 0.001%), because these compounds are very corrosive. So they must be transformed or removed, even if the total sulfur content satisfies the value in the standard.

The transformation of mercaptans to noncorrosive compounds like disulfides is called "sweeting." They are not corrosive and have a less undesirable odor than mercaptans [7–9]. Many process are developed for this aim [7–11]. Among these, the most prevalent is mercaptan oxidation by a MEROX® (Mercaptan Oxidation) process that uses a cobalt-based catalyst:

$$2R - SH + \tfrac{1}{2}O_2 \rightarrow R - S - S - R + H_2O$$

In some cases the sulfur content of jet product does not reduce, because disulfides are not separated from the product.

If the sulfur content of a straight-run kerosene fraction has to be reduced in a significant quantity, then the sulfur must be removed not only from mercaptans but also from other compounds. Technically and economically, the best process is hydrodesulphurization. The sulfur atom is then removed in the form H_2S, and in the meantime, the corresponding saturated hydrocarbon is formed. (A detailed description of hydrodesulphurization can be found in Chapter 3.) Depending on the requirements, jet fuel can be produced with sulfur content lower than 1 mg/kg (1 ppm) on a $CoMo/Al_2O_3$ ($NiMo/Al_2O_3$) catalyst, at temperatures 310–340 °C, pressure 20–40 bar, 3–4 m³/m³h liquid hourly space velocity, and 150–250 Nm³/ m³ H_2/hydrocarbon ratio.

During the heteroatom removal, not only does the desulphurization takes place but nitrogen and oxygen atoms are also removed from their compounds, forming ammonia, water, and the corresponding saturated hydrocarbon. Additionally, a partial saturation of aromatics (20–50%) takes place, depending on the applied catalyst and process parameters. This results in the reduction of particle formation during combustion.

In many countries jet fuel is passed through clay beds or other special adsorbents for removal of surface active materials (naphthenic acid, phenol, benzene sulphonic acid, etc.) before the transportation from the fuel terminals or airports [8,12,13]. Because of the increase of demand for jet fuels, crude oil distillates cannot satisfy the quantity requirements. So cracking of heavy distillates to kerosene boiling point products is necessary. Examples of these technologies are the fluid catalytic cracking (FCC) and hydrocracking (see Chapter 3). Suitably large fractions of FCC must be desulphurized, and olefins have to be saturated. The kerosene fraction of hydrocracking is an excellent Jet fuel blending component.

To improve the quality of jet fuels (combustion properties and harmful material emissions) obtained from distillation, partial or total saturation of aromatics are required. This way more preferable, mainly naphthene content jet fuel is produced (see at Table 8.4) [6].

The saturation of aromatics is carried out on $NiMo/Al_2O_3$ or Pt and/or Pd/acid support catalysts, at temperatures 280–320 °C, 30–40 bar pressure, 1.5–3.0 m^3/m^3h liquid hourly space velocity, and at 200–250 $Nm^3/m^3 H_2$/feedstock ratio. The aromatic content of jet fuel that is produced with 98–99% liquid yield is lower than 1–5%. Other quality properties are also excellent (smoke point: min. 29 mm; freezing point: −51 °C) [6].

Over the last 5 to 10 years, the issue of fuel production from alternative resources has come to the foreground [6,14–23]. As shown in Figure 8.1, some of these possibilities include production of jet from triglycerides with suitable carbon numbers and Fischer–Tropsch synthesis from different sources. Alternative fuels such as the introduction of bio-JET into the aviation were already discussed in Chapter 4 [14–16].

The properties of jet fuel produced by different methods are further improved with additives. These are important from the point of new of security.

8.2.4 Additives of Jet Fuel

In different Jet fuels the additive used and their properties are carefully specified. The additives should not interact with each other and have negative impact on other fuel quality characteristics.

The main difference between the crude oil based civil and military Jet fuels is the additives used in them. For example Jet-A1 in the United States contains antioxidant and antistatic additives, whereas the military Jet fuel must contain at least three different kinds of additives. The different types of Jet fuel additives are discussed below [28–30].

Antioxidants Aviation fuels produced by hydroprocessing lose the naturally occurring antioxidants that provide protection from oxidation. Oxidation of hydrocarbons proceeds through a free radical mechanism, and peroxide radicals

TABLE 8.4 **Properties of naphthenes and aromatics with the same carbon number**

Property	n-Penthyl-Benzene	n-Penthyl-Cyclohexane	n-Hexil-Benzene	n-Hexil-Cyclohexane	n-Hepthyl-Benzene	n-Hepthyl-Cyclohexane
Heating value, MJ/kg	34.1	36.5	34.1	36.5	34.2	36.6
Freezing point, °C	−43	−58	−42	−52	−40	−47

are formed. Peroxides are known to attack elastomers, causing embrittlement, and they also contribute to gum and particulate formation. The use of antioxidants effectively prevents peroxide degradation. Under JFSCL and Def Stan 91–91, a 17–24 mg/L dosage of an approved antioxidant must be added to the fuel blend that has been hydro processed. All the antioxidants are approved by chemistry. The use of antioxidants is optional in ASTM D1655 specification. The following are approved antioxidant additives (*dosage: 24 mg/l max*) for aviation turbine fuels.

2,6-Ditertiary butyl phenol (DBP)

2,6-Ditertiary butyl 4-methyl phenol (DBPC or BHC)

2,4-Dimethyl-6-tertiary butyl phenol

2,6-Ditertiary butyl phenol (75% min., plus 25% mixed tertiary and tritertiary-butyl phenols)

2,4-Dimethyl-6-tertiary butyl phenol (55% min., plus 15% min. 2,6- ditertiary butyl 4-methyl phenol, remainder as monomethyl and dimethyl teriary-butyl phenols)

2,4-Dimethyl-6- tertiary butyl phenol (72% min., plus 28% max. monomethyl and dimethyl tertiary butyl phenols)

Metal Deactivators Metal ions in fuel catalyze oxidation reactions. The fuel can undergo degradation, leading to the formation of gum or sludge, which contribute to poor thermal stability. Copper and zinc are the two common metals found in the jet fuel system. The metal deactivator additive (MDA) has the ability to chelate the metal ions and prevent fuel degradation. The metal deactivators in synergy with an antioxidant improve both thermal stability and oxidation stability. The use of MDA showed improved results in the JFTOT test. The following two MDAs have been found suitable for jet fuels:

N,N'-Disalicylidene-1,2-propanediamine

N,N'-Disalicylidene-1,2-cyclohexanediamine

ASTM D-1655 recommends an application of 2–5.7 mg/L max. of N,N'-disalicylidine-1,2-propane diamine metal deactivator in jet fuel.

Fuel System Icing Inhibitors Icing inhibitors deter the plugging of the fuel supply system caused by ice crystals. Most fuels contain trace amounts of dissolved or dispersed water. Water can also get into the system during storage. At low temperatures, the dissolved water may form fine droplets. Although the amounts are small, the droplets can freeze at high altitudes and cause filter plugging. Fuel system icing inhibitors were developed to protect the system against this plugging. The most widely used additive is diethylene glycol monomethyl ether (DEGME). The use of icing inhibitor is required in UK and US military jet fuels. It is optional in many civilian specifications and is very seldom used. Diethylene glycol monomethyl ether (DIGME) is recommended in a 0.10–0.15 vol% dosage.

Corrosion Inhibitors and Lubricity Improvers Corrosion inhibitors were originally added to military jet fuels to protect the fuel distribution system and aircraft engines from corrosion. Many aircraft fuel system components, especially pumps, rely on fuel to lubricate the moving parts. Hydroprocessing of fuels removes the components that provide fuel with natural lubricating properties. Because mostly military aircrafts are susceptible to lubricity problems, UK and US military specifications require corrosion inhibitors (e.g., alkyl-succinic-acid derivatives)/ lubricity additives (e.g., C8-C10 carboxylic-acid esters) in fuel. Although civilian fuel specifications do not require the use of lubricity additives and fuel produced according to ASTM D1655-04a, these additives may be supplied to fuel by an agreement between the purchaser and the supplier [27,28]. Def Stan 91–91/5 has qualified several commercial products to be used in specific concentration.

Thermal Stability Additive (Only Allowed for Use in Certain Military Jet Fuels) Jet engine fuels are often subjected to very high temperatures. Fuel-related carbon deposits have been observed in the filter screen and fuel nozzles, but not observed in piston engine aircrafts operating on Avgas. Several sulfur- and nitrogen-containing organic compounds [29,30] are responsible for the thermal instability of the fuel. Jet fuels must therefore be subjected to the Jet Fuel Thermal Oxidation Test (JFTOT) by ASTM D3241. In this procedure, the fuel is pumped over a heated tube under pressure and through a metal screen of 17 μm at 260 °C. The filter pressure drop during the test and the tube color are monitored.

Modern military jet engines require aviation fuel that has a higher thermal stability and heat sink capacity than Civilian Jet engines. The fuel is exposed to higher heat load in these jet engines.

Static Dissipater Naturally occurring polar compounds having good electrical conductivity are removed in refining processes. The refined fuels have lower electrical conductivity and have increased risk of charge generation, especially during loading or as the fuel passes through filters. To eliminate this risk, a static dissipator additive is widely used in jet kerosene. The minimum and maximum fuel conductivity requirements for Def Stan 91–91 and JP-8 are 50–450 pS/m. The addition of a static dissipator is not mandatory under ASTM D1655 specification. Currently Stadis® 450 is the only additive manufactured for use in aviation turbine fuels and approved by the major turbine manufacturers. This additive is used in a 3–5 mg/L dosage in jet fuel.

Leak Detection Additive   These are gas phase additives, such as sulfur-hexafluoride (SF_6). These are used as a part of a tracer system for fuel system leak detection at major airports. Airports occasionally run leak detection testing of hydrants, which may be carried out monthly or quarterly. Current use requires agreement by purchasers on a case to case basis.

Biocides   Application of biocides is permitted by engine and aircraft manufacturers only for turnaround period. The aircraft are refilled and fully treated with the biocide and, as a general rule, will fly on the treated fuel until it is fully used up. Fuel system icing inhibitor may also serve to inhibit fungal and bacterial growth in aircraft fuel system to a certain extent, but are not fully reliable.

REFERENCES

1. Goodger, E., Vere, R., eds. (1985). *Aviation Fuels Technology*. Macmillen, London.
2. Oliveira, E. J., Rocha, M. I., Santos, A. R. (2011). Development of an unleaded aviation gasoline. *Proceeding of 8th International Colloquim on Fuels—Conventional and Future Energy for Automobiles*, Technische Akademie Esslingen, January 19–20, Esslingen, Germany.
3. Esch, Th., Funke, H. (2011). Biogen Automobilkraftstoffe inder allgemeinen Luftfart. *Motortechn. Z.*, 72(1), 54–59.
4. Chéze, B., Gastineau, P., Chevallier, J. (2011). Forecasting world and regional aviation jet fuel demands to the mid-term (2025). *Energy Pol.*, 39, 5147–5158.
5. Lamping, M., Muether, M. (2009). Advanced diesel combustion: impact of engine hardware and fuel properties. *Proceedings of 7th Internation Colloquium Fuels*, January 14–15, Esslingen, Germany.
6. Eller, Z., Hancsók, J., Papp A. (2013). Production of jet fuel from alternative source. Proceedings of 9th International Colloquium Fuels 2013, January 15–17, Stuttgart/Ostfildern, Germany.
7. Lucas A. G., ed. (2000). *Modern Petroleum Technology*. Wiley, Chichester.
8. Hobson, G. D., ed. (1984). *Modern Petroleum Technology*. Wiley, Chichester.
9. Meyers R. A., ed. (2007). *Handbook of Petroleum Refining Processes*. McGraw-Hill, New York.
10. Vazquez, R. G. (1989). Reduce operating costs by caustic treating jet fuel streams. *Hydrocarb. Technol. Int.*, 63–65.
11. Forero, P., Suarez, F. J. (1996). Caustic treatment of jet fuel streams. *Petrol. Technol. Quart.*, (4), 43–47.
12. Cressman, P. R., Hurren, M. L., Smith, E. F., Holbrook, D. L. (1995). Caustic-free jet fuel Merox unit reduces waste disposal. *Oil Gas J.*, 93(12), 80–84.

13. Jackson, K. (1996). Filtering aviation fuels. *Petrol. Rev.*, 50(558), 24–29.

14. Gül, Ö., Rudnick, L. R., Schobert, H. H. (2008). Effect of the reaction temperature and fuel treatment on the deposit formation of jet fuels. *Energy Fuels*, 22, 433–439.

15. API. (1997). *Technical Data Book—Petroleum Refining*. API, Washington, DC.

16. Reidy, S. (2008). Renewable fuels for aviation. *Biofuels Bus.*, (June), 42–45.

17. Ryan, A. (2008). Specification and performance issues relating to biofuels in aviation, marine and heating fuels. Biofuels: Measurements and *Analysis Symposium*, April 16.

18. Nygren, E., Aleklett, K., Höök, M. (2009). Aviation fuel and future oil production scenarios, *Energy Pol.*, 37, 4003–4010.

19. Naik, C. V., Puduppakkam, K. V., Modak, A., Meeks, E., Wang, Y. L., Feng, Q., Tsotsis, T. T. (2011). Detailed chemical kinetic mechanism for surrogates of alternative jet fuels. *Combust. Flame*, 158, 434–445.

20. Outcalt, S., Laesecke, A., Freund, M. B. (2009). Density and speed of sound measurements of jet A and S-8 aviation turbine fuels. *Energy Fuels*, 23, 1626–1633.

21. Bova, T. (2007). Alternative jet fuel approach to a new horizon. 8th European Fuels Conference, March 14, Paris France.

22. Huber, M. L., Smith, B. L., Ott, L. S., Bruno, T. J. (2008). Surrogate mixture model for the thermophysical properties of synthetic aviation fuel S-8 explicit application of the advanced distillation curve. *Energy Fuels*, 22, 1104–1114.

23. Speight, J. G., ed. (2008). *Synthetic Fuels Handbook*. McGraw-Hill, New York.

24. ExxonMobil. (2008). *World Jet Fuel Specification*.

25. ASTM D-1655. (2011). *Standard Specification for Aviation Turbine Fuels*. American Society for Testing Materials. West Conshohocken, PA.

26. Adams, J. A. (1996). *Aviation jet fuel—specifications, manufacture, product quality, properties, distribution and availability*. RF7/B, October 30,–November 1, Oxford.

27. Crowther, J., Bowden, M. (1989). Fuel system lubrication—an airline's experience. IOP Tribology Group Meeting, Tribology of aviation fuel systems, London.

28. Jones, R. A. (1989). The evaluation of fuel pumps to meet the demands of fuel with poor boundary lubricating properties. IOP Tribology Group Meeting, Tribology of aviation fuel systems, London.

29. Hezlett, R. N. (1991). *Thermal Oxidation Stability of Aviation Turbine Fuels*. ASTM International, West Conshohocken, PA.

30. Acker, W. P., Hahn, R. T., Mach, T. J., Sung, R. L. (1990). Thermal stability additives for jet fuels. US Patent 4,912,873, April 3.

CHAPTER 9

Fuel Oils and Marine Fuels

In the early nineteenth century, steam produced by burning coal was the main source of power for ocean-going vessels. The refining industry started developing during those years, and initially all crude oil fractions heavier than kerosene were sold as residual fuel. Fuel oil on combustion releases a large amount of heat, which is defined as the specific energy of the fuel. This heat can be used for steam generation in steam turbines. The high pressure of the combustion gases can also be used to drive an engine, or a gas turbine. Thermal powerplants use this heat to generate steam, which then drives the turbines. For marine engines and gas turbines, mechanical energy provided by the combustion gases is used either directly for propulsion or converted into electrical energy for power plants. Thus residual fuel began to replace coal in steamships due to its low ash content and good transportable nature.

With the discovery of the diesel engine in 1892 by Rudolf Diesel these engines started gradually replacing steam engines and the first four-stroke marine diesel engine ships were operating in 1912. A series of innovations of the diesel engine made it possible to use heavy fuel oil in medium speed trunk piston engines.

In the 1920s, a thermal cracking process (see Chapter 3) was developed and widely adopted in the petroleum industry, which changed the composition of residual fuel. In this process the heavy distillate or residual fraction of crude oil is subjected to very high temperatures that cause the larger molecules to crack into smaller molecules boiling in the range of gasoline and kerosene. As a consequence of this, the residual fuel became heavier, containing larger amounts of sulfur compounds, thermal tars, and visbreaker products. Great care was required to blend these residual stocks into fuel oil to avoid sediment formation due to the incompatibility and insoluble characteristics of asphaltenes and tars. With the development of catalytic cracking, residual fuel composition further changed. Vacuum distillation came into use to provide additional clean feed for catalytic cracking. Vacuum distillates were now converted to gasoline and middle distillates in the catalytic cracker. In turn, the residue from vacuum distillation became the basic component of residual fuel oil, containing the heaviest fraction of the crude and including asphaltenes. This residue has very high viscosity and a higher amount of sulfur. The vacuum residue is often

Fuels and Fuel-Additives, First Edition. S. P. Srivastava and Jenő Hancsók.
© 2014 John Wiley & Sons, Inc. Published 2014 by John Wiley & Sons, Inc.

visbroken to reduce its viscosity (see Chapter 3). This stock is then diluted further with low-price cracked stock to yield fuel oil of the required viscosity.

In the 1930s, two-stroke marine engines became popular due to the larger size of ships. Between World War I and World War II, the share of these marine engine-driven ships increased to approximately 25% of the overall ocean-going fleet. In the mid-1950s, high alkalinity cylinder lubricants became available to neutralize the acids generated by the combustion of high-sulfur residual fuels, and the wear rates became comparable to those when distillate diesel fuel was used. Diesel ships using residual fuel oil gained in popularity and in the second half of the 1960s, motor ships overtook steamships, both in numbers and in tonnage. By the start of the twenty first century, motor ships accounted for 98% of the world fleet. These marine engines have also found their way into the power industry due to the use of low-cost residual fuels.

With the increase in crude prices, most refiners resorted to the conversion of the vacuum residue into value-added products such as gasoline and diesel fuel by the use of coking, supercritical extraction, fluid catalytic cracking (FCC), heteroatom removal, and hydrocracking, for example. These are the so-called residue conversion processes. This shift in the refining processes further resulted in the reduced application of fuel oil in power generation and other industrial applications. In the electric power industry the feedstock base remarkably shifted toward natural gas, which is less damaging to the environments.

Refiners have coped with this decrease in demand for residual fuel oil by shifting their production from lighter grades to heavier grades. This benefits both refiners and end users, since the heavier fuel oil is cheaper for end users and refiners produce more value-added products like gasoline and diesel fuel. Some refineries have eliminated residual fuel production completely by incorporating cokers or hydrocrackers units in their configuration.

Out of the streams of Figure 3.1 presented in Chapter 3, the following are used for fuel oil blending:

• Atmospheric residue
• Vacuum residue
• Heavy product of visbreaker
• Heavy gasoils of coking
• Extracts of solvent-refined base oil
• Heavy cyclic oils of distillate and residue hydrocracking
• Desulphurized residue oils
• Extracts or even resin phases produced by supercritical extractions of residues

9.1 CLASSIFICATION OF FUEL OILS

Fuel oils are classified according to their applications. For example, RFO (residual fuel oil) terminology is used for burner fuels (in combustion equipment). The term Bunker fuel oil is used for marine ship fuels. RFO has been further classified in BS 2869 into light, medium, and heavy fuel oils according to the increasing viscosity of the fuels. ASTM D 396–80 classifies fuels into NO 1 to 6 according to the increasing viscosity. French AFNOR classifies fuels according to the sulfur content of the fuels (TBTS, BTS, and HTS). Marine bunker fuel oils have been classified according to the viscosity at 50 °C and have 14 grades (IFO 30–700). (The most popular grades are, however, IFO-180 and IFO-380.)

In 2005, ISO classified fuel oils (ISO-8217-2005) into 4 distillate grades and 10 residual grades. Detailed specifications of these grades are provided in this chapter. First, we consider the quality characteristics of fuel oils in some detail.

9.1.1 Characteristics of Fuel Oils

The properties of fuel oils define fuel oil applications and classifications. Fuel oils are characterized by viscosity, flashpoint, pour point, water and sediment content, carbon residue, ash content, distillation behavior or distillation temperature ranges, specific gravity, sulfur content, heating value, and carbon–hydrogen content.

However, fuel oil specifications do not cover all these properties. *Viscosity* is an important property and indicates the oil's resistance to flow. It is significant because it indicates the ease at which oil can be pumped. Differences in fuel oil viscosities are caused by variations in the concentrations of fuel oil constituents and different refining methods.

Flashpoint is the lowest temperature to which the oil can be heated for its vapors to ignite by a flame.

Pour point is the lowest temperature at which a fuel can be stored and handled. Fuels with higher pour points can be used when heated storage and piping facilities are provided.

Water and sediment content should be as low as possible to prevent fouling the facilities. Sediment accumulates on filter screens and burner parts. Water in distillate fuels can cause tanks to corrode and emulsions to form in residual oil.

Carbon residue is obtained by a test in which the oil sample is destructively distilled in the absence of air. Higher carbon residue value is indicative of the tendency of the fuel to form deposits.

Ash is the noncombustible material in the oil. An excessive amount indicates the presence of materials that can cause high wear in pumps.

The *distillation test* shows the volatility and ease of vaporization of a fuel.

Specific gravity is the ratio of the density of a fuel oil to the density of water at a specific temperature. Specific gravities cover a range in each grade, with some overlaps between distillate and residual grades. API gravity (developed by the American Petroleum Institute) is a parameter widely used in place of specific gravity.

Air pollution considerations are important in determining the allowable *sulfur content* of fuel oils. The sulfur content is frequently limited by legislation aimed at reducing sulfur oxide emissions from combustion equipment. Sulfur in fuel oils is also undesirable because of the corrosiveness of sulfur compounds in the flue gas.

Although low-temperature corrosion can be minimized by maintaining the stack at temperatures above the dew point of the flue gas, this limits the overall thermal efficiency of the combustion equipment. *Heating value* is an important property, although ASTM Standard D 396 does not list it as one of the criteria for fuel oil classification. Heating value can generally be correlated with the API gravity.

9.1.2 Classification of Heating Fuels for Power Plants

Fuel oil is a broad term used for fractions obtained from petroleum distillation, as a distillate or a residue. These liquid fuels can be burned in a furnace or boiler for the generation of heat or used in an engine for the generation of power. In this sense, diesel is also a type of fuel oil. Fuel oils for heating are broadly classified as distillate fuel oils (lighter oils) or residual fuel oils (heavier oils). So far in the European Union there has been no unified specification for domestic and industrial heating oils, and for fuel oils. For example, there is a British standard, BS 2869, for heating oils (provided in Table 9.1), and a German standard, DIN 51603–2, for fuel oils used in Germany (Table 9.2).

Fuel oil is classified into six classes in the United States, numbered 1 to 6, according to the boiling point, composition, and purpose in ASTM D 396. The boiling point, ranges from 175 to 600 °C, and carbon chain length, 9 to 70 atoms. Viscosity also increases with number, and the heaviest oil has to be heated before it can be pumped. The price usually decreases as the fuel number increases.

Numbers 1, 2, and 3 fuel oils are referred to as distillate fuel oils. For example, No. 2 fuel oil, No. 2 distillate and No. 2 diesel fuel oil are almost the same (diesel is different in that it also has a cetane number limit that describes the ignition quality of the fuel). Distillate fuel oils are distilled from crude oil. Gasoil refers to the process of distillation.

Number 1 is similar to kerosene and is the fraction that boils off immediately after gasoline is formed. Number 1 is a light distillate intended for vaporizing-type burners. High volatility is essential to continued evaporation of the fuel oil with minimum residue.

Number 2 is the diesel fuel that trucks and some cars can run on, thus often called "road diesel." It is also similar to heating oil. Number 2 is a heavier distillate than No. 1. It is used primarily with pressure-atomizing (gun) burners that spray the oil into a combustion chamber. The atomized oil vapor mixes with air and burns. This grade is used in most domestic burners and many medium-capacity commercial-industrial burners.

Number 3 is a distillate fuel oil and is rarely used.

Number 4 fuel oil is usually a blend of distillate and residual fuel oils, such as Nos. 2 and 6; however, sometimes it is just a heavy distillate. Number 4 may be classified as diesel, distillate, or residual fuel oil. Number 4 is an intermediate fuel that is considered either a heavy distillate or a light residual. It is intended for burners that atomize oils of higher viscosity than domestic burners can handle.

Grade No. 5 (light) is a residual fuel of intermediate viscosity for burners that handle more viscous fuel than No. 4 without preheating. Preheating may be necessary in some equipment for burning and, in colder climates, for handling. Grade No. 5 (heavy) is a residual fuel more viscous than No. 5 (light) but intended for similar purposes. Preheating is usually necessary for burning and in colder climates.

TABLE 9.1 Classification of gasoil-based fuel oils (BS 2869) for agricultural, domestic, and industrial engines and boilers

Fuels	Class to BS 2869	Viscosity (mm²/s)		Flashpoint		Sulfur Content, (%)	Min. Temp. for Storage(°C)	Min. Temp. for Outflow from Storage and Handling (°C)	Application
		At 40°C	At 100°C	Abel min. (°C)	Pensky–Martens, Min.(°C)				
Kerosene	C1			43	—	0.04	Ambient	Ambient	Distillate fuel for free-standing flueless domestic appliances
Gasoil	C2	1.0–2.0		38	—	0.2	Ambient	Ambient	Similar vaporizing and atomizing burners on domestic appliances with flues
	D	1.5–5.5		56	—	0.5	Ambient	Ambient	Distillate fuel for atomizing burners in domestic and industrial use
Heavy-fuel oil	E	8.2 max.		—	66	3.5	10	10	Residual or blended fuels for atomizing burners normally requiring preheating before combustion in burner
	F	20 max.		—	66	3.5	25	30	
	G	40 max.		—	66	3.5	40	50	

TABLE 9.2 Specifications of fuel oils used in Germany (DIN 51603–2)

Characteristics	Fuel Oil L	Fuel Oil T	Fuel Oil M	Fuel Oil S	Method
	Fuel Oil Type				
Density at 15 °C, g/cm^3	≤1.1	—	—	—	DIN 51757
Density at 20 °C, g/cm^3	—	≤1.1	≤1.1	—	DIN 51757
Flashpoint, °C	≥85	≥85	≥85	≥80	DIN 51758
Kinematical viscosity					
At 20 °C, mm^2/s	≤6	≤12	—	—	DIN 51550,
At 50 °C, mm^2/s	—	—	≤40	—	DIN 51562
At 75 °C, mm^2/s	—	—	≤12	—	
At 100 °C, mm^2/s	—	—	—	50	DIN ISO 3104
At 130 °C, mm^2/s	—	—	—	20	DIN 51366
Conradson number, %	≤0.5	≤1	≤16	≤16	DIN 51551
Sulfur content, %	≤0.2	≤0.8	≤0.5	≤2.8	DIN 51400
Water content, %	≤0.3	≤0.3	≤0.3	≤0.5	DIN ISO 3733
Heating value, MJ/kg	≥38.7	≥37.8	≥38.5	≥39.5	DIN 51900
Ash, %	≤0.01	≤0.01	≤0.02	≤0.15	DIN EN 7
Crystallization point, °C	—	—	≤15	—	

Number 5 fuel oil and No. 6 fuel oil are called residual fuel oils (RFO), or heavy fuel oils. The terms heavy fuel oil and residual fuel oil are sometimes used interchangeably for No. 6. Numbers 5 and 6 are what remains of the crude oil after gasoline and the distillate fuel oils are recovered through distillation. Number 5 fuel oil is a mixture of 75–80% No. 6 oil and 25–20% of No. 2 oil. Number 6 fuel oil may also contain a small amount of No. 2 to get it to meet industry specifications.

Residual fuel oils are sometimes called light when they have been mixed with distillate fuel oil, while distillate fuel oils are called heavy when they have been mixed with residual fuel oil. Heavy gas oil, for example, is a distillate that contains residual fuel oil.

Low-sulfur residual oils are marketed in many areas to permit users to meet sulfur dioxide emission regulations. These fuel oils are produced either by refinery processes that remove sulfur from the oil (hydrodesulphurization) or by blending high-sulfur residual oils with low-sulfur distillate oils. A combination of these two procedures can also be used to get low-sulfur stocks. These oils have significantly different characteristics from regular residual oils. For example, the viscosity–temperature relationship can be such that low-sulfur fuel oils have viscosities of No. 6 fuel oils when cold, and of No. 4 when heated. Therefore normal guidelines for fuel handling and burning can be altered when using such fuels.

9.1.3   Classification of Bunker Fuels

Bunker fuel is technically any type of fuel oil used aboard ships. It gets its name from the containers on ships and in ports that it is stored in. Earlier in steamships coal was stored in coal bunkers but now they are bunker fuel tanks.

According to ASTM D 396, Bunker A is No. 2 fuel oil, bunker B is No. 4 or No. 5, and bunker C is No. 6. Since No. 6 is the most common, "bunker fuel" is often used as a synonym for No. 6 fuel oil. Number 5 fuel oil is also called Navy special fuel oil or just Navy special; No. 5 or 6 is also called furnace fuel oil (FFO). The high viscosity of bunker fuel requires heating before the oil can be pumped from a bunker tank.

Residual fuel oils have high viscosity, particularly the No. 6 oil, which requires proper handling and storage system. This fuel must be stored at around 100 °F (38 °C) and heated to 150 °F (66 °C)–250 °F (121 °C) before it can be easily pumped. BS 2869 Class G Heavy Fuel Oil is similar in behavior, requiring storage at 104 °F (40 °C) and pumping at around 122 °F (50 °C).

In the maritime field another type of classification is used for fuel oils:

- MGO (marine gas oil)—roughly equivalent to No. 2 fuel oil, made from distillate only
- MDO (marine diesel oil)—a blend of heavy gas oil that may contain very small amounts of black refinery feed stocks, but with a low viscosity of up to 12 mm^2/s, so it does not need to be heated for use in internal combustion engines
- IFO (intermediate fuel oil)—a blend of gasoil and heavy fuel oil, with less gasoil than marine diesel oil
- MFO (marine fuel oil)—same as HFO
- HFO (heavy fuel oil)—pure or nearly pure residual oil that is roughly equivalent to No. 6 fuel oil

The characteristics of No. 6 fuel oil are listed in Table 9.3. No. 4 fuel oil has kinetic viscosity of 20 mm^2/s at 40 °C and No. 5 fuel oil has kinetic viscosity of 40 mm^2/s at 50 °C.

Marine diesel oil contains some heavy fuel oil, unlike regular diesels. Also marine fuel oils sometimes contain waste products such as used motor oil.

TABLE 9.3 Characteristics of No. 6 Bunker fuel oil

Properties	Typical Values
Kinematical viscosity at 50 °C, mm^2/s	500
Density, kg/m^3	985
Flashpoint, °C	60
Energy content, MJ/kg	43
Chemical composition	
Carbon, %	86
Sulfur, %	2.5
Ash content, %	0.08
Vanadium, mg/kg	200

Marine fuels were traditionally classified by their kinematic viscosity. This is a mostly valid criterion for the quality of the oil as long as the oil is made only from atmospheric distillation. Today, almost all marine fuels are based on fractions from other more advanced refinery processes, and viscosity says little about the quality as a fuel. CCAI and CII are two indexes that describe the ignition quality of residual fuel oil, and CCAI is especially often calculated for marine fuels. Despite this, marine fuels are still quoted on the international bunker markets with their maximum viscosity (according to the ISO 8217) due to the fact that marine engines are designed to use different viscosities of fuel. The following fuels are generally quoted for marine application:

- IFO 380—intermediate fuel oil with a maximum viscosity of 380 mm²/s at 50 °C
- IFO 180—intermediate fuel oil with a maximum viscosity of 180 mm²/s at 50 °C
- LS 380—low-sulfur (<1.5%) intermediate fuel oil with a maximum viscosity of 380 mm²/s at 50 °C
- LS 180—low-sulfur (<1.5%) intermediate fuel oil with a maximum viscosity of 180 mm²/s at 50 °C
- MDO—marine diesel oil
- MGO—marine gas oil

The first British standard for fuel oil came in 1982. The latest standards are ISO 8217 from 2005 and 2010. The ISO standard describes four types of distillate fuels and 10 types of residual fuels. Over the years the standards have become stricter on environmentally important parameters such as sulfur content. The latest standard also banned the addition of used lubricating oil to the residual fuels.

Some parameters of marine fuel oils [1] according to ISO 8217 (3rd ed., 2005) are provided in Table 9.4, and marine residual fuels (10 grades) are described in Table 9.5.

In 2010, ISO 8217 was modified. Tables 9.6 and 9.7 show the advisory data [3]. These were partly new quality parameters (e.g., hydrogen sulfide content, acid number, total sediment by hot filtration, and oxidation stability in the case of marine distillate fuels; hydrogen sulfide content, and acid number in the case of marine residual fuels) and partly new limits (e.g., cloud point in the case of DMX; grade and total sediment by hot filtration in the case of RMA grade).

Importantly, the aluminum and silicon (Al+Si) content was significantly decreased, which can be less than 25–60 mg/kg instead of the earlier 80 mg/kg. The vanadium and ash content of most fuel oils was lowered as well. The acid number of marine distillate fuels is max. 0.5 mg KOH/g and that of marine residual fuels is less than 2.5 mg KOH/g. (These values are used instead of the former strong acid number.) The oxidation stability of fatty acid methyl esters (FAME) is specified as well in the case of marine distillate fuels. The lubricity test is compulsory in products having less than 0.05% sulfur content.

TABLE 9.4   Marine distillate fuels—ISO 8217:2005

Properties	DMX*	DMA	DMB	DMC
Density at 15 °C, max., kg/m³	—	890.0	900.0	920.0
Viscosity at 40 °C, max., mm²/s	5.5	6.0	11.0	14.0
Viscosity at 40 °C, min., mm²/s	1.4	1.5	—	—
Water, max., % v/v	—	—	0.3	0.3
Sulfur content,[a] max., % (m/m)	1.0	1.5	2.0	2.0
Aluminum + silicon content,[b] max., mg/kg	—	—	—	25
Flashpoint,[c] min., °C	43	60	60	60
Pour point, summer, max., °C	—	0	6	6
Pour point, winter, max., °C	—	−6	0	0
Cloud point, max., °C	−16	—	—	—
Calculated cetane index, min.	45	40	35	—

Note: DMX is used for equipment such as emergency generators and not normally used in the engine room.
[a]Maximum sulfur content is 1.5% in the designated areas (wef 1-07-2010 1% sulfur is max.).
[b]The aluminum + silicon value is used to check for remains of the catalyst after catalytic cracking. Most catalysts contain aluminum or silicon, and the catalyst remains can damage the engine.
[c]The flashpoint of all fuels used in the engine room should be at least 60 °C

The sulfur content reduction of distillate and residue marine fuels is a continual issue [4–12]. In ISO 8217:2010, the sulfur content of residue marine fuels is defined according to the relevant statutory requirements. The sulfur content of the distillate marine fuels may be very low (≤10–500 mg/kg), whereas for residue marine fuels, 0.5–1.5% sulfur content is generally the target value.

In March 2009, the United States and Canada announced the formation of a North American Emission Control Area (ECA), which will require marine fuel sulfur content to be reduced to 0.1% by 2015. Open-sea sulfur content specifications are to be reduced to 0.5% by 2020 [13].

Starting in 2010, the total fuel oil consumption is being annually decreased by about 0.4 million b/d, but the marine fuel consumption has been growing at an annual rate of about ca. 2.3% [13]. In due course, technology and costs may force shippers to switch from residual- to distillate-based bunker fuel to meet the new specifications. Already demand for distillate bunker fuel is growing at an annual rate of 2–3%. Consequently demand for residual bunker fuel is expected to decrease by about 0.7 million bpd by 2020.

9.2   PRODUCTION OF FUEL OILS

As fuel oil consumption decreases worldwide, that of other white products will increase. Consequently, in the modern refineries, it is now practical to use such systems of technologies that produce fuel oils only in the required quantities in order to satisfy more stringent specifications. The biggest problem of fuel oils remains that of the sulfur, nitrogen, and metals content. This has been the focus of research in the last 10 to 15 years. The quality specifications for fuels oils and the stricter emissions

TABLE 9.5 Marine residual fuels—ISO 8217:2005

Property					Grades					
	RMA 30	RMB 30	RMD 80	RME 180	RMF 180	RMG 380	RMH 380	RMK 380	RMH 700	RMK 700
Density at 15°C, max., kg/m³	960	975	980	991	991	991	991	1010	991	1010
Viscosity at 50°C, max., mm²/s	30.0	30.0	80.0	180	180	380	380	380	700	700
Water, max. v/v%	0.5	0.5	0.5	0.5	0.5	0.5	0.5	0.5	0.5	0.5
Sulfur,[a] max., %	3.5	3.5	4.0	4.5	4.5	4.5	4.5	4.5	4.5	4.5
Aluminum+silicon,[b] max., mg/kg	80	80	80	80	80	80	80	80	80	80
Flashpoint[c] min., °C	60	60	60	60	60	60	60	60	60	60
Pour point, summer, max., °C	6	24	30	30	30	30	30	30	30	30
Pour point, winter, max., °C	0	24	30	30	30	30	30	30	30	30

Note: The use of waste lubricants as a blend component in marine fuels is now prohibited. ISO 8217–2005 also specifies that if the levels of calcium, zinc, and phosphorous exceed the given limits concurrently, the fuel is deemed to contain waste lubricants. CIMAC provides guidelines for marine fuels and lubricants [2].

[a]Maximum sulfur content is 1.5% in designated areas (since 1-07-2010 1% is max.).

[b]The combined aluminum and silicon value is used to check for remains of the catalyst after catalytic cracking. Most catalysts contain aluminum or silicon and the catalyst remains can damage the engine.

[c]The flashpoint of all fuels used in the engine room should be at least 60°C.

TABLE 9.6 Marine distillate fuels

Properties	DMX	DMA	DMZ	DMB
Density at 15 °C, max., kg/m³		900.0		
Viscosity at 40 °C, min., mm²/s		1.400–2.000		
Microcarbon residue max., %		0.3[a]		
Microcarbon residue, max., %	—	—	—	0.3
Sulfur content, max., % (m/m)		1.0–2.0		
Ash, max., % m/m		0.01		
Flashpoint, min., °C		60.0		
Pour point, summer, max., °C		0–6		
Pour point, winter, max., °C		−6–0		
Calculated cetane index, min.		45–35		
Acid number, max., mg KOH/g		0.5		
Lubricity, corrected wear scar diameter, (wsd 1.4 at 60 °C), max., μm		520		
Hydrogen sulfide [5], max., mg/kg	2.0	2.0	2.0	2.0

[a]Residue at 10%.

specifications for power plant and marine shipping have already led to the development of fuel oils having high heteroatom content (4–5%), and also to low-sulfur fuel oil (LSFO) and ultra-low sulfur fuel oil (ULSFO) [14–16].

The high sulfur, nitrogen, and metal present in residues lower fuel quality (mainly due to the sulfur and nitrogen content), especially the fuel produced using cracking technologies. These fuels require quality improver technologies to release more hydrogens in the hydrogenation process [17, 18]. The basic nitrogen compounds (e.g., pyridine, quinoline, acridin, and alkyl derivates) neutralize the acidic catalysts used in catalytic cracking. The main properties of vacuum residues produced from different crude oils that affect cracking are summarized in Table 9.8.

Over the last decade, refineries have turned to improving the quality and quantities of fuel oils produced by residue-processing technologies. With the decreasing demand for fuel oils, more refineries have started to use technologies that convert heavy fuel oils and other residues into light products.

Residue-processing technologies can be classified according to the following [19–23]:

Noncatalytic Processes
- **Extraction processes**:
  ○ Conventional
  ○ Supercritical extraction
- **Visbreaking**:
  ○ Conventional
  ○ Aquaconversion/emulsification of fuel
  ○ Hydro

TABLE 9.7 Marine residual fuels

Property	Grades										
	RMA[1]	RMB	RMD	RME	RMG				RMK		
	10	30	80	180	180	380	500	700	380	500	700
Viscosity at 50°C, max., mm²/s					10.00–700.0						
Density at 15°C, max., kg/m³					920–1010						
Microcarbon residue, max., %					2.50–20.00						
Aluminum+silicon, max., mg/kg					25–60						
Sodium, max., mg/kg					50–100						
Ash, max., %					0.040–0.150						
Vanadium, max., mg/kg					50–450						
Water, max., % v/v					0.3–0.5						
Pour point, summer, winter, max., °C					0–30						
Sulfur, max., % m/m					Statutory requirement						
Total sediment, aged, max., % m/m					0.10						
Acid number, max., mg KOH/g					2.5						

Note: Flashpoint: min 60°C; the fuel must be free from used lubricating oil and be considered to contain ULO when either one of the following conditions is met: calcium >30 and zinc >15, or calcium >30 and phosphorus >15.

TABLE 9.8 Main properties of vacuum residues from different crude oils that affect cracking

Feedstock	Yield Relative to Crude Oil (%)	Density (g/cm³)	H/C Atom Ratio	Asphaltene Content (%)	Conradson Number (%)	Sulfur Content (%)	Nitrogen Content (%)	Nickel Content (mg/kg)	Vanadium Content (mg/kg)
Njord Blend	6.5	0.9384	—	<0.5	5.9	0.32	0.22	3.1	0.5
Norne Blend	11.5	0.9600	—	<0.5	9.6	0.58	0.24	7.9	2.4
Troll Blend	11.5	0.9678	—	<0.5	9.6	0.55	—	4.6	3.9
Aasgard Blend	5.0	0.9688	—	0.9	11.0	1.2	0.32	1.6	21
Statfjord	13.7	0.9745	—	0.7	12.6	0.71	0.46	8.9	10.8
Gach Saran	32	1.0085	1.49	6.1	17.5	3.27	0.62	72	227
Venezuela Tia Juara	33	1.0092	1.48	12.5	21.3	2.90	0.60	66	669
Grane	33	1.0159	—	3.9	16.8	1.44	0.70	18	53
Middle East, Kuwait	21	1.0187	1.48	4.9	19.8	4.77	0.70	25	103
Bachaquero	59	1.0238	1.45	10.3	19.3	3.34	0.65	82	614
Russia, Arlansk	35	1.0298	1.44	7.8	20.5	4.53	0.72	127	335
Khafji	31	1.0404	1.42	12.6	23.1	5.56	0.47	50	152
Mexico, Maya	45	1.0427	1.41	18.1	24.5	4.98	0.78	100	487
Agbami	—	1.0458	—	6.3	—	0.47	0.48	105	11
Canada, Lloydminster	43	1.0489	1.42	13.8	23.9	5.72	0.62	117	230
Basrah Heavy	37	1.0512	1.38	6.1	25.6	6.14	0.61	71	211
Orinoco	57	1.0520	1.40	16.9	36.2	4.30	0.99	147	626

- **Coking**:
  - Delayed
  - Flexi
  - Fluid
- **Combined-cycle gasification**

Catalytic Processes
- **Hydrogenation catalytic processes**:
  - *Residue heteroatom removal* of metal, sulfur, and nitrogen
  - *Residue hydrocracking* of fix beds, slurry phases, and ebulatted beds
- **Catalytic process without hydrogen**:
  - Residue fluid catalytic cracking (RFCC)

In most refineries, more residue-processing technologies are being used, but these technologies are basically coking, residue hydrocracking, and residue FCC. Using residue hydrocracking and FCC, the distillate yield of the refinery can vary between 78% to 100% from crude oil, which contains about 31% residue [20]. The total product yield of all refineries has been found to be based on these technologies in the processing of fuel oil. The lowest fuel oil yield was observed in refineries whose coker units are set up beside a residue FCC or residue hydrocracking plant.

In terms of quality, the products of refineries differ. In the heteroatom residue hydrocracking plants, the fuel oil obtained does meet the required low sulfur and nitrogen percentages, but with the modern residue fluid hydrocracking process, the sulfur- and nitrogen-removing efficiency is raised to between 55% and 95% [21–32].

9.3 FUEL OIL STABILITY AND COMPATIBILITY

Residual fuel oil storage stability is affected by the large presence of asphaltene sediments. In heavy fuel oil tanks, stratification takes place as a result. Asphaltenes are insoluble in n-heptane but soluble in toluene. Fuel oils are required by ISO 10307–2 to keep the total potential sediment down to 0.10 m/m% max. Stratification in heavy fuel oil tanks is minimized when this specification is met. Asphaltenes are high molecular weight molecules in crude oil that contain organically bound vanadium and nickel. They also contain a fairly high percentage of sulfur and nitrogen. Asphaltenes have a predominantly aromatic structure and are polar molecules; they are kept in colloidal suspension by their outer molecular structure. Thermally cracked asphaltene molecules lose some of their outer structure (depending on the severity of the thermal cracking process). A milder process like visbreaking also affects this outer molecular structure. As this happens, asphaltenes start coagulating and form a sludge. Thus the blending of heavy fuel oil must be carried out in a very controlled manner. The choice of blending stock and the refinery process used in producing the fuel oil will greatly affect the blending process. Paraffinic cutter stocks can make the fuel unstable. Therefore aromatic cutter stocks (e.g., heavy- and/or light-cycle oil) are preferred. When the mixing of two fuels does not cause asphaltene coagulation, they are said to be compatible with each other. Two heavy fuels with different compositions (e.g., an atmospheric heavy fuel from paraffinic crude, and the other from a visbreaker operation) can be incompatible with each other. Dispersants can therefore be used to control the compatibility problem in heavy fuel oils.

Because of the special nature of residual fuels, it is necessary to have a fuel-handling system with steam-heated storage tanks and lines with purifiers/clarifiers and fine filters before the fuel goes into the engine.

9.4 ADDITIVES FOR RESIDUAL FUELS

Residual fuel oils are unprocessed products obtained from the heavy fraction of crude oil after valuable products such as gasoline, diesel, kerosene, and aviation turbine fuels are recovered by a combination of distillation and secondary refining processes. This processing increases the percentage of asphaltenes, which leads to stability problems. Blending this residue with a commercial residual fuel oil is presently a complicated process requiring more of aromatic cutter stocks to stabilize the fuel. The residual fuels as supplied to marine ships also contain water, sediments, and catalyst fines. These must be removed on board before using the fuel in the engines. There may also be microbial contamination, which has to be controlled by the addition of a biocide. Residual fuels may contain some of the following additives for improved performance [33–36]:

- Pre-combustion conditioning treatment additives such as demulsifiers and dispersants
- Combustion improvers
- Ash modifiers and anti slagging additives
- H_2S scavengers → oil-soluble sulfide derivatives
- Biocides

The pre-combustion additives are required during the purification process of the fuel. These additives mainly remove water and disperse any separated asphaltenes. Dispersants or sludge inhibitors may contain aromatic compounds or alkyl naphthalenes for solubilizing the asphaltenes. Improving combustion of marine residual fuel to reduce emissions is a developing area of research [37]. Presently, there are large numbers of commercial products available for this application, but their claims are difficult to verify. This is mainly due to the very complex chemical nature of fuel.

Combustion improvers used in marine fuels are usually organometallic compounds, containing iron or other metals. The industry has been widely using an iron-based component as an oxidation catalyst to accelerate the combustion of residual fuel oils. This results in a cleaner combustion chamber, cleaner exhaust valves, turbocharger nozzle rings and blades, and so on. Other benefits of black smoke reduction and energy efficiency are also associated with the improved combustion.

Anti-slagging additives are sometimes used to reduce corrosion on the fire side of the oil-fired boilers. Corrosion is caused by the sulfur, vanadium, and sodium present in the residual fuel. At temperatures higher than 600 °C, sodium and vanadium cause metal corrosion. During combustion, sulfur is converted to sulfur dioxide, which is catalyzed by vanadium pentaoxide to sulfur trioxide, the main compound responsible for corrosion. Magnesium naphthenate [38] has been reported to be beneficial, since burning magnesium oxide reacts with vanadium penta oxide to form a noncorrosive magnesium vanadate.

REFERENCES

1. ISO 8217. (2005). Petroleum products—fuels (class F)—specifications of marine fuels.
2. CIMAC. (2006). *Recommendations Regarding Fuel Quality for Diesel Engines.* volume 21.
3. Choudhuri, R. (2010). Update on proposed 4th edition of ISO 8217. Technical Committe for Bunkering, Singapore.
4. Gregory, D. (2003). Impact of proposed sulphur legislation. 4th European Fuels Conference, March 18, Rome.
5. Brun, B. (2003). Market and refining impacts of sulphur reduction in marine bunker fuel. 9th World Fuels Conference Europe, May 21, Brussels.
6. Robinson, N. (2003). The EU's marine fuel sulphur proposal. Harts World Fuel Conference, May 21, Brussels.
7. Higgins, T. (2004). Bunker fuel sulfur reduction on horizon. *World Refin.,* 14(8), 4.
8. Peckham, J. (2005). Refiners, shipping companies encouraged by European Council position on bunker fuel options. *Global Refin. Fuels Rep.,* 9(3), 1–12.
9. Wright, T. L. (2006). Tanker group: marine distillate not fuel oil from 2010. *Hydrocarb. Proc.,* 85(12), 13.
10. Gregory, D. (2007). Merchant shipping. 8th Annual European Fuels Conference, March 15, Paris.
11. Andersen, L. K. (2010). Opportunities and challenges with future biofuels in shipping. European Biodeisel 2010 Conference, Barcelona.
12. Vautrain, J. (2009). New regs require lower bunker fuel sulfur levels. *Oil Gas J.,* 106(44), 46–49.
13. Gillis, D., Yokomizo, G., van Wees, M., Rossi, R. (2010). Residue conversion options to meet marine fuel regulations. *Petrol. Technol. Quart.,* (3), 39–51.
14. Plain, C., Duddy, J., Kressman, S., Lo Coz, O., Tasker, K. (2004). Options for Resid Conversion, Axens IFP Group Technologies.
15. Plain, C., Benazzi, E., Guillaume, D. (2006). Residue desulphurisation and conversion, *Petrol. Technol. Quart.,* 2, 57–63.

16. Humfrey, J. (2005). *Current and Future Refinery Trends*. Shell Marine Products, EMEA.

17. Wang, Z., Que, G., Liang, W. (1998). Distribution of major forms of sulphur in typical Chinese sour vacuum residues and VRDS residues. Symposium on Chemical Analysis of Crude Oils for Optimizing Refinery Yields and Economics, 215th National Meeting, American Chemical Society, March 29–April 3, Dallas.

18. Que, G., Li, N. (1998). Separation and characterization of nitrogen compounds in petroleum vacuum residues. Symposium on Chemical Analysis of Crude Oils for Optimizing Refinery Yields and Economics, 215th National Meeting, American Chemical Society, March 29–April 3, Dallas.

19. Molyneux, R. A. (1996). Residue upgrading—technology options to maximise profitability. European Refining Technology Conference, October 28–30, London.

20. Gillis, D. B., Houde, E. J. (1996). Cost effective residue upgrading options. 13th Petroleum Conference, October 21–24, Cairo.

21. Fujita, K., Abe, S., Inoue, Y., Plantenga, F. L., Leliveld, B. (2001). New developments in resid hydroprocessing. Ecological and Economic Challenges for the Transportation Sector and Related Industries in the Next Decade. June 11–13, Nordwijk aan Zee, The Netherlands.

22. Hobson, G. D., ed. (1984). *Modern Petroleum Technology*. Wiley, Chichester.

23. Meyers R. A., ed. (2007). *Handbook of Petroleum Refining Processes*. McGraw-Hill, New York.

24. Marafi, A., Stainslaus, A., Hauser, A., Fukase, S., Matsushita, K., Al-Barood, A., Absi-Halabi, A. (2001). Effect of operating severity on products quality during atmospheric residue hydroteating over an industrial MoO_3/Al_2O_3 HDM catalyst. *Fuel Chem. Div. Prepr.*, 46(1), 238–239.

25. Ross, J., Kressman, S., Harlé, V., Tromeur, P. (2000). Meeting ON-SPEC products with residue hydroprocessing. IFP Refining Seminar 2000, November 13, Rome.

26. Shi, B., Lin, D., Wang, L., Que, G. (2001). Synergism between hydrogen donors and dispersed catalysts in residue HC. Symposium on General Papers Presented before the Division of Petroleum Chemistry. 222nd National Meeting, American Chemical Society, August 26–30, Chicago.

27. Ali, S. A., Biswas, M. E., Yoneda, T., Miura, T., Hamid, H., Iwamatsu, E., Al-Suaibi, H. (1998). A novel catalyst for heavy oil hydrocracking. Science and Technology in Catalysis, July 19–24, Tokyo.

28. Tonks, G. (2005). *Market Prospect for Low Sulphur Fuels*. Shell Marine Products, EMEA, London.

29. Mosconi, J.-J. (2003). A strategy for market evolution in diesel and heavy fuels. DGRM/STD, May.

30. Stockle, M., Knight, T. (2008). The impact of low sulphur bunkers on refinery configuration. ERTC 13th Annual Meeting, November 17–19. Vienna.

31. Stockle, M., Knight, T. (2008). Impact of low sulphur bunkers on refinery configuration. *Petrol. Technol. Quart. Catal.*, 27–31.

32. Geuking, W. B. (2006). Reducing the sulphur content of residual marine fuels. *Concawe Rev*, 15(1), 22–24.

33. Anon. (2007). Fuel additives for whiter whites. *Insight*, 33, 36.

34. *Residual Fuel Oil Technology*. (2012). Oryxe Energy.

35. Clariant commercial presentation. (2011). *Scavtreat H₂S Scavengers for Heavy Fuel Oil and Resids*, Slovnaft Refinery, October 26, Bratislava.

36. Anon. (2007). Marine fuel additives established, effective and expanding. *Insight*, 33, 14–15.

37. Bastenhof, D. (1997). Large diesel engine-fuels and emission—an outlook on tomorrow, *International Symposium on Fuels and Lubricants Proceeding* (eds. B. K. Basu, and S. P. Srivastava), New Delhi, 219–224.

38. Cummings W. M. (1977). Fuels and lubricant additives—1. *Lubrication* 32 (1), 12.

Acidity/acid number   The necessary amount of potassium-hydroxide, in milligrams, used to neutralize total acid content (water soluble and insoluble) of a single gram of petroleum product.

(1) *Total acidity method* is a measure of the amount of KOH needed to neutralize all or part of the acidity of a petroleum product. It is a measure of the combined organic and inorganic acidity. The acids in the sample are extracted in neutral alcohol and then titrated against standard alcoholic potassium hydroxide under heated conditions to get an indication of the corrosive properties of the product.

(2) *Inorganic acidity* is a measure of the mineral acid present.

(3) *Organic acidity* is obtained by deducting the inorganic acidity from the total acidity

Additive   Chemicals added in minor proportions to a parent substance to create, enhance, or suppress a certain property or properties in the parent material. Additives are also referred to as improvers. Any substance added to a base fuel to change its properties, characteristics, or performance is an additive.

Aniline point   The lowest temperature at which equal volumes of aniline and hydrocarbon fuels or lubricant base stock are completely miscible. An earlier method of measuring the aromatic content of a hydrocarbon blend to determine its solvency is now out of use.

A high aniline point indicates that the fuel is highly paraffinic. In the case of aromatics, the aniline point is low. The aniline point also gives an indirect measurement of the calorific value of the product.

Antifoam agent   An additive used to suppress the foaming tendency of petroleum products. Usually silicone or a polymer is used to break up surface bubbles by reducing the interfacial tension.

Antistatic additive   An additive that increases the conductivity of a hydrocarbon fuel or lubricant to improve the dissipation of electrostatic charges during high-speed dispensing of fuels or rotating equipments, thus reducing the fire/explosion hazard.

Fuels and Fuel-Additives, First Edition. S. P. Srivastava and Jenő Hancsók.
© 2014 John Wiley & Sons, Inc. Published 2014 by John Wiley & Sons, Inc.

API gravity   The acronym for the American Petroleum Institute. API is a special function of relative density (specific gravity) at 60/60 °F, represented by

$$API = \frac{141.5}{SG} - 131.5$$

$$SG = \frac{141.5}{131.5 + API}$$

where

API = degrees API

SG = specific gravity (60 °F/60 °F)

An accurate determination of the gravity of petroleum and its products is necessary for the conversion of measured volumes to volumes at the standard temperature of 60 °F.

Gravity is a factor governing the quality of crude oils. However, the gravity of a petroleum product is an uncertain indication of its quality. Correlated with other properties, gravity can be used to give the approximate hydrocarbon composition and heat of combustion

Ash/ash (sulfated)   Inorganic residue remaining after ignition of combustible material. It is determined by treating the residue with sulfuric acid and evaporating it to a dryness expressed as % by mass.

Antiknock   Resistance to detonation or "pinging" of spark-ignited engines.

Antiknock agents   Chemical compounds that, when added in small amounts to the fuel charge of an internal-combustion engine, have the property of suppressing, or at least strongly depressing, a knocking noise.

Antioxidants   Chemicals added to gasoline, jet, diesel fuels, and other products to inhibit oxidation.

Aromatic hydrocarbons   Hydrocarbons characterized by the presence of the benzene ring.

Asphaltenes   Coal-like substances in the residues of distilled crude oil. Asphaltenes are rich in the undesirable elements (e.g., sulfur, nitrogen, nickel, and vanadium) that occur in fuel oils.

ASTM   The acronym for the American Society for Testing and Materials. The test procedures and specifications developed by the ASTM for petroleum products and lubricants are used worldwide.

Bactericide   Additive to inhibit bacterial growth in fuel systems where water or moisture ingress take place.

Base number   Amount of acid (perchloric or hydrochloric) required to neutralize petroleum product's basicity and expressed as mg KOH/gram of samples.

Bitumen   Brown or black viscous residue from the vacuum distillation of crude petroleum, or from propane extraction of shortened atmospheric residue. It also occurs in nature as asphalt "lakes" and "tar sands."

Blending   The process of mixing two or more components or hydrocarbon streams to obtain a final product.

Blow-by   In internal combustion engines, leakage of combustion gases from combustion chamber, past the rings, into the crankcase is called blow-by gases. The passage of unburned fuel and combustion gases through the piston rings of IC engines results in fuel dilution and contamination of the crankcase oil.

Boiling range   The spread of temperatures over which a hydrocarbon product starts to boil or distill vapors and proceeds to complete evaporation. The boiling range is determined by test procedures for specific petroleum products.

Calorific value   The quantity of heat released on combustion. Calorific value is a measure of the heat-producing ability of a fuel. A weighed quantity of the fuel sample is burned in oxygen in a bomb calorimeter under controlled conditions. The calorific value is calculated from the weight of the sample and the rise in temperature. It can also be calculated from the formula:

$Qv = 12400 - 2100d^2$

where Qv = Calorific value, gross cals/g

$\qquad d^2$ = density at 15 °C

Carbon residue   Amount of residue left after the burning/pyrolysis of an oil. This residue provides some indication of the relative coke-forming tendencies of the fuel, and its amount can be determined by the Conradson method or the Ramsbottom method.

Catalyst   A substance used to accelerate or retard a chemical reaction without itself undergoing significant chemical change or change in volume during the process.

Catalytic converter   An integral part of vehicle emission control systems. Oxidizing converters remove hydrocarbons and carbon monoxide (CO) from exhaust gases, while reducing converters control nitrogen oxide (NO_x) emissions. Both use noble metal (platinum, palladium, or rhodium) catalysts that can be "poisoned" by the sulfur in the fuel. Lead in fuel is now banned and P is restricted.

Cetane index   An empirical method for predicting the cetane number of diesel fuel by a calculation based on API gravity and the mid-boiling point. A value is calculated from the physical properties of the diesel fuel to predict its cetane number.

Cetane number   The measure of the ignition quality of a diesel fuel. The higher the cetane number, the easier a high-speed, direct-injection engine will start, and the less "white smoking" and "diesel knock" will occur after startup. The cetane number of diesel fuels is determined in a single cylinder CFR engine by comparing the ignition delay characteristics of the fuel with that of the reference fuels. For this purpose, normal cetane (100 CN) and hepta methyl nonane or alpha methyl naphthalene, which have a CN of 0, are used.

Cetane number improver   An additive (iso-propyl- or 2-ethyl-hexyl-nitrate) that increases the cetane number of a diesel fuel.

Cloud Point   Cloud point is the temperature at which a cloud or haze of wax crystals appears at the bottom of the test jar when the gasoil is cooled under prescribed conditions. It gives a rough idea of the temperature above which the gasoil can be safely handled without any fear of congealing or filter clogging.

Compression ratio   The ratio of the volume of combustion space at bottom dead center to that at top dead center in an internal combustion engine.

Copper strip corrosion   A qualitative measure of the tendency of a petroleum product to corrode a copper strip. A cleaned and smoothly polished copper strip is immersed in the sample, which is then maintained at the specified temperature for the specified length of time. This strip is removed from the sample, washed with sulfur-free petroleum spirit and examined for evidence' of etching, pitting or discoloration. It is then compared with ASTM copper strip corrosion standard color code to measure the degree of corrosion. This test serves as a measure of possible corrosion of copper, brass, or bronze parts of the fuel system.

Corrosion   Detrimental change in the size or characteristics of material under exposure or use. Corrosion usually results from chemical action, either regularly and slowly as in rusting or rapidly as in metal pickling. Corrosion tests are carried out on petroleum products to determine whether corrosive sulfur compounds or other corrosive compounds are present.

Corrosion inhibitor   An additive that protects metal surfaces from chemical attack by water or other contaminants.

Cracking   The process by which an organic compound is split into two or more compounds of lower molecular weight. The cracking process has become increasingly important in the petroleum industry as a means of breaking down the heavier components of petroleum into gas, naphthas, and distillates, thereby increasing the yield of gasoline, jet fuel and the gasoil obtained from crude oils. The cracking process may be carried out with heat and pressure (thermal cracking) or in presence of a catalyst (catalytic cracking), or with hydrogen and catalyst together (hydrocracking).

Crude oil   Crude petroleum. A naturally occurring mixture, consisting predominately of hydrocarbons and organic compounds containing sulfur, nitrogen, oxygen, and traces of metallic constituents, which is capable of being removed from the earth in a liquid state. Crude petroleum is commonly accompanied by varying quantities of extraneous substances such as water, inorganic matter, and gas. Basic types of crudes are asphaltic, naphthenic, paraffinic, or intermediate, depending on the relative proportion of these types of hydrocarbons present.

Density   The mass of the liquid per unit volume. Density is used for calculating the mass when volume of the bulk is known (*Volume × Density = Mass*).

Detergent dispersant   A substance incorporated in fuels that gives them the property of cleaning and keeping clean the fuel system and engine parts and preventing deposits where they would be harmful.

DIN   Deutsche Industrie Norm. The German Industry Standard.

Distillation   The basic test used to characterize liquid fuels. For any petroleum product, boiling takes place over a range of temperatures rather than at a single temperature. This range is of great importance in fuels.

Emissions (mobile sources)   Emissions of exhaust gases that may be regarded as pollutants (CO, NO_x, hydrocarbons, particulate matters, CO_2, etc.).

Emissions (stationary sources)   Emissions from stationary power plants or engines. Fuel composition can influence emissions of sulfur oxides, nitrogen oxides, metals, halogenic compounds, and other particulates from these sources.

Emulsifier   Chemical additive that promotes the formation of a stable emulsion of petroleum products with water. Emulsifiers are composed of polar surface active compounds of different chemicals.

Engine deposits   Hard or persistent accumulation of sludge, varnish, and carbonaceous residues due to blow-by of unburned and partially burned fuel, or the partial breakdown of the lubricant.

Engler distillation   A small-scale laboratory test used to determine the boiling range of a hydrocarbon stream.

Exhaust gas recirculation (EGR)   System to reduce automotive emission of nitrogen oxides (NO_x). It routes exhaust gases into the intake manifold where the air–fuel mixture is diluted to reduce peak combustion temperatures, thereby reducing the tendency to form NO_x.

Fatty acid   An organic acid of mainly unsaturated (olefinic double-bond) hydrocarbon chain structure originally derived from fats and fatty oils.

Flash point   The lowest temperature at which, under specified test conditions, a petroleum product vaporizes rapidly enough to form above its surface an air-vapor mixture that gives a flash when ignited by a small flame. The flash point of a petroleum product is an indication of the risk of fire or explosion associated with its use or storage. It can be determined in a closed and open cup. The flash point is an indicator of the fire and explosion hazards associated with a petroleum product.

Friction   Resistance to motion of one object over another. Friction depends on the smoothness of the contacting surfaces, as well as the load applied.

Fuel ethanol   Ethanol (ethyl alcohol, C_2H_5OH) without denaturants, but containing some impurities and water.

Gasoline   A volatile mixture of liquid hydrocarbons, containing small amounts of additives and suitable for use as a fuel in spark-ignition, internal combustion engines.

Gasoline–ethanol blend   A spark-ignition automotive engine fuel containing denatured fuel ethanol in a base gasoline.

Gums   The amount of nonvolatile heptane-insoluble residue left when the sample is evaporated in a jet of hot air at $160\,°C$. For jet fuels, the evaporation is carried out in a jet of super heated steam at $232\,°C$.

Exhisting gums   The amount of gum formed after the sample is aged in an oxidation stability bath and evaporated under specified conditions.

Potential gums   The gum contained in gasoline that may accumulate in the tank, fuel line, pump, where it tends to interfere with the action of moving parts, injectors, and obstruct the flow of fuel.

Hydrocarbons   Any chemical compound made up exclusively of carbon and hydrogen atoms. Hydrocarbons form the principal constituents of petroleum.

Hydrogenation   The process of adding hydrogen to the hydrocarbon molecule.

Hydrolytic stability   Ability of additives and certain biodiesel to resist chemical decomposition (hydrolysis) in the presence of water.

Induction period   A period under given conditions in which a petroleum product does not absorb oxygen at a substantial rate to form gum.

Inhibitor   An additive that improves the performance of a petroleum product by controlling/ inhibiting undesirable chemical reactions, i.e. oxidation inhibitor, rust inhibitor, corrosion inhibitor etc.

Kinematic viscosity   Measure of a fluid's resistance to flow under gravity at a specific temperature (usually $40\,°C$ or $100\,°C$).

Knock   The sound of "ping" associated with autoignition of a portion of the fuel–air mixture ahead of the advancing flame front in the combustion chamber of a gasoline engine.

Lead   Commonly used name for tetraethyl or tetramethyl lead (TEL), an additive earlier used in gasoline to improve octane ratings. Use of TEL has now been discontinued due to environmental restrictions.

Liquified petroleum gases   LPG, a bottled gas. Industry term for any material composed predominately of the following hydrocarbons or mixtures of them: propane, propylene, butanes, and butylenes. LPG is recovered from crude oil, and natural and refinery gases.

Lubricity   The ability to lubricate.

Metal deactivators   Organic compounds that suppress the catalytic action of heavy metal compounds sometimes contained in hydrocarbon distillates. Metal compounds (e.g., copper derivatives) tend to promote gum formation.

MMT   Methylcyclopentadienyl-manganese-tricarbonyl. MMT is a gasoline antiknock additive.

Neutralization number   The number of milligrams of potassium hydroxide required to neutralize one gram of a sample. It indicates the acidity of a petroleum product.

Octane number   Term used to indicate numerically the relative antiknock value of automotive and aviation gasolines having a rating below 100. The octane number is based on a comparison with the reference fuels, isooctane and normal-heptane that have been assigned octane numbers of 100 and zero, respectively, in the knock-rating scale. The octane number of a fuel is the volume percent of isooctane in a blend of the two reference fuels that matches the knocking tendencies of the fuel sample when tested under specified conditions. Under mild conditions, the engine measures Research Octane Number (RON); under severe conditions Motor Octane Number (MON). An Antiknock Index (AKI) is also used, which is the average of RON and MON: $(R+M)/2$.

Octane number, aviation method   Octane number of aviation gasolines, determined by a test method that indicates the knock characteristics at a lean fuel–air ratio, at lean or cruise rating.

Octane number, motor method   Octane number of automotive gasolines determined by a test method that indicates the knock characteristics under severe conditions: at high temperatures, high speed, and/or high load.

Octane number, research method   Octane number of automotive gasolines determined by a test method that indicates the knock characteristics under mild conditions: temperatures and speed approximating ordinary driving conditions.

Octane requirement (OR)   The lowest octane number reference fuel that will allow an engine to run without knocking under standard conditions of service. This is a characteristic of each individual vehicle.

Octane requirement increase (ORI)   As results of deposits accumulate in the combustion chamber, the octane requirement of engine increases compared to that of the new engine. The value is the octane requirement increasing (ORI).

Oxidation   The process by which oxygen combines with other substances. The oxidation reaction in petroleum products may lead to gum and resin formation, which is of importance in the utilization of gasolines, jet fuels, and gasoils, particularly those that contain unsaturated compounds.

Oxidation inhibitor   A substance added in small quantities to a petroleum product to increase its oxidation resistance (reduce the rate of oxidation), thereby increasing service or storage life. These are also called antioxidants.

Oxidation stability   Resistance of a petroleum product to oxidation, and therefore a measure of its potential service or storage life.

Oxygenates   An oxygen-containing, ashless organic compound, such as alcohol or ether (MTBE, TAME, ETBE), that can be used as a gasoline-blending component.

Paraffinic   A type of petroleum fluid derived from paraffinic crude oil and containing a high proportion of straight chain saturated hydrocarbons. Presence of high amount of paraffins lead to cold flow problems.

Pensky–Martens closed tester   Laboratory apparatus used to determine the flashpoint and fire points of fuels.

Performance number   PN. An arbitrary scale, normally used to denote knock characteristics of aviation gasolines having an octane rating above 100.

Pour point   The lowest temperature at which a hydrocarbon product will pour when chilled without disturbance under specified conditions.

Pour point depressant   A compound that, when added to a wax-containing product such as diesel fuel and lubricants, reduces the solid point of the product. The additive apparently functions by modifying the crystal structure of wax that separates at low temperatures.

PPM Parts per million, usually by weight.

Pre-ignition Ignition of the fuel–air mixture in a gasoline engine before the spark plug fires. Often caused by incandescent fuel or deposits in the combustion chamber. It wastes power and may damage the engine.

Reference fuels Standardized laboratory engine fuels and blends that are used in determining the octane numbers of engine gasolines and the cetane numbers of diesel fuels, and for other properties.

Refining A series of processes to convert crude oil and its fractions into finished petroleum products, including thermal cracking, catalytic cracking, alkylation, hydrocracking, hydrogenation, hydrodesulphurization, and isomerization.

Reid vapor pressure RVP. A test for determining the vapor pressure of volatile hydrocarbon products (gasolines) under controlled conditions.

Road octane A numerical value based on the relative antiknock performance in an automobile of a test gasoline as compared with specified reference fuels. Road octanes are determined by operating a car over a stretch of level road or on a chassis dynamometer under conditions simulating those encountered on the highway.

Smoke point A test made of kerosene and jet fuels that indicates the highest point to which the flame can be turned before it will smoke. The smoke point is the maximum flame height, in millimeters, at which the fuel will burn without smoking under specified conditions in a smoke point apparatus. This is an important test for kerosene and indicates the aromatic content of kerosene. Higher smoke point is better for domestic use.

Specific gravity The ratio of the weight of a given volume of material to the weight of an equal volume of some standard substance. In a hydrocarbon product, the standard reference material is distilled water, and the temperature of both the hydrocarbon product and the water is 60 °F.

Syncrude Unconventional crudes such as those derived from tar sands, oil shale and coal liquefaction, and the product mixture of Fischer–Tropsch synthesis.

Tar sands A mixture of 84–88% sand and mineral-rich clays, 4% water, and 8–12% bitumen, heavy oil. Bitumen is a dense, sticky, semisolid substance that is about 83% carbon. The word tar is a correct description because the tar or heavy oil separated from the sand is a highly aromatic, high-sulfur material more closely related to a tar derived from coal than to crude oils. Tar sand is also known as oil sands and heavy oils. The extensive Athabasca tar sands in Canada have received considerable attention for decades, but costs have always proved a deterrent for commercial exploitation on a large scale.

Viscosity The measure of the internal friction or resistance to flow of a fluid. In determining viscosities of liquid hydrocarbon products, values are often expressed as the number of seconds in time required for a certain volume of the liquid under test to pass through a standard orifice under prescribed conditions.

The viscosity of a liquid is the measure of its resistance to flow. It is expressed either in Saybolt seconds or in mm²/s (kinematic viscosity). Viscosity is an important characteristic of fuels and lubricants.

Index

Fuels and Fuel-Additives, First Edition. S. P. Srivastava and Jenő Hancsók.
© 2014 John Wiley & Sons, Inc. Published 2014 by John Wiley & Sons, Inc.